CULTURAL HERITAGE AND TOURISM IN AFRICA

Cultural Heritage and Tourism in Africa examines the multiple and diverse manifestations of cultural heritage-based tourism in Africa from a regional, social science, and sustainability perspective.

This book delivers a comprehensive treatise on the interdependent concepts of cultural heritage and tourism. Heritage is one of the most pervasive tourism assets worldwide and lies at the foundations of tourism in many localities, including Africa. However, despite its salience, there has not been a systematic examination of Africa's heritage resources, markets, policies, practices, successes, and challenges in a tourism framework, despite the continent's immense heritage value. This book reviews the different types of heritages that pervade the cultural environment of Africa and comprises its vast heritagescapes. It also examines the increasing potential for the growth of heritage tourism throughout the entire continent. The contributions in this volume delve into current thinking about space and place and their effects on heritage, mobilities, globalization, colonialism and indigeneity, conflict, identity and nation-building, connections with other regions through migration and the slave trade, and a greater emphasis on the ordinary heritage of Africa, which has long been ignored by tourism scholars and industry representatives. The chapters herein are authored by Africa specialists, most being from Africa, offering a truly African perspective. The chapters are conceptually rigorous and empirically rich with examples from all regions of the African continent.

This unparalleled interdisciplinary glimpse at cultural heritage and tourism in Africa delivers strong value and is a vital resource for all students and researchers of tourism, cultural studies, heritage studies, geography, anthropology, sociology, history, and global studies.

Dallen J. Timothy is Professor of Community Resources and Development, and Senior Sustainability Scientist at Arizona State University, Arizona, USA. He is also a Senior Research Associate at the University of Johannesburg, South Africa; Visiting Professor at Hunan Normal University, Guangxi University, and Luoyang Normal University, China; and Guest Professor in the Erasmus Mundus European Master in Tourism Management program based at the University of Girona, Spain. He is the founding editor of the *Journal of Heritage Tourism* and currently serves on the editorial boards of 25 international journals. He is commissioning or co-commissioning editor over four book series with Routledge and Channel View Publications, and has ongoing research projects in Africa, North America, Asia, Europe, and the Middle East on topics related to heritage, international borders and tourism, religious tourism, and community empowerment.

Routledge Cultural Heritage and Tourism Series

Series editor: Dallen J. Timothy, Arizona State University, USA

The Routledge Cultural Heritage and Tourism Series offers an interdisciplinary social science forum for original, innovative and cutting-edge research about all aspects of cultural heritage-based tourism. This series encourages new and theoretical perspectives and showcases ground-breaking work that reflects the dynamism and vibrancy of heritage, tourism and cultural studies. It aims to foster discussions about both tangible and intangible heritages, and all of their management, conservation, interpretation, political, conflict, consumption and identity challenges, opportunities and implications. This series interprets heritage broadly and caters to the needs of upper-level students, academic researchers, and policy makers.

Cultural Heritage and Tourism in Japan
Takamitsu Jimura

Medieval Imaginaries in Tourism, Heritage and the Media
Jennifer Frost and Warwick Frost

Children, Young People and Dark Tourism
Edited by Mary Margaret Kerr, Philip R. Stone and Rebecca H. Price

Tourism and Development in the Himalaya
Social, Environmental and Economic Forces
Edited by Gyan Nyaupane and Dallen J. Timothy

Cultural Tourism and Cantonese Opera
Jian Ming Luo

Screen Tourism and Affective Landscapes
The Real, the Virtual, and the Cinematic
Edited by Erik Champion, Jane Stadler, Christina Lee and Robert Moses Peaslee

Cultural Heritage and Tourism in Africa
Edited by Dallen J. Timothy

For more information about this series, please visit https://www.routledge.com/Routledge-Cultural-Heritage-and-Tourism-Series/book-series/RCHT

CULTURAL HERITAGE AND TOURISM IN AFRICA

Edited by Dallen J. Timothy

Routledge
Taylor & Francis Group
LONDON AND NEW YORK

Cover image: © Domingo Leiva / Getty Images

First published 2023
by Routledge
4 Park Square, Milton Park, Abingdon, Oxon OX14 4RN

and by Routledge
605 Third Avenue, New York, NY 10158

Routledge is an imprint of the Taylor & Francis Group, an informa business

British Library Cataloguing-in-Publication Data
A catalogue record for this book is available from the British Library

ISBN: 978-0-367-72223-4 (hbk)
ISBN: 978-0-367-72224-1 (pbk)
ISBN: 978-1-003-15395-5 (ebk)

DOI: 10.4324/9781003153955

Typeset in Bembo
by SPi Technologies India Pvt Ltd (Straive)

CONTENTS

FIGURES

TABLES

CONTRIBUTORS

David Adeloye completed his PhD in tourism from the Department of Tourism, University of Otago, New Zealand. His research interests include tourist behaviour, risk perception, risk management, destination image, and tourism development. David has undertaken research in Nigeria addressing tourism and security. He is currently working as an independent researcher based in Dunedin, New Zealand, working in the field of community wellbeing.

Emmanuel Akwasi Adu-Ampong is an Assistant Professor in the Cultural Geography Chair Group, Wageningen University, the Netherlands, and a Senior Research Associate at the School of Tourism and Hospitality, University of Johannesburg, South Africa. He currently holds a Dutch Research Council (NWO) Veni Grant (VI. Veni.201S.037) for 2021–2024, researching the geographies of slavery heritage tourism in the Ghana–Suriname–Netherlands triangle. In this project, he seeks to synthesize the relational notions of place and embodied performativity in cultural geography with transformative cross-cultural encounters in critical tourism studies and ideas about the social construction of heritage in critical heritage studies. His other research interests are in sustainable tourism development, tourism policy and planning, cultural heritage management and innovations in qualitative research methodologies.

Gebeyaw Degarege is a Lecturer in the Faculty of Environment, Society and Design at Lincoln University, New Zealand, with research interests in sustainable tourism planning and management, food tourism, food security, food sovereignty, poverty, sustainable livelihoods and sustainable development. Gebeyaw has undertaken tourism policy and food security research in Ethiopia and is currently leading a project in New Zealand that considers the relationship between hunting and food security.

Ikechukwu O. Ezeuduji is a Professor at the University of Zululand, South Africa, obtained both his Master of Science (2005) and PhD (2008) from the BOKU-University of Natural Resources and Life Sciences, Vienna, Austria. He thereafter obtained a Master of Business Administration (2013) from The Open University, Milton Keynes, UK, and a Postgraduate Diploma in Higher Education (2019) from the University of KwaZulu-Natal, South Africa. He has more than ten years of teaching experience in higher education institutions. He has widely published in broader areas of tourism development (rural and events tourism), tourism management (strategic tourism management and tourism entrepreneurship), and tourism marketing (brand essence, brand competitiveness, brand image and brand loyalty).

Edson Gandiwa is a Director of Scientific Services at the Zimbabwe Parks and Wildlife Management Authority. He has a PhD in Wildlife Conservation and Management from Wageningen University, the Netherlands. He has widely published in the areas of wildlife conservation and management.

Ermias Kifle Gedecho completed his PhD in hotel and tourism management at the Hong Kong Polytechnic University in September 2022 and is now an assistant professor at the Kotebe University of Education. He has a broader research interest in tourism, hotel, and event management: religious tourism, diaspora festivals, quality of life, well-being, diaspora tourism, hotel data privacy management, hotel performance, local food experience, values, and tourist behaviour.

Liza Gijanto is an Associate Professor of Anthropology and Coordinator of Museum Studies at St. Mary's College of Maryland, USA. She is the co-editor of the Society of Africanist Archaeology's biannual bulletin *Nyame Akuma* and the associate editor for African contributions for *Historical Archaeology*. As an archaeologist focused on West Africa and the Atlantic slave trade, she has conducted fieldwork in The Gambia and participated in projects in Liberia. Her book *Life of the Trade* (2016) examines the impact of the transatlantic trade on everyday life in the Gambian commercial centre of Niumi.

Brent Lovelock is a Professor and Head of the Department of Tourism at the University of Otago. His research focuses on sustainable, regenerative, and ethical tourism, community-based tourism, and the relationships between tourism and natural systems. He is co-Director of the centre for Recreation Research at the Otago Business School and is currently leading a project that explores nature-based tourism and recreation and its connection with native and introduced flora and fauna. He has undertaken research in Africa on coffee tourism (Ethiopia, Rwanda) and is collaborating on research that explores introduced invasive animal species and food security in New Zealand.

Joseph E. Mbaiwa is the Director of the Okavango Research Institute (ORI), University of Botswana. He is also a Professor of Tourism studies. He holds a PhD in Recreation, Park & Tourism Sciences from Texas A&M University (USA). Joe Mbaiwa has over 20 years' experience in tourism and natural resource management research and consultancy experience in Botswana. He is widely published on issues of tourism development, rural livelihoods, and conservation.

Kokel Melubo is an Associate Professor in the Department of Tourism Management at the College of African Wildlife Management, Mweka, Tanzania. He is a graduate of the University of Dar es Salaam, Tanzania (MSc in Natural Resources Management), and the University of Otago (PhD in Business Tourism). His research interests include indigenous tourism, protected area management, mountaineering, tour guiding, corporate social responsibility, and sustainable tourism.

Ishmael Mensah is a Professor of tourism and hospitality management and Ghana Director of the Confucius Institute at the University of Cape Coast. He is also an External Expert at Hunan City University. He holds a PhD in tourism from the University of Cape Coast as well as post-graduate certificates in Hospitality Administration and Event Planning & Tradeshow Management from Georgia State University where he was a fellow under the Ghana Tourism Capacity Development Initiative. He has published widely in high-ranking academic journals in the areas of roots tourism, environmental management, special events management, service quality, and community-based tourism.

Goemeone Mogomotsi is an Associate Professor (International Environmental Policy) at the Okavango Research Institute, University of Botswana. He is also serving on secondment as a Senior Assistant to the Vice Chancellor, University of Botswana. He has published in natural resources use and policy, water law, waste management, legal aspects of religious freedom and financial services regulation. He is an admitted attorney, conveyancer, and notary public in the Courts of Botswana.

Patricia Mogomotsi is an Associate Professor (Natural Resources Economics) at the University of Botswana, Department of Economics. She holds a PhD in Economic and Management Science, with Tourism Management from the North-West University, South Africa. Her research interests include natural resources management, tourism economics, water institutions and regulations, waste management, and climate change adaptation among others.

Naomi Moswete is Senior Lecturer in the Department of Environmental Science, University of Botswana. Her research interests include Human geography, tourism as a strategy for rural development, community-based tourism, transboundary conservation areas–ecotourism nexus, parks–people relationships, heritage management and cultural tourism. She is an editor in CABI Tourism Cases and sectional editor in *Parks and Recreation Administration Journal*.

Chiedza Ngonidzashe Mutanga is a Senior Research Fellow in Sustainable Tourism at the Okavango Research Institute (ORI), University of Botswana. She holds a PhD in Tourism and Wildlife Conservation from Chinhoyi University of Technology, Zimbabwe. Her main area of specialization lies in tourism development and wildlife resource conservation which include wildlife tourism, ecotourism, community-based natural resource management and protected area tourism.

Siyabulela Nyikana holds a PhD in Tourism & Hospitality from the University of Johannesburg. He is currently a Senior Lecturer at the University of Johannesburg in the School of Tourism and Hospitality. His research interests lie in tourism development and management generally, but especially as relating to the use of niche tourism sectors such as religion, business, sport, and events to fast-track tourism development and growth in emerging economies in Africa.

Roselyne N. Okech is Professor of Tourism Studies at Grenfell Campus, Memorial University of Newfoundland, Canada. Her research interests are in the fields of cultural heritage development and ecotourism management. She has written numerous articles on the subject appearing in reputable international journals including *Tourism Review International* and *Asia-Pacific Journal of Tourism Research* among others. She has won several international awards including the Outstanding Research Award in Costa Rica and Female Researchers Award in Croatia. She has also been featured in TV stations in Kenya and Montenegro regarding her teaching and research. Professor Okech has engaged in guest lectures (Kenya, Rwanda, The Netherlands, Hong Kong PolyU, and India) covering various tourism aspects. She was recently named a Fellow with the Carnegie African Diaspora Fellowship Program.

Jarkko Saarinen is a Professor of Human Geography at the University of Oulu, Finland, and Distinguished Visiting Professor (Sustainability Management) at the University of Johannesburg, South Africa, and Visiting Professor at the Department of Civil and Industrial Engineering, Sustainable Destination Development, Uppsala University, Sweden. His research interests include sustainable development, sustainable tourism, tourism–community relations, nature and heritage conservation, and climate change adaptation studies.

João Sarmento is an Associate Professor at the Geography Department, University of Minho, Portugal, and a researcher at the Communication and Society Research Centre in the same university. He graduated in Geography from the University of Porto, Portugal, and received his doctorate in Geography from the University College Cork, Ireland. He has published extensively in the fields of Cultural Geography, Colonial/Postcolonial Studies, Tourism Studies and Urban Studies, in journals such as *Environment & Planning D.*, *Journal of Historical Geography*, *Social & Cultural Geography*, *Tourism Geographies*, and *European Urban and Regional Studies.* He is the author of *Fortifications, Postcolonialism and Power. Ruins and Imperial Legacies*

(Routledge, 2016). His research mostly focuses on Africa, and attempts to connect heritage, memory, landscape, violence, and space.

Gladys B. Siphambe is currently the acting Projects Coordinator and a Heritage Manager at the Department of National Museum and Monuments (DNMM), Government of Botswana. She holds a PhD in Tourism Management from The University of Botswana. Gladys Siphambe has 30 years' experience in cultural and heritage tourism management and research in Botswana. She has worked extensively with Communities on issues of tourism development, rural livelihoods, and conservation.

Lesego S. Stone is a Senior Lecturer (Tourism Management) at the Department of Tourism and Hospitality Management, University of Botswana, Botswana. Her research interests are in sustainable tourism development with specific reference to nature-based tourism, tourism marketing, community-based tourism, and community participation in tourism.

Moren Tibabo Stone is an Associate Professor in the Department of Environmental Science, University of Botswana. He holds a PhD in Community Resources and Development (Environment and Tourism) from Arizona State University, Arizona, USA. He teaches environment and tourism courses. His research interests are sustainable tourism development with specific reference to nature-based tourism, protected area tourism, ecotourism, conservation, and rural community livelihoods.

Philip R. Stone is Executive Director of the Institute for Dark Tourism Research (iDTR) at the University of Central Lancashire, UK. He is an internationally recognized scholar in the field of "dark tourism" and "difficult heritage" and has published extensively about the subject. Philip is also a media consultant on dark tourism, with clients including the *BBC* (British Broadcasting Corporation), *CNN*, *The New York Times*, *The Guardian*, and *The Washington Post*. Philip is the founder and Editor-in-Chief of the *International Journal of Dark Tourism Studies*. His latest books include the first-ever tourist guidebook dedicated to dark tourism – *111 Dark Tourism Places in England You Shouldn't Miss* (Emons, 2021), and C*hildren, Young People and Dark Tourism* (Routledge, 2023). His forthcoming book is *The Future of Dark Tourism: Enlightening New Horizons* (Channel View Publications).

Dallen J. Timothy is Professor of Community Resources and Development, and Senior Sustainability Scientist at Arizona State University, Arizona, USA. He is also a Senior Research Associate at the University of Johannesburg, South Africa; Visiting Professor at Hunan Normal University, Guangxi University, and Luoyang Normal University, China; and Guest Professor in the Erasmus Mundus European Master in Tourism Management program based at the University of Girona,

Spain. He is the founding editor of the *Journal of Heritage Tourism* and currently serves on the editorial boards of 24 international journals. He is commissioning or co-commissioning editor over four book series with Routledge and Channel View Publications, and has ongoing research projects in Africa, North America, Asia, Europe, the Middle East on topics related to heritage, international borders and tourism, religious tourism, and community empowerment.

Clinton D. van der Merwe is a Senior Lecturer and teaches Geography Education in the Department of Humanities Education at the Faculty of Education, of the University of Pretoria. Clinton has over 25 years of teaching experience, having taught at the Oprah Winfrey Leadership Academy for Girls (OWLAG) – as the Head of Department: Humanities (and the Acting Head of Academics). Before that, Clinton was a Geography Lecturer and Methodologist at the Wits School of Education; a councillor on the South African Geographical Names Council (SAGNC) by Ministerial Appointment; and a National Examiner for Geography at the Department of Basic Education (DBE). Clinton is the Editor-in-Chief of the *Journal of Geography Education in Africa* (*JoGEA*) and was instrumental in starting the South African National Geography Olympiad (SANGO). Clinton holds a PhD in Heritage Tourism from the University of Johannesburg and is an avid parkrunner.

Gustav Visser studied at Stellenbosch University and the London School of Economics. He joined the University of the Witwatersrand in 2000 as a post-doctoral fellow after which he was appointed lecturer in the Department of Geography at the University of the Free State in 2002 and where he became a Professor in 2009. In 2015, he was appointed Professor of Geography at Stellenbosch University. His main research interest concerns identity-based consumption and urban morphological change, best expressed in his work on the tourism and development nexus in urban areas. This interest has been published in over 130 scholarly journal articles, book chapters, and books.

Aaron Yankholmes is Associate Professor in the Department of Marketing, Nottingham Business School (NBS) Nottingham Trent University in the UK, having worked previously for the University of Huddersfield (UK), University of South Wales (UK), Breda University of Applied Sciences (The Netherlands), Bournemouth University (UK), Institute for Tourism Studies (Macau SAR, China), and Macau University of Science and Technology (Macau, Macau SAR, China). Aaron holds a PhD in Tourism Management from The Hong Kong Polytechnic University. One of his primary research interests centres around cultural heritage tourism and its management and marketing.

1

AFRICA'S HERITAGE AND TOURISM

Dallen J. Timothy

Introduction

Heritage tourism is one of the fastest-growing sectors of the tourism industry, and it is far more encompassing in its reach than many observers acknowledge. The term 'heritage' derives from the Late Latin term *hereditare* and the Old French word *eritage*, both indicating an inheritance from prior generations. A common working definition of heritage is whatever we inherit from the past, utilize and value in the present, and hope to pass on to future generations (Graham et al., 2000; Timothy, 2021a). Thus, the concept of heritage is multidimensional, complex, and exceptionally inclusive. Tourism is a key force in the heritagization of the past. It is one of the primary users of the past (together with education and science, government, commerce, and others), it draws upon heritage values and projects certain values upon heritage, and it is seen by many as a tool for protecting heritage for today and for the future (Timothy & Boyd, 2003). Heritage tourism, therefore, indicates the use of the past as a tourism asset. Although heritage may be broadly interpreted as being cultural and natural, the focus of this book is cultural heritage.

This introductory chapter sets the tone for the book by examining the meanings and practices of heritage and heritage tourism, and how it manifests generally and more specifically in Africa. The chapter first discusses the natural and cultural environmental setting(s) of the continent and how various types of heritage derived from the uniqueness of Africa and the manifestation of heritage there. Following this discussion, the chapter examines broad trends in tourism throughout the region and then hones in on cultural heritage-based tourism in Africa and outlines the main contributions of each chapter of the book.

DOI: 10.4324/9781003153955-1

Understanding Cultural Heritage-Based Tourism

Cultural heritage encompasses the tangible and intangible record of human activity on the earth. It may be of ancient origins or have recent roots. It may be living and vibrant, or it may be frozen in time and part of the archaeological record. Heritage may be extraordinary and lavish or rather ordinary and mundane, and its scale of recognition (i.e., touristic appeal) may be global or very local and personal (Timothy, 1997, 2018). Although many observers and scholars erroneously view heritage tourism as a form of use that focuses on tangible elements of the distant past, it is far more inclusive than that. Living cultures, faith practices, languages, artforms, music, culinary traditions and recipes, folklore and stories, family traditions, and migrant narratives are all part of the intangible heritage, whereas historic buildings, archaeological sites, battlefields and sites of disasters, industrial remains, farms, villages, agricultural landscapes, museum pieces, stadiums, theaters, and railways are a few examples of tangible heritage that appeal to tourists.

People choose to experience cultural heritage for a variety of reasons, and it frequently provides the primary motive of a journey (serious cultural tourists), or it may function as a secondary attraction once in a destination for other purposes (casual cultural tourists) (Stebbins, 1996; Timothy, 2021a). McKercher (2002) suggests that cultural tourists' motivations and engagement with heritage may be seen on a continuum of intense engagement (purposeful), to less engagement (casual and incidental), and engaging only when an individual happens upon a heritage attraction (serendipitous cultural tourist). Serious or purposeful heritage tourists often plan their journeys with the intention of visiting a historical locale, with the heritage becoming the main motive of the journey. Where heritage is an 'anchor attraction' of the destination, it usually becomes an iconic symbol of the place, often used as a logo for tourism marketing and the main draw to a destination. In these cases, global-scale heritage is a major attraction, whereas local or regional heritage does not attract the same level of attention (Adie & Hall, 2017; Lopez-Guzman & Gonzalez Santa-Cruz, 2016; Škrabić Perić et al., 2021; Timothy, 1997).

Research shows that people visit heritage sites for many different reasons (Chhabra, 2010; Hall & Zeppel, 1990; Jewell & Crotts, 2002; Poria et al., 2003; Prayag & Del Chiappa, in press). These include educational purposes or life enrichment, such as learning about different cultures and stretching imaginations; immersing oneself in something different from the home environment; leisure pursuits; participating in formal education programs (field trips); satisfying curiosity; developing bragging rights; or exploring personal connections to a place.

Given the wide range of assets utilized in heritage tourism and the multitudes of motivations for visiting, heritage tourism clearly reaches beyond old buildings and historic sites. In fact, it is so pervasive that it includes all, or certain elements of, other types of tourism, blurring the lines between categories of tourism—typologies that are often devised by scholars and industry practitioners as a means of making tourism easier to study and market. There are many types of tourism, but dark tourism, archaeotourism, agritourism, pilgrimage and religious tourism, culinary tourism, sport tourism, and ethnic and indigenous tourism are several categories that have very strong heritage underpinnings and in fact are salient manifestations of heritage tourism (Timothy, 2020). Even volunteer tourism, shopping tourism, and sun-sea-sand (SSS) tourism have strong heritage elements when volunteers assist in archaeological digs or national parks, retail tourists shop in historic marketplaces or buy traditional artworks, and beach resort-based tourism (SSS) through time becomes part of the heritagescape of coastal areas, such as in many Caribbean and Mediterranean island destinations. In fact, tourism itself and the landscapes it creates may become part of a destination's heritage milieu, as Weaver (2011) noted in the case of Las Vegas, USA, and Gold Coast, Australia. Heritage may derive from indigenous pasts, colonial pasts in settler societies, or from a mix of these. Every place has heritage, or latent heritage, because every place has a past that can be utilized for present purposes. Thus, heritage as a tourism asset is widespread and complex, and the memories, identities, and values associated with it manifest in many different ways.

Africa and Tourism

The second largest continent on Earth, Africa is home to 54 sovereign states (independent countries) and several non-state territories and disputed regions (Tables 1.1 and 1.2). Africa's largest country demographically is Nigeria with nearly 220 million residents, with the smallest country in area and population being the Seychelles with 96,387 inhabitants and comprising only 550 km^2 in physical territory. Algeria is physically the largest country on the continent at over 2.3 million km^2.

Regional geographers and other social scientists often divide Africa into two major realms: Sub-Saharan Africa and North Africa, with North Africa frequently being grouped with the Middle East into the region known as the Middle East and North Africa (MENA), owing to both areas' common natural and cultural characteristics (Timothy et al., 2022). Despite North Africa's demographic and physiographic differences from Sub-Saharan Africa, the decision was taken to focus on all of Africa in this volume.

The physical geography of Africa is comprised of several important features. These include major deserts, the most famous of which are the Sahara, the Kalahari, and the Namib. The Sahara is the world's largest hot desert, covering approximately 9 million km^2, or approximately one-quarter of the continent's total area (National Geographic, 2022). Just south of the Sahara Desert is the Sahel—a semi-arid transition zone where the desert meets the grasslands and savannas, and the Berber and Arab cultures of the north meet the Black African cultures of the south. Overgrazing and severe droughts have caused the Sahelian region to become drier and more inhospitable to pastoralism and agriculture through the process of desertification, or the southward spread of the Sahara into the grasslands, which is exacerbated by global climate change (Timothy et al., 2022).

TABLE 1.1 The sovereign states of Africa and their basic characteristics

Sovereign state/ country	*Capital city*	*Area/size (sq. km)*	*Population (2020 est.)*	*GDP per capita (2019 est., USD $)*	*Major ethnic group(s)*
Algeria	Algiers	2,381,740	43,576,691	11,511	Arab-Berber
Angola	Luanda	1,246,700	33,642,646	6,670	Ovimbundu, Kimbundu, Bakongo
Benin	Porto-Novo	112,622	13,301,694	3,287	Fon, Adja, Yoruba
Botswana	Gaborone	581,730	2,350,667	17,767	Tswana, Kalanga, Basarwa
Burkina Faso	Ouagadougou	274,200	21,382,659	2,178	Mossi, Fulani, Gurma
Burundi	Gitega	2,178	12,241,065	752	Hutu, Tutsi, Twa
Cabo Verde	Praia	4,033	589,451	7,172	Creole/Mestiços, Mandyako, Fulani
Cameroon	Yaoundé	475,440	28,524,175	3,642	Bamileke-Bamu, Beti/Bassa, Biu-Mandara
Central African Republic	Bangui	622,984	5,357,984	945	Baya, Banda, Mandjia
Chad	N'Djamena	1,284,000	17,414,108	1,580	Sara, Kanembu, Arab
Comoros	Moroni	2,235	864,335	3,060	Antalote, Cafre, Makoa
Congo, Democratic Republic	Kinshasa	2,344,858	105,044,646	1,098	Monga, Luba, Kongo
Congo, Republic	Brazzaville	342,000	5,417,414	3,673	Kongo, Teke, Mbochi
Cote d'Ivoire	Yamoussoukro	322,463	28,088,455	5,213	Akan, Voltaique, Mande
Djibouti	Djibouti	23,200	938,413	5,535	Somali, Afar, Arabs
Egypt	Cairo	1,001,450	106,437,241	11,763	Egyptian, Nubian
Equatorial Guinea	Malabo	28,051	857,008	18,558	Fang, Bubi, Mdowe
Eritrea	Asmara	117,600	6,147,398	1,600	Tigrinya, Tigre, Saho
Eswatini	Mbabane	17,364	1,113,276	8,622	Swazi, Zulu, Tsonga
Ethiopia	Addis Ababa	1,104,300	110,871,031	2,221	Oromo, Amhara, Tigray
Gabon	Libreville	267,667	2,284,912	14,950	Fang, Shira-Punu, Nzabi-Duma

(*Continued*)

TABLE 1.1 (Continued)

Sovereign state/ country	*Capital city*	*Area/size (sq. km)*	*Population (2020 est.)*	*GDP per capita (2019 est., USD $)*	*Major ethnic group(s)*
Gambia, The	Banjul	11,300	2,221,301	2,223	Mandinka, Fulani, Wolof
Ghana	Accra	238,533	32,372,889	5,413	Akan, Mole-Dagbon, Ewe
Guinea	Conakry	245,857	12,877,894	2,562	Fulani, Malinke, Susu
Guinea-Bissau	Bissau	36,125	1,976,187	1,989	Fulani, Balanta, Mandinga
Kenya	Nairobi	580,367	54,685,051	4,330	Kikuyu, Luhya, Kalenjin
Lesotho	Maseru	30,355	2,177,740	2,704	Sotho
Liberia	Monrovia	111,369	5,214,030	1,428	Kpelle, Bassa, Grebo
Libya	Tripoli	1,759,540	7,017,224	15,174	Arab-Berber
Madagascar	Antananarivo	587,041	27,534,354	1,647	Merina, Betsimisaraka, Betsileo
Malawi	Lilongwe	118,484	20,308,502	1,060	Chewa, Lomwe, Yao
Mali	Bamako	1,240,192	20,137,527	2,322	Bambara, Fulani, Sarakole
Mauritania	Nouakchott	1,030,700	4,079,284	5,197	Haratin, Arab-Berber, Halpulaar
Mauritius	Port Louis	2,040	1,386,129	22,870	Indo-Mauritians, Creole, Chinese
Morocco	Rabat	716,550	36,561,813	7,515	Arab-Berber
Mozambique	Maputo	799,380	30,888,034	1,281	Makhuwa, Tsonga, Lomwe
Namibia	Windhoek	824,292	2,678,191	9,637	Ovambo, Kavangos, Herero
Niger	Niamey	1,267,000	23,605,767	1,225	Hausa, Djerma, Tuareg
Nigeria	Abuja	923,768	219,463,862	5,136	Hausa, Yoruba, Igbo
Rwanda	Kigali	26,338	12,943,132	2,227	Hutu, Tutsi, Twa
São Tomé and Príncipe	São Tomé	964	213,948	3,970	Creole/Mestiços, Angolares, Forros
Senegal	Dakar	196,722	16,082,442	3,395	Wolof, Fulani, Serer
Seychelles	Victoria	455	96,387	29,223	Creole, British, French
Sierra Leone	Freetown	71,740	6,807,277	1,718	Temne, Mende, Limba

(*Continued*)

TABLE 1.1 (Continued)

Sovereign state/ country	*Capital city*	*Area/size (sq. km)*	*Population (2020 est.)*	*GDP per capita (2019 est., USD $)*	*Major ethnic group(s)*
Somalia	Mogadishu	637,657	12,094,640	888	Somali, Bantu, Bajuni
South Africa	Cape Town & Pretoria	1,219,090	56,978,635	12,482	Black, Coloured, White
South Sudan	Juba	644,329	10,984,074	1,600	Dinka, Nuer, Bari
Sudan	Khartoum	1,861,484	46,751,152	3,958	Sudanese Arabs, Fur, Beja
Tanzania	Dodoma	947,300	62,092,761	2,660	Sukuma, Nyamwezi, Chaga
Togo	Lomé	56,785	8,283,189	1,597	Ewe, Kabye, Tem
Tunisia	Tunis	163,610	11,811,335	10,756	Arab-Berger
Uganda	Kampala	241,038	44,712,143	2,187	Baganda, Banyankole, Basoga
Zambia	Lusaka	752,618	19,077,816	3,470	Bemba, Tonga, Chewa
Zimbabwe	Harare	390,757	14,829,988	1,371	Shona, Ndebele, Venda

Source: Central Intelligence Agency (2022).

TABLE 1.2 Non-sovereign state territories in Africa

Non-state territory	*Sovereign state*	*Area*	*Population (2019, 2020 est.)*	*Explanation*
Canary Islands	Spain	7,493 km^2	2,207,000	An integral part of Spain, it is an autonomous community of Spain. It lies 100 km west of the Moroccan Atlantic coast. The Canaries are part of the EU and Schengen Area, but they are outside of Europe's VAT area, making some prices lower than in mainland Europe.

(*Continued*)

TABLE 1.2 (Continued)

Non-state territory	*Sovereign state*	*Area*	*Population (2019, 2020 est.)*	*Explanation*
Ceuta	Spain	18.5 km²	85,144	Ceuta is part of Spain's remnant territories on the North African coast. It is attached to Morocco by an 8 km land border but is an integral part of Spain and is classified as a Spanish 'autonomous city'. Its sovereignty is disputed by Morocco.
Islas Alhucemas	Spain	4.6 ha²	25–30 military personnel	Three islands—Peñón de Alhucemas, Isla de Tierra, and Isla de Mar—are part of Spain's remnant territories on the North African coast. They lie only 300 m from the Moroccan coast and their sovereignty is disputed by Morocco.
Islas Chafarinas	Spain	52.5 ha²	30–50, mostly military personnel	Three islands—Isla del Congreso, Isla Isabel II, and Isla del Rey—are part of Spain's remnant territories on the North African coast. They lie 3.3 km from the Moroccan coast and their sovereignty is disputed by Morocco.
Madeira	Portugal	801 km²	253,945	This Portuguese archipelago is 520 km from the Atlantic coast of Africa, but it is part of the African Plate. An integral part of Portugal, it is an autonomous region and belongs to the EU, the Schengen Area, and the European VAT Area.

(Continued)

TABLE 1.2 (Continued)

Non-state territory	*Sovereign state*	*Area*	*Population (2019, 2020 est.)*	*Explanation*
Mayotte	France	374 km^2	270,372	An integral part of France, this French overseas department and region is geographically one of the Comoros islands off the east coast of Africa. When the Comoros voted for independence in 1974, Mayotte voted to remain in France. Its sovereignty is disputed by the Comoros.
Melilla	Spain	12.3 km^2	84,689	Melilla is part of Spain's remnant territories on the North African coast. It is attached to Morocco by a 10.5 km land border but is an integral part of Spain and is classified as a Spanish 'autonomous city'. Its sovereignty is disputed by Morocco.
Pelagie Islands	Italy	25.5 km^2	6,556	Three islands: Lampedusa, Linosa & Lampione. Located in the Mediterranean Sea on the Tunisian continental shelf. Part of the Agrigento province of Italy.
Peñón de Vélez de la Gomera	Spain	1.9 ha^2	Unknown/ military secret	Part of Spain's remnant territories on the North African coast. It is a small peninsula connected to Morocco by a narrow border and is home to Spanish military personnel. Its sovereignty is disputed by Morocco.

(*Continued*)

TABLE 1.2 (Continued)

Non-state territory	*Sovereign state*	*Area*	*Population (2019, 2020 est.)*	*Explanation*
Reunion	France	2,512 km^2	858,450	Reunion is an overseas department and region of France, 550 km east of Madagascar. It is an integral part of France and is classified as an EU outermost region.
Sahrawi Arab Democratic Republic	Disputed territory, majority controlled by Morocco	266,000 km^2 (53,200 km^2 under control of the SADR)	567,402	SADR is a de facto state with only partial recognition. It claims all of Western Sahara as its national territory but controls only 20% of the area. Morocco occupies 80% of the territory and disputes any Sahrawi claims to sovereignty over the region. Some 30,000 people live in the SADR-controlled areas, and 173,600 Sahrawi refugees live in camps in Algeria.
Socotra	Yemen	3,796 km^2	44,670	Socotra is a governorate of Yemen and lies some 232 km off the Horn of Africa. It consists of seven islands and is under the military control of the UAE and Saudi Arabia during the Yemeni Civil War.
Somaliland	Somalia	176,120 km^2	3,500,000	Although part of Somalia's sovereign territory, Somaliland declared independence in 1991 and functions as a de facto state outside the effective control of Somalia. Its independence remains unrecognized by the community of sovereign states and the UN.

Source: Compiled by the author from multiple sources.

Highland areas and mountain ranges are also found in limited areas of Africa. In the far north, the Atlas Mountains stretch from southwestern Morocco across northern Algeria and into Tunisia at heights exceeding 4,000 m, providing water to enable agriculture to prosper, as well as substantial snowfall, creating opportunities for skiing at resorts in Algeria and Morocco (Timothy, 2019b). The Ethiopian Highlands (1,500–4,550 m) is Africa's largest cluster of mountains, which formed as a result of the tectonic activity of the Great Rift Valley, where the Arabian Plate and the African Plate (Somali and Nubian plates) are separating in the East African Rift (EAR). As part of the same rift system, a number of volcanoes formed, the most famous of these being Mt Kilimanjaro (5,895 m), Mt Kenya (5,199 m), and other highland areas of Tanzania. The Drakensberg Mountains in southern Africa (up to 3,482 m), provide a dramatic landscape in South Africa and Lesotho (Adams et al., 1999).

As elsewhere, water resources are a critical part of subsistence livelihoods and tourism in Africa. Oases (desert springs or wells) in the Sahara provide sustenance for much of North Africa's population and have supported the development of villages, towns, and small cities throughout history, particularly along historic trade routes (Mattingly & Sterry, 2013). Another prominent water feature is the African Great Lakes of the EAR. The dividing effect of EAR plate tectonics formed cracks and basins that have filled with water over many millennia, creating dozens of lakes that have sustained life in eastern Africa for thousands of years. The largest and best known of these are Lake Victoria (Kenya, Tanzania, Uganda), Lake Tanganyika (Burundi, DR Congo, Tanzania, Zambia), and Lake Malawi (Malawi, Mozambique, Tanzania)—three of the largest fresh-water lakes in the world (Adams et al., 1999; Odada et al., 2003).

Rivers have also played a crucial role in Africans' survival and mobility, as well as the European colonization of the continent. The Nile River, the world's longest river (approximately 6,650 km), is the lifeblood for millions of people in the Sahel and Sahara in northeastern Africa. The Nile is formed from three main tributaries: the White Nile (originating at Lake Victoria in Uganda), the Blue Nile (originating in the Ethiopian Highlands), and the Red Nile (Atbara) (originating in the Ethiopian Highlands) (Said, 2013). Other important rivers include the Congo River (4,700 km in DR Congo, Congo, and Angola), Niger River (4,184 km in Guinea, Mali, Niger, Benin, and Nigeria), Senegal River (1,086 km in Senegal, Mauritania, Mali, and Guinea), Volta River (1,500 km in Ghana, Côte d'Ivoire, and Burkina Faso), Limpopo River (1,750 km in South Africa, Botswana, Zimbabwe, and Mozambique), Orange River (2,432 km in Lesotho, South Africa, and Namibia), and the Zambezi River (2,574 km in Zambia, Angola, Namibia, Botswana, Zimbabwe, and Mozambique) (Adams et al., 1999; Hoag, 2013).

Much of the continent is comprised of grasslands, or savannas, totaling more than 13 million km^2 from the Sahel to the southernmost tip of the continent (National Geographic, 2022). Historically, rainforests covered much of Africa, particularly West and Central Africa around the Gulf of Guinea and in the Congo Basin. Today, however, relatively little rainforest remains owing to human activities, such as deforestation, farming, and urban development (Timothy et al., 2022). Africa's rainforests are home to thousands of unique and endemic animal and plant species, and the savanna lands are home to most of the best-known African fauna, including elephants, giraffes, zebras, lions, wildebeests, rhinos, hyenas, and leopards. These and other endemic animal species are one of the most significant tourist attractions in eastern and southern Africa, and their management has led to the establishment of many game preserves that provide safari sightseeing and trophy hunting opportunities (Lindsey et al., 2007; Munthali, 2007; Muntanga et al., 2017; Norton, 1996).

Although this is not a complete depiction of all physiographic regions, topographic features, or fluvial and biotic systems of Africa, the above discussion does indicate the diversity of the continent's ecosystems and physical features, which have determined how people have lived and survived for millennia and how cultures have evolved. The physical geography of Africa has both limited contact between people and enabled it, and it is a significant determiner of human behavior in pre-colonial, colonial, and post-colonial times. Africa's natural landscapes and physiography have determined its people's cultural traditions, built environments, human evolutions, and more recently the patterns and growth of tourism.

Humans and their hominid ancestors have lived in Africa for millions of years, as evidenced by the fossilized remains of Lucy, a 3.18-million-year-old *Australopithecus* human ancestor discovered in Ethiopia in 1974, and the Taung Child (2.8-million-year-old *Australopithecus africanus*), which was discovered in South Africa in 1924 (Finneran, 2013; Naidu, 2008). As ideas emerged, technology evolved, mobilities increased, and humans adapted to local environments and incursions by other humans, communities, organized societies, and empires began to emerge. The Egyptian kingdoms and empires (ca. 3100 BC–330 BC); the Kingdom of Kush (ca. 1069 BC–350 AD) in today's Sudan; the Kingdom of Aksum (1st–10th centuries AD) in today's Ethiopia, Eritrea, Djibouti and Sudan; the Mali Empire (13th–16th centuries AD) with Timbuktu being a major remnant of that state; and the Zimbabwe Empire (13th–15th centuries AD) in todays' Zimbabwe, Botswana, and Mozambique are a few of the best documented ancient civilizations (Asante, 2019; Hall & Stefoff, 2006; O'Connor & Reid, 2003). These ancient civilizations and later empires, including the Kingdom of Benin, which lasted until 1897, and the Ashanti Empire (1701–1901), all testify of the ingenuity and sophisticated social order of various African societies in Ancient, Post-classical, and Modern times.

Africa today is home to diverse cultures, races, faiths, and ethno-linguistic groups. The religious map of Africa is fairly straightforward, with the majority of the population of North Africa and large parts of West and East Africa adhering to Islam, although every country in the realm has a notable Muslim presence (Levtzion & Pouwels, 2000). A slightly higher number of Africans (approximately 45%) belong to various Christian denominations, some of which originated in Africa. Most large Christian sects are represented on the continent, with Angola, Rwanda, the Seychelles, Zambia, and São Tomé and Príncipe reporting the highest percentages of their populations adhering to Christianity (Bongmba, 2015; Central Intelligence Agency, 2022). Traditional African religion (animism and ancestor worship) dominates much of Sub-Saharan Africa and is frequently blended with Christian and Muslim beliefs into unique manifestations that can only be found in this part of the world. Vodun (voodoo), a uniquely African set of beliefs, is widely practiced in West Africa and was spread to the Caribbean and other parts of the New World by enslaved people centuries ago and has become an important element of living heritage inside and outside of Africa (Dorsey, 2005; Goldberg, 1983; Hernandez-Ramdwar, 2013).

Other religions are prominent in certain areas. For example, Mauritius and South Africa are home to large Hindu populations. Ethiopia has strong Jewish roots with deep connections to the Holy Land, and small Jewish minorities can be found scattered across North Africa (especially Tunisia and Morocco), West Africa (e.g., Ghana, Sierra Leone, Nigeria, Cameroon, and Gabon), Southern Africa, and East Africa (e.g., Kenya and Uganda) (Miles, 2019; Timothy, 2021b). Some regions are also home to small minorities of Buddhists, Sikhs, Jains, and Zoroastrians. The religious diversity of Africa has produced a number of important religious pilgrimages, spiritual destinations, faith-based events, and many other manifestations of religious culture that draw believers as pilgrims and curiosity-seekers to see and experience sacrosanct sites and events.

Having between 1,000 and 2,000 languages and dialects, Africa is home to around one-third of the world's languages (Harvard University, 2022). Most of the region's languages belong to one of five language families: Niger-Congo, Nilo-Saharan, Afroasiatic, Khoisan, and Austronesian. The most widely spoken languages of the five phyla groups are as follows:

- Niger-Congo: Swahili, Yoruba, Igbo, Fula
- Nilo-Saharan: Kanuri, Nubian, Dinka, Luo, Teso
- Afroasiatic: Arabic, Berber, Amharic, Hausa
- Khoisan: Sandawe, Taa, Khoekhoe, !Kung
- Austronesian: Malagasy

Africa's history of colonialism also means that several European languages pervade the business and education environment and are a means of communicating in multi-ethnic and multi-linguistic states. The most common are French, English, Portuguese, and Spanish. The colonial map of Africa parallels the use of these lingua francas in countries such as Senegal and Burkina Faso (French), Ghana and Kenya (English), Cabo Verde and Mozambique (Portuguese), and Equatorial Guinea (Spanish) (Heine & Nurse, 2000).

The number of ethno-linguistic groups in Africa is diverse and immense. Ethnic identity is closely aligned with the language families noted above. The Afroasiatic group includes large numbers of Arab, Berber, Tigrayan, Amharan, Somali, and Hausa people in North Africa and the Horn of Africa (Heine & Nurse, 2000; Timothy, 2019a). The Austronesian group is comprised of Madagascar's Malagasy population. The Khoisan people are concentrated mainly in southern Africa and include the !Xóõ, Kung, Sandawe, Nama, and San populations. In the Nilo-Saharan group, the most populous are the Maasai, Dinka, Kanuri, Luo, and Nuer in the Sahara, Sahel, and East Africa. Some of the largest of the Niger-Congo ethnic groups are the Shina, Zulu, Yoruba, Kongo, Igbo, Fula, and Akan in the western, southern, and eastern parts of the continent (Shoup, 2011).

The colonial history of Africa is both unique and typical of European overseas exploration and colonization. Although European settlements had existed in North Africa for centuries, the main thrust of European colonization and imperialism started with Portuguese and Spanish settlements in areas such as today's Cape Town (South Africa), Cabo Verde, São Tomé and Príncipe, the Canary Islands, and the northern coastal cities of Melilla and Ceuta. Many of these initial 14th- and 15th-century settlements were previously uninhabited and became fortressed centers of trade for the Spanish and Portuguese (Martin-Márquez, 2008; Minter, 1972).

Throughout the 18th and early 19th centuries, the largest European empires—Great Britain, France, Portugal, and Spain—had established coastal outliers in Africa, largely for transporting slaves to the New World, but few had penetrated the continent's interior. Between 1880 and 1914, the colonization of most of Africa accelerated in the 'Scramble for Africa' or the 'Partition of Africa', which was negotiated between Europe's most powerful states and entailed the occupation of approximately 90% of the continent except for modern-day Liberia, Ethiopia, and part of Somalia—the latter two eventually being occupied by Italy in the 1930s (Adu Boahen, 1987).

Portugal's African holdings were geographically dispersed, with Mozambique in southeastern Africa, Angola in the southwest, São Tomé and Príncipe in Central Africa, and Cape Verde and Guinea Bissau in West Africa. Spain's possessions were concentrated in central and northwestern Africa, including the Canary Islands, parts of Morocco, Western Sahara, and Equatorial Guinea. The British and French held the largest territories on the continent, with the British controlling much of western, southern, and eastern Africa, and Egypt, as well as parts of Libya following the Second World War. France's African colonies were widespread but were particularly concentrated in North, West, and Central Africa, as well as tiny Djibouti on the Horn of Africa and almost all of the African islands in the Indian Ocean (Adu Boahen, 1987).

In addition to the largest metropolitan empires, Germany, Belgium, and Italy claimed territories in the late 19th and early 20th centuries (Ben-Ghiat & Fuller, 2005; Conrad, 2012; Davidson, 1994). Germany's African possessions included German East Africa (today Burundi, Rwanda, Tanzania, and parts of Mozambique), German South West Africa (today Namibia), and German West Africa (today parts of Cameroon, Togo, Chad, Central African Republic, Congo, and Gabon). Belgium's main colony was Congo Free State and Belgian Congo (today's DR Congo), and it received Ruanda-Urundi (today Rwanda and Burundi) from Germany after the First World War. Italy's colonial holdings from the 1890s until its defeat in WWII included Libya, Eritrea, Somalia, and eventually Ethiopia briefly from 1936 until 1947 (Ben-Ghiat & Fuller, 2005).

The primary motives for the European colonization of Africa were the extraction of natural resources (including gold, agricultural products, and slaves), competition with other metropoles for world dominance, to explore unknown areas, and to spread Christianity. Lingua francas, transportation infrastructure, educational establishments, governance models, urban design, architectural styles, and heritage protective legislation are all leftovers of the colonial era (Demissie, 2012; Fredholm, 2016; Timothy & Boyd, 2003). Nonetheless, much of the colonial rule in Africa was brutal toward the native populations and extremely exploitative socially and materially. Many observers suggest that historical colonialism continues to underlie much of Africa's underdevelopment and many of its countries' inability to elevate beyond colonial dependency and political instability (Alemazung, 2010; Langan, 2018; Táíwò, 2010). In the scramble for Africa, European metropoles drew largely arbitrary state boundaries, often dividing people who had theretofore not been divided, while simultaneously conjoining together into states different tribes and socio-ethnic groups and imposing upon them a nationality they knew nothing about, disregarding the tribal and communal identities of most African societies (Cusack, 1999; Fouéré & Hughes, 2015; Lefebvre, 2011). These geopolitical impositions laid the foundation for conflict, ethnic discord, and national identity crises that have yet to be resolved and have raised challenges to Africa's heritage milieu. It has also raised difficult questions about the importance of heritage and which heritage (or whose heritage) deserves to be commemorated, preserved, and marketed for tourism (Adie, 2021; Baillie & Sørensen, 2021; Bolaane, 2014; Boswell & O'Kane, 2011; De Jong & Rowlands, 2016; Mariz, 2016; Ndoro et al., 2018). Colonial heritage and heritage tourism are frequently used in Africa as mechanisms to build solidarity and a common national memory where the notions of 'nationality' and statehood did not previously exist (Cusack, 1999; Fouéré & Hughes, 2015; Keitumetse, 2016; Makuvaza, 2014; Ndoro et al., 2018; Okech, 2008; Peterson et al., 2015; Schmidt, 2003; Schmidt & Pikirayi, 2016).

Africa is rich in natural resources, and its people have survived for millennia on hunting and gathering, farming, and fishing. Some people still practice hunting and gathering, but subsistence agriculture is now the continent's most widespread economic activity. Mining, oil drilling, and manufacturing dominate certain parts of the continent, and in many countries, remittances from emigres who live and work abroad are a significant economic engine. The tertiary and IR 4.0 economic sectors, including tourism, other services, and technology, are becoming increasingly prominent throughout Africa. This is especially notable with the growth of special interest forms of tourism and the desire of many modern travelers to avoid tourist hotspots, to stay away from the well-trodden tourist paths, and the growing influence of the 'slow tourism' movement wherein people seek deeper and more authentic experiences in unspoiled destinations (Fullagar et al., 2012; Lessmeister & Scherle, 2008; Ramkissoon, 2015).

Economically, Africa is home to some of the world's poorest countries, although industrial projections seem to suggest that more countries in the realm will achieve 'middle income' status by 2030 (de la Torre & Rigolini, 2020). Between 2013 and 2019, Africa was on a steep upward development trajectory with a growth rate of 5.6% per annum, and the African Development Bank indicated that the region was one of the fastest-growing economies until its growth was severely curtailed by the COVID-19 pandemic, which affected Africa disproportionately (Anyanwu & Salami, 2021; Bwire et al., 2022; Tichaawa & Rogerson, 2022). In 2021, Africa rebounded strongly with a real economic growth rate of 6.9% (African Development Bank, 2022). In 2021, economic growth was highest in North Africa (11.7%) and East Africa (4.8%). Despite notable improvements, the continent's economies continue to reel from the ongoing pandemic, corruption and political instability, and climate change.

These geographical and sociodemographic characteristics illustrate a part of the world whose ecosystems, physical geography, and cultural geography are extremely diverse, rich, and as yet relatively untapped by tourism (Bolaky, 2016; Dieke, 2000; Novelli, 2016; Novelli & Adu-Ampong, 2021). The region's climates, topographies, and natural resources, as well as its ethno-linguistic composition, colonial history, and economic development, have determined how Africans live, worship, earn livelihoods, celebrate, eat, and play. These conditions have resulted in a rich cultural heritage that has much more potential for tourism than what is currently being utilized (Akama & Sterry, 2002; Boswell, 2011; Dhaher et al., 2021; Drummond et al., 2021; Gravari-Barbas & Jacquot, 2014; Gumede & Ezeuduji, 2021; Olukole, 2014; Saarinen et al., 2014; Saarinen & Rogerson, 2015; Seyfi & Hall, 2021; Shackley, 1997; Shano, 2014; Spenceley & Nelson, 2013).

In addition to Africa's sovereign states, most of the realm's non-state territories also have thriving tourism sectors or are interested in pursuing the industry for socioeconomic development. Mass tourism thrives in the Canary Islands and Madeira. Smaller scale but important tourism thrives in Ceuta, Mayotte, Melilla, and Reunion. On Yemen's Socotra Island, a budding tourism sector is currently operated by the Emiratis because of Yemen's ongoing civil war, and Somaliland, a breakaway 'de-facto state' or proto-state, which functions independently of Somalia, has worked hard to develop its tourism sector, although its lack of recognition as an independent country by the global community raises many challenges (Jeffrey, 2015). Even the Sahrawi Arab Democratic Republic hosts a limited number of tourists (students, solidarity and social tourists, artists, and journalists), largely through its government in exile in the refugee camps of southwestern Algeria, but also in the parts of Western Sahara it still controls (Jablonska et al., 2016).

Although Africa is the second largest continent and one of the world's great physical and anthropic realms, it is oxymoronically one of the least visited regions of the world. Nonetheless, Africa is extremely varied in its tourism appeal and levels of development. Some African states have thriving tourism industries, such as South Africa, Botswana, Lesotho, Kenya, Tanzania, Seychelles, Mauritius, The Gambia, Morocco, Tunisia, and Egypt, whereas others struggle to attract visitors or maintain even rudimentary levels of international visitation (e.g., Burundi, Chad, Djibouti, Equatorial Guinea, Niger, and Sierra Leone). The most visited countries in Africa in terms of international arrivals are South Africa, Egypt, Morocco, and Tunisia, with major secondary destinations including Ghana, Eswatini, Kenya, Algeria, Mauritius, Côte d'Ivoire, Namibia, Uganda, Tanzania, Zimbabwe, and

Zambia (Table 1.3). Obviously, the proximity of North Africa to its European market base and the proliferation of low-cost air carriers have created more opportunities for Morocco, Tunisia, and Egypt, whereas high travel costs, long distances from market origins, relative inaccess, poverty, and a lack of adequate infrastructure within many Sub-Saharan states have restricted tourism growth, compared to the north. Tourism is an important export economic sector in many African countries, even those whose arrival numbers are low by global standards. Table 1.4 shows the top 20 countries whose economies are most dependent on tourism per percentage of the total share of exports.

TABLE 1.3 International tourist arrivals by country in 2019 and 2020

Country	*International Arrivals (2019)*	*International Arrivals (2020)*
Algeria	2,371,000	591,000
Angola	218,000	64,000
Benin	337,000	354,000
Botswana	n/a	n/a
Burkina Faso	143,000	67,000
Burundi	n/a	n/a
Cabo Verde	758,000	180,000
Cameroon	1,021,000	n/a
Central African Republic	87,000	n/a
Chad	81,000	10,400
Comoros	45,100	7,000
Congo, Democratic Republic	n/a	n/a
Congo, Republic	n/a	n/a
Cote d'Ivoire	2,070,000	668,000
Djibouti	n/a	n/a
Egypt	13,026,000	3,677,000
Equatorial Guinea	n/a	n/a
Eritrea	n/a	n/a
Eswatini	1,226,000	345,000
Ethiopia	812,000	518,000
Gabon	n/a	n/a
The Gambia	620,000	246,000
Ghana	1,130,307	355,108
Guinea	n/a	n/a
Guinea-Bissau	n/a	n/a
Kenya	2,049,000	568,000
Lesotho	1,142,000	n/a
Liberia	n/a	n/a
Libya	n/a	n/a
Madagascar	486,000	87,000
Malawi	n/a	n/a
Mali	217,000	75,000
Mauritania	n/a	n/a
Mauritius	1,418,000	316,000
Morocco	13,109,000	2,802,000

(Continued)

TABLE 1.3 (Continued)

Country	*International Arrivals (2019)*	*International Arrivals (2020)*
Mozambique	2,033	959,000
Namibia	1,651,000	187,000
Niger	192,000	85,000
Nigeria	n/a	n/a
Rwanda	n/a	n/a
São Tomé and Príncipe	34,900	10,700
Senegal	n/a	n/a
Seychelles	428,000	125,000
Sierra Leone	71,000	26,000
Somalia	n/a	n/a
South Africa	14,797,000	3,887,000
South Sudan	n/a	n/a
Sudan	n/a	n/a
Tanzania	1,527,000	621,000
Togo	876,000	482,000
Tunisia	9,429,000	2,012,000
Uganda	1,543,000	473,000
Zambia	1,266,000	502,000
Zimbabwe	2,294,000	639,000

Sources: Statista (2022a); World Tourism Organization (2022).

TABLE 1.4 Tourism as total share of exports, 2019

Country	*Tourism share of total export trade (%)*
Cabo Verde	71
São Tomé and Príncipe	65
The Gambia	53
Comoros	45
Mauritius	39
Seychelles	38
Rwanda	29
Tanzania	29
Uganda	27
Madagascar	23
Sudan	22
Morocco	21
Kenya	16
Togo	16
Tunisia	14
Senegal	11
Eritrea	11
Ethiopia	10
Cameroon	10
Zambia	10

Source: Statista (2022b).

For some countries, recent arrivals data and economic statistics are unavailable, which illustrates a longstanding problem in certain parts of the region for understanding tourism's magnitude for research and marketing purposes: inconsistent and unreliable tourism statistics (Chaula, 2019; Rogerson, 2007). As Table 1.3 indicates, the COVID-19 pandemic severely inhibited international arrivals between 2019 and 2020, as countries closed their borders to arrivals or limited arrivals severely, and source countries restricted which countries their citizens could visit. As of late 2022, most countries had re-opened, and tourist arrivals have started to grow once again, although it may take many years for some African states to achieve their pre-COVID levels of tourism.

Heritage Tourism in Africa

Africa's large diversity of indigenous ethno-linguistic groups, ancient civilizations, industrial development and varied socioeconomic histories, exit of emigrants and arrival of immigrants, and its colonial past and present have created an extremely rich and varied patchwork of living cultures and traditions that continue to thrive today despite the pressures and perils of modernization, globalization, and technology. Rural environments provide a stage for nature-based tourism and ethnic and indigenous tourism, and cities largely function as tourism gateways and are home to colonial architecture and urban heritage (Demissie, 2012). National museums and ethnic/cultural villages (open-air museums) abound to protect the past and teach of pre-colonial, colonial, and post-colonial eras and in creating narratives that might help develop national identities in states that were created by outsiders during the 'scramble for Africa' for their own administrative convenience (Cusack, 1999). In both urban and rural contexts, cultural heritage-based tourism is a foundational product for most African countries, even those whose beaches are the primary tourist draw (Gijanto, 2011; Soper, 2007).

The prevalence of nomadic and hunter-gatherer societies often meant that relatively few permanent settlements were developed anciently. Thus, many of the heritage assets of Africa are intangible (Keitumetse, 2012; Manwa et al., 2016; Mustafa, 2019), or are represented by specific types of material remains such as rock art, burial tombs, ancient industrial areas, and sites of discovery of humankind's ancestors (Duval & Smith, 2013, 2014). With some notable exceptions, a number of which have become UNESCO World Heritage Sites, relatively few built environmental features remain from Prehistoric or Classical times, compared to the ancient environments in other world regions (Willems & Comer, 2011).

Although Africa has a great deal of heritage tourism potential, only a handful of countries have tapped into this lucrative area of tourism. Nevertheless, there are many manifestations of cultural heritage that form the heritagescapes of Africa. These are the focus of the chapters that follow. Each chapter examines specific types of heritage tourism, or issues surrounding heritage tourism that require further investigation.

Contributions in This Book

Following this introductory chapter, Moren Stone, Lesego Stone, Goemeone Mogomotsi, and Patricia Mogomotsi (Chapter 2) examine issues and challenges surrounding intangible cultural heritage, particularly with regard to indigenous people as purveyors of living heritage. Ownership of culture and cultural appropriation in relation to the failure of laws to protect living culture creates substantial challenges that suppress ownership and access to development capital, and result in underdeveloped infrastructure, stakeholder conflicts, and changes to living culture wrought by tourism. In Chapter 3, Kokel Melubo delves into issues specifically related to indigenous tourism as a manifestation of living heritage. Much of his discussion centers on justice issues, including native land rights and intellectual property/cultural rights, which are often ignored in indigenous tourism settings where tourism is developed from the top down by outside stakeholders. Melubo provides several examples of this and such issues as cultural appropriation, commodification, denial of indigenous access, and other challenges that indigenous people face in Africa. Despite its potential to create jobs and access to earnings, especially for women, indigenous culture-based tourism requires a delicate balancing act that will enable local people to maintain and protect their cultures while also benefitting socioeconomically through cultural tourism. These issues are especially pertinent in Sub-Saharan Africa where much of the heritage tourism product is based upon elements and manifestations of indigenous heritage. Like the two chapters before it, Chapter 4 (Ikechukwu Ezeuduji) deals primarily with issues of intangible heritage as they manifest in cultural events and festivals. These sorts of happenings have existed for centuries in Africa but only in recent decades have they become part of the mainstream tourism product of many countries and are now seen as opportunities to preserve culture, develop new heritage elements, and empower communities by sharing their cultures with outsiders. Heritage events enable local residents to showcase what is special about their home communities, which enables a more robust economy but also helps fill seasonal lulls in tourism, strengthen a sense of place, and create a more visible brand in the tourism sector nationwide.

Although Africa is not especially well known globally as a pilgrimage destination, it certainly has a longstanding religious tourism sector that dates back hundreds, or perhaps even thousands, of years. Chapter 5 (by Ermias Gedecho and Siyabulela Nyikana) outlines the role of pilgrimage and other forms of religious tourism in Africa, suggesting a wide range of opportunities and important challenges facing the region. Religious tourism/pilgrimage is often regarded as less lucrative than other forms of tourism, so investments and development efforts are often lacking compared to other tourism development efforts. Traditional African religions form the foundation of much living and built culture throughout the continent, and therefore emit much potential appeal for tourism development. Religious tourism based on imported religions, such as Islam and Christian denominations, has an important role to play in tourism, and certain sacrosanct sites associated with these faiths have become salient destinations in several countries throughout the region.

The original African sects of Christianity, including Egypt's Coptic Christians and the Ethiopian Orthodox Tewahedo Church, both of which predate European colonization, are an especially important part of the heritagescapes of those countries and have stimulated pilgrimage travel for centuries.

In Chapter 6, Gustav Visser examines some of the regional disparities throughout Africa in terms of urban development and urban heritage, with a notable lack of built heritage outside of urban environments. Capital cities and other urban agglomerations are usually bastions of colonial heritage because these were developed as administrative centers during the colonial period. Nonetheless, many of Africa's cities have been influenced by post-colonial changes and urban renewal, often through the effects of heritage tourism. In the context of rural tourism, Joseph Mbaiwa and Gladys Siphambe (Chapter 7) investigate regional disparities in the rural parts of Africa, suggesting that rural landscapes, agricultural areas, villages, events and festivals, open-air museums and cultural villages, natural areas, and handicrafts are among the most pervasive manifestations of rural heritage tourism in Africa. Rural heritage-based tourism, according to Mbaiwa and Siphambe, has a great deal of potential to help alleviate poverty and provide supplemental incomes to agrarian societies throughout the realm, even though it faces many challenges, including inaccess, poor infrastructure, and security fears.

Related to rural tourism, Brent Lovelock, Gebeyaw Degarege, and David Adeloye (Chapter 8) describe the dynamics of agri-heritage as a tourism asset, suggesting that agritourism has yet to be well developed in Africa, despite its rapid growth in other agriculturally rich rural areas of the world. Agricultural landscapes and production systems, particularly as regards indigenous agrarian knowledge and practices, are crucibles of ancient knowledge and practices that indicate human survival and dealings with nature. Thus, they have important stories to tell and have the potential to become increasingly important tourism assets the way they have in many other parts of the world. Traditional food growing systems and special products are making significant headway in parts of Africa, including coffee tourism in Ethiopia and cocoa tourism in Côte d'Ivoire. Following Lovelock, Degarege, and Adeloye's contribution, Chapter 9 (Roselyne Okech and Dallen Timothy) provides a descriptive and conceptual foundation for the growth of culinary heritage in Africa. Although African foods generally, with the exception of Ethiopian, Egyptian, and Moroccan cuisine, have not become widespread 'ethnic foods' across the globe, there is more potential for this branding and place-making through culinary heritage. This is especially true among various African diasporic communities. Native foods of Africa have health, ecological, and sociocultural benefits, and through the heritagization process, culinary traditions can play a salient role in destination growth, employment, and empowerment as the tourism marketplace continues to expand toward sustainability, slow food, deep cultural tourism, and authentic experiences.

The transatlantic slave trade is the focus of the next chapter by Aaron Yankholmes (Chapter 10). His essay tells the story of the transatlantic slave trade from the 16th to the 19th centuries and how its memorials and memorialization have become a substantial tourism asset for a handful of West African states, especially Ghana, but also to a lesser extent, Senegal and The Gambia. Despite slaves having been taken from territories that today are widespread across West and Central Africa, the development of slave heritage and the tourism sector that embraces it has been marginal for most countries of the region. Slave heritage is both tangible (memorials and slave forts) and intangible (narratives, stories, and emotional geographies), drawing people from every corner of the globe but most notably from the African diaspora to what they consider their ancestral homeland, which in most cases is represented today by Ghana. The designation of slave sites in Ghana and elsewhere as UNESCO World Heritage is wrought with controversy, for there is no consensus on the global (and political) narratives of slavery reflecting the local values and narratives of heritage and tourism in West Africa today. Slavery-based heritage tourism is personal, national, and global and exudes extremely emotive responses among different audiences.

Beyond the slavery-induced African diaspora, Emmanuel Akwasi Adu-Ampong and Ishmael Mensah (Chapter 11) look at African diaspora-based tourism more holistically to include the phenomena of return travel, roots, and genealogy tourism, visiting friends and relatives tourism (VFR), and homeland travel—all associated with many different types of diasporas, including economic, refugee, and forced migration. All of these tourism types are manifestations of personal or familial heritage with strong connections with, and a collective consciousness between, people's present place and past place, which may weaken or strengthen with successive generations. Several countries in Africa (e.g., Ethiopia, Ghana, Senegal, and Zimbabwe) have been successful in marketing to their overseas diasporas, drawing them 'home' for clan reunions and to partake of the cultural heritage of their motherlands.

Industrial heritage is increasing in popularity throughout the world in post-Fordist economies, but in Africa, industrial heritage is one of the most neglected forms of tourism, according to Chapter 12 (Clinton van der Merwe and Dallen Timothy). Africa is rich in natural resources, which have been used by ancient peoples, but more intently by Africa's colonial overlords. With the decline in industrial output in some sectors, destinations are becoming increasingly aware of the potential value of exploiting their extractive or manufacturing past for tourism. The chapter examines both derelict and currently functioning industrial landscapes as tourism assets and the challenges industrial destinations face in trying to grow this element of human heritage for tourism development.

In Chapter 13, Philip Stone examines the diverse painful pasts of Africa. He looks at African dark tourism and develops a typology of dark events and attractions with varying degrees of lightness and darkness. On the lighter end of the scale, dark heritage includes tombs and burial grounds, funerary rituals and death events, with epidemics and natural disasters, dictatorships and corruption, racism and apartheid, and slavery being darker. Murders and assassinations, warfare and terrorism, and genocide manifest the darkest of the dark heritages on the continent. Based on empirical evidence, this particular combination of dark phenomena, Stone argues, is somehow uniquely African.

Much of the dark heritage alluded to by Stone derives directly or indirectly from European colonialism, which is the focus of João Sarmento's contribution. In Chapter 14, Sarmento deconstructs the traditional view that Africa's heritage derives from three spheres of influence: indigenous, Islamic, and Christian, or the tripartite spatio-temporal framework of pre-colonial, colonial, and post-colonial (Teye, 2009). Africa's past is far too complex to be boiled down to characterizations such as these. He describes the impacts of European colonization on Africa's patrimony, including its desecration and plundering, the misunderstanding of Africa's heritage values by the Europeans (including UNESCO's concept of universal value), and how the states created by the European metropoles have divided shared heritages and caused domestic problems as nations within state boundaries have tried, sometimes unsuccessfully, to annex heritage as a means of developing a modern national identity. Colonialism in Africa lies at the root of much discontent on the continent and has been both a barrier to, and a facilitator of, heritage protection and cultural tourism development.

In Chapter 15, Liza Gijanto examines the regional disparities in the archaeological record, with some parts of Africa being favored over others in digs and research. Some regions of the continent have a richer array of archaeological sites, which is apparent with the attention they receive, most of the work being undertaken by European and North American scientists with funding provided from abroad. Although archaeological remains are not as prominent in Sub-Saharan Africa as they are in the north and other world regions, there is much potential for the growth of archaeotourism.

Chiedza Mutanga and Edson Gandiwa (Chapter 16) focus on Africa's natural heritage and how it too manifests in cultural terms and functions as a crossover between pure nature and mixed cultural heritage. Many of Africa's protected areas and game preserves originated during the colonial control of the continent, as Europeans enacted legislation to protect natural areas and wildlife for their own recreational, economic, and prize-hunting needs. Like the discussion in Chapter 14, the chapter elucidates the incongruity between European colonial values and African values with regard to land, flora, and fauna. Many of the metropolitan values regarding nature have been adopted by African states in the post-colonial era as a way of protecting wildlife and natural areas for tourism purposes.

Chapter 17 by Jarkko Saarinen and Naomi Moswete examines the effects of climate change on the cultural heritage of Africa. Increasingly intense weather events, coastal flooding and erosion, and extended droughts in some areas with continued desertification in the Sahel will all continue to affect tangible heritage (directly) and intangible heritage (indirectly) if warming is not mitigated effectively.

The concluding chapter by Dallen Timothy summarizes many of the issues learned throughout the book and speculates on some of the future needs in heritage tourism research in Africa.

Africa is one of Earth's greatest cultural and natural treasures. With a few exceptions, cultural heritage-based tourism on the continent remains underdeveloped, and its heritages have been afforded much less academic and industry attention than the manifestations of cultural heritage in other parts of the world (Ndoro et al., 2018; Rogerson & Rogerson, 2011; Rogerson & Visser, 2011; Timothy, 2021a). Nevertheless, the cultural heritage of Africa is immense, diverse, and in need of protection and sustainable use; it has the potential to be an increasingly salient part of the region's tourism development efforts. With this realization, this book emerged as a tool for consolidating what is known about cultural heritage-based tourism in Africa. This collection of essays aims to enlighten scholars and practitioners about the assets, opportunities, and challenges that exist for heritage tourism development in a pan-African context and to provide a framework of support for additional research and sustainable management of the continent's cherished cultural heritage.

References

Adams, W.M., Goudie, A., & Orme, A.R. (1999). *The physical geography of Africa*. Oxford: Oxford University Press.

Adie, B.A. (2021). Touring 'our' past: World Heritage tourism and post-colonialism in Morocco. In C.M. Hall & S. Seyfi (Eds.), *Cultural and heritage tourism in the Middle East and North Africa: Complexities, management and practices*, pp. 102–114. London: Routledge.

Adie, B.A., & Hall, C.M. (2017). Who visits World Heritage? A comparative analysis of three cultural sites. *Journal of Heritage Tourism, 12*(1), 67–80.

Adu Boahen, A. (1987). *African perspectives on colonialism*. Baltimore, MD: Johns Hopkins University Press.

African Development Bank (2022). *African economic outlook, 2021*. Abidjan: African Development Bank.

Akama, J., & Sterry, P. (Eds.) (2002). *Cultural tourism in Africa: Strategies for the new millennium*. Arnhem, Netherlands: Association for Tourism and Leisure Education.

Alemazung, J.A. (2010). Post-colonial colonialism: An analysis of international factors and actors marring African socio-economic and political development. *The Journal of Pan African Studies, 3*(10), 62–84.

Anyanwu, J.C., & Salami, A.O. (2021). The impact of COVID-19 on African economies: An introduction. *African Development Review, 33*(1), 1–16.

Asante, M.K. (2019). *The history of Africa: The quest for eternal harmony*. London: Routledge.

Baillie, B., & Sørensen, M.L.S. (Eds.) (2021). *African heritage challenges: Communities and sustainable development*. Singapore: Springer.

Ben-Ghiat, R., & Fuller, M. (Eds.) (2005). *Italian colonialism*. New York: Palgrave Macmillan.

Bolaane, M. (2014). San cross-border cultural heritage and identity in Botswana, Namibia and South Africa. *African Study Monographs, 35*(1), 41–64.

Bolaky, B. (2016). Tourism for economic development in Africa. *Journal of Research in Business, Economics and Management*, 7(4), 1222–1248.

Bongmba, E.K. (Ed.) (2015). *The Routledge companion to Christianity in Africa*. London: Routledge.

Boswell, R. (2011). Managing heritage in Antananarivo, Madagascar. *Journal of Contemporary African Studies, 29*(4), 455–472.

Boswell, R., & O'Kane, D. (2011). Introduction: Heritage management and tourism in Africa. *Journal of Contemporary African Studies, 29*(4), 361–369.

Bwire, G., Ario, A.R., Eyu, P., Ocom, F., Wamala, J.F., Kusi, K.A., Ndeketa, L., Jambo, K.C., Wanyenze, R.K. & Talisuna, A.O. (2022). The COVID-19 pandemic in the African continent. *BMC Medicine, 20*, 167.

Central Intelligence Agency (2022). CIA World Factbook. Online: https://www.cia.gov/the-world-factbook/

Chaula, J. (2019). Opportunities and challenges for integrating statistical and spatial data in East African Countries. *American Academic Scientific Research Journal for Engineering, Technology and Sciences, 59*(1), 42–48.

Chhabra, D. (2010). Student motivations: A heritage tourism perspective. *Anatolia, 21*(2), 249–269.

Conrad, S. (2012). *German colonialism: A short history*. Cambridge: Cambridge University Press.

Cusack, I. (1999). Hispanic and Bantu inheritance, trauma, dispersal and return: Some contributions to a sense of national identity in Equatorial Guinea. *Nations and Nationalism, 5*(2), 207–236.

Davidson, B. (1994). *Modern Africa: A social and political history*. London: Routledge.

De Jong, F., & Rowlands, M. (Eds.) (2016). *Reclaiming heritage: Alternative imaginaries of memory in West Africa*. London: Routledge.

de la Torre, A., & Rigolini, J. (2020). *MIC forum: The rise of the middle class*. Washington, DC: World Bank.

Demissie, F. (Ed.) (2012). *Colonial architecture and urbanism in Africa: Intertwined and contested histories*. Aldershot: Ashgate.

Dhaher, N., Seyfi, S., & Hall, C.M. (2021). Cultural heritage and tourism in Tunisia: Evolution, challenges, and perspectives. In C.M. Hall & S. Seyfi (Eds.), *Cultural and heritage tourism in the Middle East and North Africa: Complexities, management and practices*, pp. 87–101. London: Routledge.

Dieke, P.U.C. (2000). *The political economy of tourism development in Africa*. New York: Cognizant.

Dorsey, L. (2005). *Voodoo and Afro-Caribbean Paganism*. New York: Citadel Press.

Drummond, J., Drummond, F., & Rogerson, C.M. (2021). Latent opportunities for heritage tourism in South Africa: Evidence from Mahikeng and surrounds. *African Journal of Hospitality, Tourism and Leisure, 10*(5), 1591–1609.

Duval, M., & Smith, B.W. (2013). Rock art tourism in the uKhahlamba/Drakensberg World Heritage site: Obstacles to the development of sustainable tourism. *Journal of Sustainable Tourism, 21*(1), 134–153.

Duval, M., & Smith, B.W. (2014). Seeking sustainable rock art tourism: The example of the Maloti-Drakensberg Park World Heritage Site. *The South African Archaeological Bulletin, 69*, 34–48.

Finneran, N. (2013). Lucy to Lalibela: Heritage and identity in Ethiopia in the twenty-first century. *International Journal of Heritage Studies, 19*(1), 41–61.

Fouéré, M.A., & Hughes, L. (2015). Heritage and memory in East Africa today: A review of recent developments in cultural heritage research and memory studies. *Azania: Archaeological Research in Africa, 50*(4), 542–558.

Fredholm, S. (2016). Assets in the age of tourism: The development of heritage planning in Ghanaian policy. *Journal of Contemporary African Studies, 34*(4), 498–518.

Fullagar, S., Markwell, K., & Wilson, E. (Eds.) (2012). *Slow tourism: Experiences and mobilities*. Bristol: Channel View Publications.

Gijanto, L.A. (2011). Competing narratives: Tensions between diaspora tourism and the Atlantic past in The Gambia. *Journal of Heritage Tourism, 6*(3), 227–243.

Goldberg, A. (1983). Identity and experience in Haitian voodoo shows. *Annals of Tourism Research, 10*(4), 479–495.

Graham, B., Ashworth, G.J., & Tunbridge, J.E. (2000). *A geography of heritage: Power, culture & economy*. London: Arnold.

Gravari-Barbas, M., & Jacquot, S. (2014). World Heritage, tourism and sustainable development in Africa: Discourses, approaches and challenges. *Via Tourism Review*, *4*(5), 2–4.
Gumede, T.K., & Ezeuduji, I.O. (2021). Managing heritage tourism brand in South Africa: A synthesis of literature. *African Journal of Hospitality, Tourism and Leisure*, *10*(4), 1302–1320.
Hall, C.M., & Zeppel, H. (1990). Cultural and heritage tourism: The new grand tour. *Historic Environment*, 7(3/4), 86–98.
Hall, M., & Stefoff, R. (2006). *Great Zimbabwe*. Oxford: Oxford University Press.
Harvard University (2022). Introduction to African languages. Online: https://alp.fas.harvard.edu/introduction-african-languages#:~:text=population%20census%20statistics.-,Languages%20of%20Africa,languages%20of%20Africa%20by%20country
Heine, B., & Nurse, D. (2000). *African languages: An introduction*. Cambridge: Cambridge University Press.
Hernandez-Ramdwar, C. (2013). African traditional religions in the Caribbean and Brazil: Models of religious tourism and impacts of commodification. *Journal of Heritage Tourism*, *8*(1), 81–88.
Hoag, H.J. (2013). *Developing the rivers of West and East Africa: An environmental history*. London: Bloomsbury.
Jablonska, J., Jaremko, M., & Timčák, G.M. (2016). Social tourism, its clients and perspectives. *Mediterranean Journal of Social Sciences*, 7(3), 43–52.
Jeffrey, J. (2015). Somaliland entices tourists: The quasi-independent region of Somalia is trying to market itself as a tourist destination. *African Business*, *419*, 66–68.
Jewell, B., & Crotts, J.C. (2002). Adding psychological value to heritage tourism experiences. *Journal of Travel & Tourism Marketing*, *11*(4), 13–28.
Keitumetse, S. (2012). Conversation piece: Intangible cultural heritage in Botswana. In M.L. Stefano, P. Davis, & G. Corsane (Eds.), *Safeguarding intangible cultural heritage*, pp. 149–152. Woodbridge: Boydell.
Keitumetse, S.O. (2016). *African cultural heritage conservation and management: Theory and practice from Southern Africa*. Cham, Switzerland: Springer.
Langan, M. (2018). *Neo-colonialism and the poverty of 'development' in Africa*. New York: Palgrave Macmillan.
Lefebvre, C. (2011). We have tailored Africa: French colonialism and the 'artificiality'of Africa's borders in the interwar period. *Journal of Historical Geography*, *37*(2), 191–202.
Lessmeister, R., & Scherle, N. (2008). "Off the beaten track?" Special interest tourism in the Maghreb countries. *Geographische Rundschau*, *4*(3), 4–9.
Levtzion, N., & Pouwels, R.L. (Eds.) (2000). *The history of Islam in Africa*. Athens, OH: Ohio University Press.
Lindsey, P.A., Frank, L.G., Alexander, R., Mathieson, A., & Romañch, S.S. (2007). Trophy hunting and conservation in Africa: Problems and one potential solution. *Conservation Biology*, *21*(3), 880–883.
Lopez-Guzman, T., & Gonzalez Santa-Cruz, F. (2016). International tourism and the UNESCO category of intangible cultural heritage. *International Journal of Culture, Tourism and Hospitality Research*, *10*(3), 310–322.
Makuvaza, S. (Ed.) (2014). *The management of cultural World Heritage Sites and development in Africa: History, nomination processes and representation on the World Heritage List*. Heidelberg: Springer.
Manwa, H., Moswete, N., & Saarinen, J. (2016). *Cultural tourism in Southern Africa*. Bristol: Channel View Publications.
Mariz, V. (2016). The understanding of the touristic value of Portuguese overseas monuments: The case of Angola (1959–1974). *Journal of Spatial and Organizational Dynamics*, *4*(2), 157–168.

Martin-Márquez, S. (2008). *Disorientations: Spanish colonialism in Africa and the performance of identity*. New Haven, CT: Yale University Press.

Mattingly, D.J., & Sterry, M. (2013). The first towns in the central Sahara. *Antiquity*, *87*(336), 503–518.

McKercher, B. (2002). Towards a classification of cultural tourists. *International Journal of Tourism Research*, *4*(1), 29–38.

Miles, W.F. (2019). Who is a Jew (in Africa)? Definitional and ethical considerations in the study of sub-Saharan Jewry and Judaism. *The Journal of the Middle East and Africa*, *10*(1), 1–15.

Minter, W. (1972). *Portuguese Africa and the West*. New York: Monthly Review Press.

Munthali, S.M. (2007). Transfrontier conservation areas: Integrating biodiversity and poverty alleviation in Southern Africa. *Natural Resources Forum*, *31*(1), 51–60.

Mustafa, M.H. (2019). Intangible heritage and cultural protection in the Middle East. In D.J. Timothy (Ed.), *Routledge handbook on tourism in the Middle East and North Africa*, pp. 57–70. London: Routledge.

Mutanga, C.N., Vengesayi, S., Chikuta, O., Muboko, N., & Gandiwa, E. (2017). Travel motivation and tourist satisfaction with wildlife tourism experiences in Gonarezhou and Matusadona National Parks, Zimbabwe. *Journal of Outdoor Recreation and Tourism*, *20*, 1–18.

Naidu, M. (2008). Creating an African tourist experience at the Cradle of Humankind World Heritage Site. *Historia*, *53*(2), 182–207.

National Geographic (2022). Africa: Physical geography. Online: https://education.nationalgeographic.org/resource/africa-physical-geography

Ndoro, W., Chirikure, S., & Deacon, J. (Eds.) (2018). *Managing heritage in Africa: Who cares?* London: Routledge.

Norton, A. (1996). Experiencing nature: The reproduction of environmental discourse through safari tourism in East Africa. *Geoforum*, *27*(3), 355–373.

Novelli, M. (2016). *Tourism and development in Sub-Saharan Africa: Current issues and local realities*. London: Routledge.

Novelli, M., Adu-Ampong, E.A., & Ribeiro, M.A. (Eds.) (2021). *Routledge handbook of tourism in Africa*. London: Routledge.

O'Connor, D., & Reid, A. (Eds.) (2003). *Ancient Egypt in Africa*. London: Routledge.

Odada, E.O., Olago, D.O., Bugenyi, F., Kulindwa, K., Karimumuryango, J., West, K., Ntiba, M., Wandiga, S., Aloo-Obudho, P., & Achola, P. (2003). Environmental assessment of the East African Rift Valley lakes. *Aquatic Sciences*, *65*, 254–271.

Okech, R. (2008). Tourism and globalization of cultural heritage in Africa: Mind the difference? *Tourism Culture & Communication*, *8*(3), 195–205.

Olukole, T.O. (2014). Geographical information systems database of cultural heritage resources of Osogbo and their tourism potential. *African Journal of Hospitality, Tourism and Leisure*, *3*(2), 1–10.

Peterson, D.R., Gavua, K., & Rassool, C. (Eds.) (2015). *The politics of heritage in Africa: Economies, histories, and infrastructures*. New York: Cambridge University Press.

Poria, Y., Butler, R., & Airey, D. (2003). The core of heritage tourism. *Annals of Tourism Research*, *30*(1), 238–254.

Prayag, G., & Del Chiappa, G. (in press). Nostalgic feelings: Motivation, positive and negative emotions, and authenticity at heritage sites. *Journal of Heritage Tourism*. DOI: 10.1080/1743873X.2021.1874000

Ramkissoon, H. (2015). Authenticity, satisfaction, and place attachment: A conceptual framework for cultural tourism in African island economies. *Development Southern Africa*, *32*(3), 292–302.

Rogerson, C.M. (2007). Reviewing Africa in the global tourism economy. *Development Southern Africa, 24*(3), 361–379.

Rogerson, C.M., & Rogerson, J.M. (2011). Tourism research within the Southern African Development Community: Production and consumption in academic journals, 2000–2010. *Tourism Review International, 15*(1–2), 213–224.

Rogerson, C.M., & Visser, G. (2011). African tourism geographies: Existing paths and new directions. *Tijdschrift voor Economische en Sociale Geografie, 102*(3), 251–259.

Saarinen, J., Moswete, N., & Monare, M.J. (2014). Cultural tourism: New opportunities for diversifying the tourism industry in Botswana. *Bulletin of Geography, Socio-economic Series, 26*(26), 7–18.

Saarinen, J., & Rogerson, C.M. (2015). Setting cultural tourism in southern Africa. *Nordic Journal of African Studies, 24*(3&4), 207–220.

Said, R. (2013). *The River Nile: Geology, hydrology and utilization*. Amsterdam: Elsevier.

Schmidt, P. (2003). Heritage and tourism in Eritrea: Issues and alternative directions. *Journal of Eritrean Studies, 2*(1–2), 59–76.

Schmidt, P.R., & Pikirayi, I. (Eds.) (2016). *Community archaeology and heritage in Africa: Decolonizing practice*. London: Routledge.

Seyfi, S., & Hall, C.M. (2021). Cultural heritage tourism in the MENA: Introduction and background. In C.M. Hall & S. Seyfi (Eds.), *Cultural and heritage tourism in the Middle East and North Africa: Complexities, management and practices*, pp. 1–33. London: Routledge.

Shackley, M. (1997). Tourism and the management of cultural resources in the Pays Dogon, Mali. *International Journal of Heritage Studies, 3*(1), 17–27.

Shano, T.M. (2014). Developing heritage and cultural tourism in Lesotho: The case of Ha Kome cave village. Unpublished master's thesis, University of Pretoria.

Shoup, J.A. (2011). *Ethnic groups of Africa and the Middle East: An encyclopedia*. Santa Barbara, CA: ABC-CLIO.

Škrabić Perić, B., Šimundić, B., Muštra, V., & Vugdelija, M. (2021). The role of unesco cultural heritage and cultural sector in tourism development: The case of EU countries. *Sustainability, 13*(10), 5473.

Soper, A.K. (2007). Developing Mauritianness: National identity, cultural heritage values and tourism. *Journal of Heritage Tourism, 2*(2), 94–109.

Spenceley, A., & Nelson, F. (2013) Heritage tourism in Africa. In R. Staiff, R. Bushell, & S. Watson (Eds.), *Heritage and tourism: Place, encounter, engagement*, pp. 251–273. London: Routledge.

Statista (2022a). Number of international tourist arrivals in Ghana from 2016 to 2020. Online: https://www.statista.com/statistics/1230202/number-of-international-tourists-arriving-in-ghana/#:~:text=Number%20of%20tourist%20arrivals%20in%20Ghana%202016%2D2020&text=In%202020%2C%20Ghana%20registered%20355%2C108,%2D19)%20in%20the%20country

Statista (2022b). Tourism as a share of total exports in Africa as of 2019, by leading country. Online: https://www.statista.com/statistics/1232700/tourism-as-share-of-total-exports-in-africa-by-country/

Stebbins, R.A. (1996). Cultural tourism as serious leisure. *Annals of Tourism Research, 23*, 945–948.

Táíwò, O. (2010). *How colonialism preempted modernity in Africa*. Bloomington: Indiana University Press.

Teye, V.B. (2009). Tourism and Africa's tripartite cultural past. In D.J. Timothy & G.P. Nyaupane (Eds.), *Cultural heritage and tourism in the developing world: A regional perspective*, pp. 165–185. London: Routledge.

Tichaawa, T.M., & Rogerson, C.M. (2022). Africa tourism in change: Past and present. *Tourism Review International, 26*(1), 1–7.

Timothy, D.J. (1997). Tourism and the personal heritage experience. *Annals of Tourism Research, 34*(3), 751–754.

Timothy, D.J. (2018). Making sense of heritage tourism: Research trends in a maturing field of study. *Tourism Management Perspectives, 25*, 177–180.

Timothy, D.J. (2019a). The Middle East and North Africa: A dynamic cultural realm. In D.J. Timothy (Ed.), *Routledge handbook on tourism in the Middle East and North Africa*, pp. 24–35. London: Routledge.

Timothy, D.J. (2019b). Tourism trends and patterns in MENA: A resource perspective. In D.J. Timothy (Ed.), *Routledge handbook on tourism in the Middle East and North Africa*, pp. 36–54. London: Routledge.

Timothy, D.J. (2020). Heritage consumption, new tourism and the experience economy. In M. Gravari-Barbas (Ed.), *A research agenda for heritage tourism*, pp. 203–217. Cheltenham: Edward Elgar.

Timothy, D.J. (2021a). *Cultural heritage and tourism: An introduction*, 2nd Edn. Bristol: Channel View Publications.

Timothy, D.J. (2021b). Tourism and the multi-faith heritage of the Middle East and North Africa: A resource perspective. In C.M. Hall & S. Seyfi (Eds.), *Cultural and heritage tourism in the Middle East and North Africa: Complexities, management and practices*, pp. 34–53. London: Routledge.

Timothy, D.J., & Boyd, S.W. (2003). *Heritage tourism*. Harlow: Prentice Hall.

Timothy, D.J., Lew, A.A., & Hall, C.M. (2022). *World regional geography: Human mobilities, tourism destinations, sustainable environments*, 3rd Edn. Dubuque, IA: Kendall Hunt.

Weaver, D.B. (2011). Contemporary tourism heritage as heritage tourism: Evidence from Las Vegas and Gold Coast. *Annals of Tourism Research, 38*(1), 249–267.

Willems, W.J., & Comer, D. (2011). Africa, archaeology, and world heritage. *Conservation and Management of Archaeological Sites, 13*(2–3), 160–173.

World Tourism Organization (2022). UNWTO Global tourism dashboard, country profile—inbound tourism. Online: https://www.unwto.org/tourism-data/country-profile-inbound-tourism

2

INTANGIBLE HERITAGE AS A CULTURAL ASSET FOR AFRICAN TOURISM

Moren Tibabo Stone, Lesego S. Stone, Goemeone Mogomotsi, and Patricia Mogomotsi

Introduction

Cultural tourism has grown rapidly in recent decades (Smith & Richards, 2013; Timothy, 2021a). Cultural tourism (or heritage tourism) incorporates sociocultural, economic, and environment aspects and traditions of societies (Moswete et al., 2015). Culture can refer to both people and their ordinary social characteristics reflected in their traditions and day-to-day patterns of behaviour, which differentiate them from others, as well as their exceptional representations of creative and artistic endeavours (Robinson & Picard, 2006). Cultural heritage-based tourism has the potential to encourage local communities to take pride in their culture and contributes to strengthening community identity (Mitchell & Ashley, 2010; Timothy, 2020). This form of tourism relies on local products, artefacts, and the living practices of the host community. Therefore, simply put, it is a form of tourism that is built around people utilizing cultural resources (Okech, 2009). Compared to other continents, Africa, especially Sub-Saharan Africa, is known for its vast environmental and cultural diversity (Ezeuduji, 2017; Ndoro et al., 2018). The beautiful and varied cultures that form both the tangible and intangible heritages of the region make Africa a diverse environment that cannot be substituted by any other part of the world (Ankomah & Crompton, 1990; Nic Eoin & King, 2013). Cultural tourism that utilizes intangible heritage, therefore, is driven by people and their unique ways of life (Ezeuduji, 2017).

The word "heritage" in its wider denotation is concomitant with the word "inheritance". Heritage is something transferred from one generation to another, whereas the concept of tourism is really a form of modernity and present-centredness (Nuryanti, 1996). Heritage is what we inherit from the past and use and value in the present day (Timothy, 2021a). In contemporary society, heritage is often treated as an economic commodity, especially for tourism (Graham, Ashworth &

DOI: 10.4324/9781003153955-2

Tunbridge, 2000). Cultural heritage management and tourism have often been at odds, as cultural heritage managers sometimes find it hard to accept treating valuable (and sometimes fragile) human inheritances as a tourism product to be used for economic gain (Gillot, 2020; Ho & McKercher, 2004).

While this might be the case in scholarly and many professional circles, the cultural resources in Africa provide the region with substantial tourism development opportunities. Visiting rural Africa to experience the living cultures of the native populations is considered a highlight of tourists' journeys and an important strategy within the management portfolio of sustainable rural development (Ezeuduji & Rid, 2011). Intangible heritage-based tourism development aims to improve rural lives through the participation of rural people in tourism. Nevertheless, the expected benefits from intangible heritage-oriented tourism may not always be delivered to rural communities. Hence, the motivation for community participation by people who are the custodians of these important cultural resources may be low, minimizing community development opportunities through cultural tourism.

The purpose of this chapter is to explore the problems associated with intangible heritage-based tourism in Africa. This form of tourism has generated increasing research interest in individual African countries, but there have not been any major academic attempts to create an overview of this area of cultural tourism or to critically assess the field more widely. Concepts, issues, and challenges discussed in this chapter are general and exploratory in nature, but they can serve as a guide for future studies in each African country. Although important intangible heritage exists throughout the continent, the empirical elements of this chapter focus mainly on southern Sub-Saharan Africa – a region endowed with many intangible heritage assets and cultural attractions.

Understanding Intangible Heritage and Tourism

Heritage includes material forms of culture such as monuments, historical or architectural remains, and artefacts on display in museums, but intangible heritage refers to immaterial forms of culture such as philosophies, traditions, and art in all their manifestations, celebrations of great events or personalities in history, distinctive ways of life, literature, folklore, language, religion, tales and myths, history, music, songs, dances, rituals, customs, and lifestyles (Herbert, 1989; Smith & Akagawa, 2008; Stefano & Davis, 2017; Timothy, 2021a; UNWTO, 2008; Zeppel & Hall, 1992). UNESCO's (2003) Convention for the Safeguarding of the Intangible Cultural Heritage defines intangible cultural heritage (ICH) as the "practices, representations, expressions, knowledge, skills – as well as the instruments, objects, artefacts and cultural spaces associated therewith – that communities, groups, and, in some cases, individuals recognize as part of their cultural heritage" [Article 2.1]. Intangible heritage tourism is predominantly concerned with exploring immaterial (intangible) remnants of the past, but ICH is not necessarily a fixed or static representation of the past, especially in the context of tourism (Smith, 2006).

Heritage is a complex and multifaceted concept. It is constantly reconstructed and reinterpreted in an attempt to meet the specific demands of tourists and reflects the sociocultural changes of contemporary society (Park, 2013). Intangible heritage represents some of the strongest manifestations of cultural authenticity and facilitates a deeper understanding of a destination's culture, which in turn strengthens the competitiveness of ICH within the broader arena of cultural heritage tourism (Gonzalez, 2008; Kim et al., 2019). ICH creates an opportunity for countries and communities to develop alternative and diverse product offerings away from the more expected mainstream tourism products (Giudici et al., 2013), although there are several high-profile cases in Africa of living cultures having been an important tourism product for many years already, including that of the Maasai in eastern Africa, the Zulu and Swazi in southern Africa, and the Berbers of North Africa. In the context of Sub-Saharan Africa, ICH-based tourism provides an avenue for communities to preserve their living heritage, (re)educate international visitors about Africans' ancient civilizations and their role in developing scientific knowledge, and to share their painful experiences with colonialism and slavery. As a result, it is the aim of this chapter to evaluate the role of intangible heritage in contemporary tourism. This is an especially important consideration as ICH can assist in diversifying the tourism sector in Africa.

Historical Impediments to Indigenous Participation in Heritage Tourism

Intangible heritage in Africa is not only the province of indigenous people; some countries' mainstream and colonial populations also possess and express important elements of ICH. Nonetheless, ICH in the region is usually connected to indigenous people, tribal people, or certain ethnic groups. The literature on indigenous participation in heritage tourism highlights a number of impediments that shape development consequences and are entrenched in historical inhibitions affecting communities' participation and success in heritage tourism development. These include, among others, lack of ownership of heritage resources, lack of availability and access to development capital, and inadequate physical tourism infrastructure (Hinch & Butler, 2007; van Veuren, 2001; Zeppel, 1998). These factors, as argued by van Veuren (2001), are predisposed by the colonial and/or post-colonial political and historical dynamics of different African countries, often with the sidelining of indigenous people and due to efforts geared towards bringing them into the mainstream culture, an important theme that underlies many of the challenges facing community participation in heritage tourism (Timothy, 2015).

Ownership and Access to Heritage Resources

Ownership of resources entails land ownership, as well as heritage resources. The availability or lack of resource ownership defines indigenous participation in heritage on many fronts. For instance, communities with full title deeds to their land can develop or control the physical development of tourism facilities and the spatial growth of the tourism plant. Furthermore, they stand a better chance of using their land assets to attract joint venture partnerships. However, historically, indigenous communities have repeatedly been dispossessed of their native land (Daes, 1999; Harrison & Price, 1996; van Veuren, 2001). For example, Lepp (2007) acknowledges that in Bigodi, Uganda, rural residents viewed the new idea of wealthy foreign tourists coming to their villages to participate in consuming intangible heritage tourism as outlandish because it conjured up memories of British imperialism. For this reason, residents initially feared that tourism was a plot by foreigners to steal village land. As a result, residents reacted to the idea with suspicion, anxiety, and fear (Lepp, 2007).

Thus, the development of intangible heritage tourism resources is usually slow, tedious and, on occasion, inflammatory (Lepp, 2007). Nevertheless, where indigenous communities have instituted successful land entitlements, it has created a foundation for tourism participation and socioeconomic empowerment. For example, Moswete, Saarinen, and Monare (2015) report that, through the formation of a community-based natural resource management (CBNRM) program in Botswana, local villagers formed communally owned tourism enterprises/projects to utilize cultural and heritage resources for social, political, and economic empowerment by creating CBNRM cultural tourism villages. According to the authors, the government of Botswana has realized the benefits that come with tourism, especially in remote, rural areas. The advantages and benefits of this approach include but are not limited to increasing knowledge about the concept of cultural tourism, encouraging the preservation and conservation of cultural resources (including intangible heritage) of various ethnic groups, and nurturing pride in individuals and solidarity in communities as regards their culture (Moswete, Saarinen & Monare, 2015). Other benefits include job creation, income generation, and a sense of connectedness to the resources they live with or near (Moswete et al., 2015). This is an indication that, where indigenous communities have instituted successful land claims, participation in tourism becomes increasingly viable.

In addition to ownership, providing access is vital to heritage tourism management. In Zimbabwe, authorities long maintained that the rock art in and around Domboshava belonged to the ancestors of the San people who no longer inhabited the area and, as a result, did not consider that the preservation of the site depended on the area's present inhabitants and being tolerant of their cultural practices (Chirikure & Pwiti, 2008). Because locals were excluded from using the site for rainmaking and other rituals, the rock paintings were vandalized by locals who felt if they could not benefit from the site spiritually or economically, then the government and archaeologists should not benefit from it either. This unfortunate situation eventually led to a more positive outcome. Local belief systems, traditions, and values were integrated into the use and management of the area (myths and legends were included into the interpretation of the site), rainmaking ceremonies were once again allowed, and residents were employed as tour guides and were permitted to sell curios and crafts to visitors (Chirikure & Pwiti, 2008).

Availability and Access to Development Capital

In addition to a general lack of ownership and control of heritage resources, the availability of, and access to, capital for developing sustainable heritage tourism is difficult (Lai, 2020), in terms of both human resource advancement and construction of physical infrastructure. To satisfy international cultural tourism market demands, an understanding of tourists' motivations and management skills in business operations and marketing are required (Loulanski & Loulanski, 2011). In most cases, indigenous and other rural communities rarely possess this service industry knowledge, nor do they have access to financial capital to develop human resources, marketing abilities, and intelligence about consumer trends, or to finance small and medium start-up enterprises (Ulrich-Cloher, 1998).

For instance, due to a lack of public funding for heritage tourism projects in Botswana, some of the tourism-oriented cultural villages are privately owned by individuals or families (Motse Cultural Village is a family-/private-owned enterprise), while some are community-owned. For example, Shandereka Village is a community-owned cultural village in northwestern Botswana, which showcases, protects, and preserves Hambukushu culture and is a tourist attraction (Mbaiwa, 2013; Saarinen et al., 2014). Studies by Akama and Kieti (2007) in Mombasa, Kenya, and Ezeuduji and Rid (2011) in The Gambia illustrate how large-scale, capital-intensive tourism projects have led to increasing control of these nations' tourism resources by powerful foreign interest groups, leading to the propagation of enclave tourism and economic leakage. For the countries of Africa to reduce the high percentage of overseas leakage of tourism revenues, community-based heritage tourism should be encouraged by governments through the creation of financial support frameworks, training and capacity-building programmes, and marketing assistance (van Veuren, 2001). Nevertheless, interventions by governments, NGOs, and international funding agencies are often criticized for their absence of integrated support frameworks, which further polarizes indigenous and rural communities (Kennedy & Dornan, 2009). Stakeholders' different perceptions of what constitutes community-based sustainable tourism are often downplayed in favour of reaching a common goal that is often determined by outside interests. Although extremely important in sustainable community-based tourism development, especially in the context of ICH, due to the diverse structural nature of participatory tourism development, decisions about what is to be developed and what is not, the inclusiveness of stakeholders' opinions, and diverging goals become very chaotic and difficult to accomplish equitably (Stone, 2015; Stone & Stone, 2020).

Tourism's Physical Infrastructure

Another challenging factor is the lack of physical infrastructure for tourism. Generally, heritage tourism attractions are in rural areas, which lack good road connections, have inadequate communication networks, and possess too few information centres. Nonetheless, heritage tourism enterprises such as festival and event planners, cultural tour operators, rural homestays, and cultural villages need at least a modest level of infrastructure to facilitate tourists' access and comfort, and access to necessary goods and services from outside the area (van Veuren, 2001). While there are calls for the provision of better infrastructure for tourism in rural Africa to foster the growth of heritage tourism and other types of tourism, the literature presents opposing views. Some observers argue that physical development may stimulate tourism growth and help create employment, while others opine that infrastructure improvements will lead to too much development too rapidly to the detriment of local communities, leading to exploitation by outsiders (van den Berghe, 1995). This dilemma brings uncertainties to heritage tourism development, especially to the already marginalized native and rural communities. Nonetheless, Hazell and Diao (2005) recognize that public investment in rural infrastructure and transportation is a prerequisite for growing rural economies, not just in tourism but also in agriculture and other non-farm activities. Hazell and Diao conclude that developing rural infrastructure is necessary to strengthen rural and urban linkages and to increase urban demand for what rural areas produce. Lenao's (2016) study in Botswana reports that youth in one small village expressed concerns that a lack of infrastructure (i.e., roads, telephones and cellular communication coverage, and internet) is detrimental to the development of tourism in their area.

Lepeyre (2016) reports that in the Erongo region of Namibia, the Damara tribe maintains its living cultural heritage, including traditional and spiritual beliefs, which have become an important asset for tourism. When tourism was initially considered, access to the tribal area was very limited. However, several donors and NGOs, including the European Union Namibia Tourism Development Programme (EU-NTDP) and the Namibian Community-Based Tourism Association (NACOBTA), helped fund infrastructure improvements on-site and invested significantly in technical assistance and institutional capacity building, while simultaneously protecting living culture, paintings, sacred sites, and local hospitality, ultimately boosting tourism in the region (Lepeyre, 2016).

Intangible Heritage Tourism and Cultural Appropriation

Cultural appropriation entails "the taking – from a culture that is not one's own – of intellectual property, cultural expressions or artefacts, history and ways of knowledge" (Ziff & Rao, 1997, p. 1). While cultural appropriation is an important issue, Ziff and Rao (1997) argue that acts of appropriation happen everywhere in many creative domains as cultural influences blend, merge, and synthesize. Thus, they imply that humans have been borrowing from others since the beginning of time. Cultural appropriation is perceived as a longstanding issue for indigenous people in many countries of the world. Ziff and Rao (1997, p. 8), therefore, conclude that these concerns give rise to several claims:

- Cultural appropriation can harm the appropriated community.
- It can negatively impact the integrity and identities of cultural groups.

- Cultural appropriation can impact the cultural objects themselves.
- It can either damage or transform a given cultural good or practice.
- Cultural appropriation wrongly allows someone to benefit to the material detriment of others (deprivation of material advantages).
- Current laws fail to reflect alternative conceptions of what should be treated as property or ownership in the realm of cultural goods and practices.

As a result, it is evident that in many isolated rural and remote communities in Africa, after centuries (or even millennia) of cultural evolution and building unique and strong cultural traditions, expressions, customs, and ways of life, tourism's consumption of intangibles is seen as equalling a subtle form of appropriation in these communities, especially if the heritage on display is being sold outside the communities' control or without their decision-making input (Timothy, 2015). Cultural heritage resources are frequently transformed into commodities for exchange and exploitation through modern tourism development (George, 2010). In tourism, cultural intangibles are converted into commodities or objects of exchange value for consumption by outsiders. Subsequently, through cultural commodification, intangible heritage is transformed and reconstructed into sometimes unrecognizable entities and ensconced in a consumer value system that supersedes longstanding community value systems (George, 2004). For example, tourism based on the construction and use of cultural villages has been growing in Sub-Saharan Africa (Moswete et al., 2015). A cultural village is a specific attraction that demonstrates the traditional ways of life of local people. In other parts of the world, these are frequently referred to as living museums, ethnic museums, outdoor heritage museums, ethnic parks, ethnic theme parks, or culture parks. At these purpose-built establishments, visitors learn about the culture of the people and their past and present ways of living (Timothy, 2021a; Yang, 2011; Yang et al., 2008; Zeppel, 2002). In addition to displays and built structures, the activities and attractions related to cultural villages often include guided tours, cultural shows, exhibitions, ethnic cuisine, and craft workshops, where local people can be employed as tour guides and demonstrators (Saarinen, 2007).

In Namibia, there are four types of cultural villages (Lepeyre, 2016). One model is based on the belief that tourists want "authenticity" and therefore tour operators organize township tours in urban areas and visits to real local villages that have not been set up for tourists. The second type is staged villages ('staged authenticity') where tourists come into contact with performers, who are paid to demonstrate certain elements of their culture. The third model is an interactive approach where tourists dialogue with the hosts, which may include participating in certain activities and demonstrations. The fourth type displays diverse traditional homesteads of the different ethnic groups of Namibia, which allows tourists to see everyday life, even if it is in some cases slightly staged (Lepeyre, 2016).

Cultural villages have become quite popular tourist attractions in South Africa and Botswana (Monare, 2013; Moswete et al., 2015; Saarinen, 2007; van Veuren, 2001). Tourism-oriented cultural villages are usually operated as community-based organizations (CBOs) or community trusts (CTs). There are several CBOs and CTs that have been created via a CBNRM framework, such as Xaixai and Shandereka cultural villages in northern Botswana. While these cultural villages started entirely as community projects, there has been a recent proliferation of privately owned cultural villages as they prove to be viable tourism enterprises. Some cultural villages are privately owned by individuals or families. The private ownership of cultural villages threatens the existence of community-owned establishments, as private businesses often have better business models and advanced marketing strategies. The development of privately owned cultural villages is a form of cultural appropriation by a few business elites, which seems to harm the appropriated community when they lose business to privately owned villages. This new development in the ownership of cultural villages (i.e., privately or family-owned), does not come cheap and brings about changes to cultural tourism. For instance, tourism is a newcomer in consuming cultural intangibles in Africa, and its space introduces consumer value systems that often distort the original culture through inaccurate marketing narratives, unauthentic tourism activities, and inappropriate uses of cultural resources over which a community has little or no control (George, 2010). Thus, in some ways, tourism has been known to disempower cultural communities as they lose control over local tourism and their intangible heritage (Timothy, 2007).

Through the promotion of tourism, agents of change such as tourism entrepreneurs, governments, tourism salespeople, film-makers, photographers, tour operators, and artists perceive intangible heritage as a unique product opportunity, and they are increasingly appropriating and exploiting other people's ICH for their own commercial purposes (George, 2010). These actions promote the decontextualization of the heritage with which certain communities identify, thereby separating the elements of culture from their rooted environment, placing them in a new relationship of commodification, ultimately changing their context of use and appreciation (Hafstein, 2009). ICH is rarely so explicitly separated from the communities that own it and in fact, it is the closeness of their relationship to one another (culture and community) that gives meaning and relevance (Wain, 2014). In Zanzibar, Boswell (2008) found that like in many African regions high poverty rates lead to more focus being placed on attaining basic needs and less on the documentation and preservation of intangible heritage. Timothy (2021b) notes that this is a common problem throughout Africa and other less-developed regions of the Global South, where it is difficult to expect people to preserve and protect their heritage when they are trying to put food on the table and when their basic survival is in question.

In Mauritius, the tourism industry has commodified the local Sega music to satisfy tourists' quest for the "exotic other". Despite its important place in the cultural identity of the country's Creole population, the music is now performed, often out of context, for tourists' consumption (Boswell, 2016). This raises questions about the sustainability of ICH and whether or not performing for tourists, and in some cases, altering traditional elements of culture to meet the needs of tourists, cheapens traditional heritage to the point where it is beyond repair. Ideally, the sustainable use of intangible culture would seek its preservation and enable its use as a steady stream of income by creating a competitive destination through continuous participatory community planning without sacrificing the integrity of the asset base (McElroy & de Albuquerque, 2002).

Despite these shortcomings, intangible heritage has promoted tourism and exposed local culture to tourists. For instance, in Malawi, the Lake of Stars festival promotes Malawian arts and culture, facilitating tourism and international interactions, as well as generating exposure and income for the country (Ezeuduji, 2015; Lake of Stars, 2012). The festival helps provide a positive image for Malawi and counters the dominant negative media coverage of the region (Ezeuduji, 2015). In The Gambia, staged cultural performances are hosted along the Gambia River to generate income for local communities, entertain tourists and expose African arts and cultures across the globe (Ezeuduji & Rid, 2011).

Multiple Stakeholder Management of Intangible Heritage: Interests and Conflicts

Understanding the dynamics of managing intangible heritage for tourism is important in the context of Africa. Heritage does not belong to individuals but to communities; therefore, developing heritage for tourism purposes involves many participating stakeholders. At the global level of stakeholders are several major organizations and alliances (e.g., World Monuments Fund and the Global Heritage Fund), but perhaps the best known of them is UNESCO and its World Heritage Convention, which implements a range of conventions and treaties related to heritage protection and promotion. Since the late 1950s, UNESCO has provided the dominant intellectual and policy framework for international understandings and debates about the nature and value of heritage (Smith, 2015). Smith (2015) argues that the World Heritage Convention of 1972, in particular, has not simply influenced management practices, it has also defined the ways in which heritage as a cultural phenomenon has been understood, largely from the perspective of tangible and magnificent historic places and sites. This narrow understanding was expanded, however, to be more inclusive of intangible and ordinary heritage in 2003, when UNESCO enacted its Convention for the Safeguarding of the Intangible Cultural Heritage (Khaznadar, 2012) (Table 2.1).

One of the key issues that emerged is the problem presented by the way the convention has addressed the issue of "community" (IRCI, 2012). The convention faces increasing criticism over its inability to deal meaningfully with the concept of "community", and this criticism reveals several limitations with its framing and implementation of the intangible heritage convention (Smith, 2015). For example, Manwa et al. (2016) suggest that in southern Africa (i.e., Botswana, Lesotho, Namibia, South Africa, and Zimbabwe), the issue of defining a community impedes community participation in the development of heritage tourism due to the multiplicity of cultures in overlapping areas, which makes it a challenge to decide whose culture should be included in the cultural product and therefore portrayed in advertising media.

TABLE 2.1 Examples of Africa's intangible heritage as recognized by UNESCO

Intangible Heritage	*Type of Heritage*	*Country*	*Year Inscribed*
Polyphonic singing of the Aka Pygmies	Vocal music tradition	The Central African Republic	2008
Kankurang	Initiatory rites in the Manding area	The Gambia and Senegal	2008
Rites and craftsmanship associated with the wedding costume tradition of Tlemcen	Wedding ritual	Algeria	2012
Coming forth of the masks and puppets in Markala	Ritual festival	Mali	2014
Oshituthi shomagongo	Marula fruit festival and harvest	Namibia	2015
Gada system	Traditional governance system among the Oromo people	Ethiopia	2016
Argungu	fishing and cultural festival	Nigeria	2016
Woodcrafting knowledge of the Zafimaniry	traditional handicraft	Madagascar	2008
Sega Tambour of Rodrigues Island	Song and dance performance	Mauritius	2017
Nsima	Culinary tradition	Malawi	2017
Morna	Musical practice	Cabo Verde	2019
Barkcloth making	Traditional handicraft	Uganda	2008
Gnawa	Musical performances and religious rituals	Morocco	2019
Chopi Timbila	Village orchestra music	Mozambique	2008
Budima dance	Dance	Zambia	2020

Source: UNESCO (2021).

Communities are not homogeneous; therefore, participating in intangible heritage-based tourism cannot be a one-shoe-fits-all approach. Immediate stakeholders are community members with diverse opinions, interests, and aspirations, while other stakeholders include donors, international organizations, NGOs, and regulatory agencies (Stone, 2015). While stakeholder collaboration is crucial in managing intangible heritage, it is the diverse and heterogeneous nature of stakeholders that creates problems. One resultant outcome is that collaborative partnerships simultaneously empower and disempower certain segments of the community (Stone, 2015). While the collaborative partnership approach is conceptually distinct, in practice it usually also yields unintended consequences. Collaborative efforts are borne of competing interests and conflicts that need to be managed. The diverse community itself often has fundamentally different values, constructs, and understandings of what is at stake and how tourism problems can be addressed (Lewicki, Gray & Elliot, 2003).

In Uganda, although NGOs can provide needed assistance to communities, they often do not capacitate the communities they serve. Thus, when they leave, they often leave communities unable to keep projects running (Ahebwa, Aporu & Nyakaana, 2016). Despite the initial successes of communities' performances of cultural dances and traditional ceremonies, which they know and perform well for tourists, they often face challenges in operating and competing in the business environment against private tourism interests owing to their limited business skills and knowledge (Ahebwa, Aporu & Nyakaana, 2016).

Gray (2007) cautions that assembling many collaborative stakeholders may enhance the success of partnerships in the long run, but key players who are not benefiting or feel excluded may try to block plan implementation. Therefore, intangible heritage-based tourism development empowers communities unequally. As argued by Bramwell and Lane (2000), the complexity of engaging diverse stakeholders in projects makes it difficult to involve them all equally. This is so because tourism planning is a complex process, where choices influence, and depend on, each other, thereby influencing stakeholders' perceptions in different ways (Hedlund et al., 2012). Thus, Cusick, McClure, and Cox (2010) argue that the tourism sector is interested in expanding opportunities for intangible heritage-based tourism because the market has economic potential, yet its role is still marginal despite the rhetoric of support from the industry and state agencies.

The pursuit of tourism based on living culture and intangible heritage needs to be unpacked further and identified so that the creation and development of new and diverse heritage products can be implemented to capitalize on the multitudes of heritage opportunities in Africa.

Discussion and Conclusion

This chapter explores the development of intangible heritage tourism in Africa and shows that through the adoption of tourism development in many rural and remote areas where indigenous traditions, habits, and local cultures hold longstanding meanings and values over many generations, tourism-induced changes are becoming progressively apparent (George, 2010) and seem almost inevitable. The changes are driven by commoditizing intangible heritage assets into products that can be packaged and sold to tourists. Hewison (1988, p. 240) argues that

> we already have a changed language in which we talk about the arts as we no longer discuss them as expressions of imagination or creativity, we talk about "product"; we are no longer moved by the experiences the arts have to offer, we "consume" them, culture has therefore, become a commodity.

Although the value of heritage assets is far more complex than that of most goods and services (Timothy, 2021a), exploiting heritage for tourism consumption, if not managed properly and cautiously, may commercialize, trivialize, and standardize the cultural meanings and values of intangible heritage (Ho & McKercher, 2004).

A community's cultural inheritance should provide the community with a competitive advantage and matchless asset base, one that distinguishes it from other communities that might appeal to tourists. African cultures rely strongly on orality and ephemeral phenomena, providing one of the key challenges for the emerging cultural and heritage tourism sector (Marschall, 2008). However, with the increasing commodification of ICH to provide new and unique products in an effort to satisfy the needs of tourists, many smaller rural and remote communities face new challenges formerly considered outside their local domain (George, 2010). Questions of ownership of intellectual property, traditional knowledge, copyrights, and forms of protection relating to a community's ICH are unresolved issues in the touristification process.

Inadequate financial resources and unequal power relations among stakeholders (e.g., government, NGOs, donors, and local community members) discourage widespread community participation; nevertheless, participatory planning and development is fundamental to achieving community empowerment and self-reliance (Kim et al., 2019).

Although what constitutes a community is contested, the concepts of indigeneity and ethnicity are contested too. For example, Barnard (2006) and Kuper (2003), discuss in detail how highly contested and debated the concept of "indigenous" is. Conceptually, ethnicity refers to "a socio-culturally distinct group of people who share a common history, culture, language, religion, and way of life" (Yang et al., 2008, p. 752). What is clear is that a community is not a static and isolated group of people (Carlsson, 2000). Rather, it is more useful to think of community as a dynamic multidimensional, cross-scalar, sociopolitical unit or network that changes over time (Stone & Nyaupane, 2014). Therefore, the authors conclude that the quest for a singular or monolithic approach to culture and tourism in multifaceted cultural communities, such as those in Africa, frequently yields animosity, resistance, and resentment in community-based projects. This can be especially poignant in situations where intangible, living heritage is the main product of tourism.

In light of the problems engulfing ICH and its tourism use, questions arise about whether or not intangible heritage can be a driver for promoting sustainable tourism. Smith (2015) points out that this disconnect is caused by deficient guiding frameworks to help execute intangible culture-based tourism. The UNESCO Convention for the Safeguarding of the Intangible Cultural Heritage has significantly raised international and community awareness of the legitimacy of the concept of intangible heritage, although in raising this awareness, the convention has not yet provided a framework that privileges the community orientation of intangible heritage (Smith, 2015).

This chapter demonstrates that impediments to the development of intangible heritage-based tourism include a lack of ownership of heritage resources, inaccess to development capital, insufficient physical infrastructure, intangible cultural appropriation dynamics, and multiple stakeholders managing ICH, which leads to multiple interests and conflicts. Indications are that intangible heritage-based tourism has both positive and negative impacts. The negative impacts or impediments that hinder the attainment of more positive outcomes should be recognized and mitigated in the development of intangible heritage tourism in Africa. Both positive and negative impediments impact the sustainability of tourism development in Africa and need an adaptive management approach that endeavours to reconcile the two. ICH as a sustainable tourism resource should aim to meet the economic and social needs of the communities whose heritage is being utilized while at the same time maintaining and transmitting the cultural and social values inherent within intangible heritage. To achieve this, it is necessary for practitioners to be cognizant of the potential negative impacts of tourism as tourism unfolds and develops.

We have adopted an inclusive approach to understanding the crossovers between ICH and tourism, covering broad elements of intangible heritage and the challenges and opportunities that accompany it. However, differences do exist among the various elements of intangible heritage and what it means within each society. Therefore, a focus on each element of immaterial heritage will provide more explicit and specific findings that can contribute to the longevity and sustainable use of ICH as a sustainable tourism resource.

References

Ahebwa, W.M., Aporu, J.P., & Nyakaana, J.B. (2016). Bridging community livelihoods and cultural conservation through tourism: Case study of Kabaka Heritage Trail in Uganda. *Tourism and Hospitality Research*, *16*(2), 103–115.

Akama, J.S., & Kieti, D. (2007). Tourism and socio-economic development in developing countries: A case study of Mombasa resort in Kenya. *Journal of Sustainable Tourism*, *15*(6), 735–748.

Ankomah, P.K., & Crompton, J.L. (1990). Unrealized tourism potential: The case of sub-Saharan Africa. *Tourism Management*, *11*(1), 11–28.

Barnard, A. (2006). Kalahari revisionism, Vienna and the Indigenous people debate. *Social Anthropology*, *14*(1), 1–16.

Boswell, R. (2008). *Challenges to identifying and managing intangible cultural heritage in Mauritius, Zanzibar and Seychelles*. Dakar: Council for the Development of Social Science Research in Africa.

Boswell, R. (2016). Making the past pay? Intangible (cultural) heritage in South Africa and Mauritius. In M.L. Stefano, & P. Davis (Eds.), *Bottom of Form The Routledge companion to intangible cultural heritage*, pp. 121–134. London: Routledge.

Bramwell, B., & Lane, B. (2000). Collaboration and partnerships in tourism planning. In B. Bramwell, & B. Lane (Eds.), *Tourism collaboration and partnerships: Politics, practice and sustainability*, pp. 1–19. Clevedon: Channel View Publications.

Carlsson, L. (2000). Policy networks as collective action. *Policy Studies Journal*, *28*, 502–520.

Chirikure, S., & Pwiti, G. (2008). Community involvement in archaeology and cultural heritage management: An assessment from case studies in Southern Africa and elsewhere. *Current Anthropology*, *49*(3), 467–485.

Cusick, J., McClure, B., & Cox, L. (2010). Representations of ecotourism in the Hawaiian Islands: A content analysis of local media. *Journal of Ecotourism*, *9*(1), 21–35.

Daes, E.A. (1999). *Indigenous people and their relationship to land*. New York: United Nations.

Ezeuduji, I.O. (2015). Strategic event-based rural tourism development for sub-Saharan Africa. *Current Issues in Tourism*, *18*(3), 212–228.

Ezeuduji, I.O. (2017). Change management for sub-Saharan Africa's rural tourism development. *Current Issues in Tourism, 20*(9), 946–959.

Ezeuduji, I.O., & Rid, W. (2011). Rural tourism offer and local community participation in The Gambia. *TOURISMOS: An International Multidisciplinary Journal of Tourism, 6*(2), 187–211.

George, E.W. (2004). *Commodifying local culture for tourism development and community sustainability: The case of one rural community in Atlantic Canada*. Unpublished doctoral dissertation, University of Guelph, Canada.

George, E.W. (2010). Intangible cultural heritage, ownership, copyrights, and tourism. *International Journal of Culture, Tourism and Hospitality Research, 4*(4), 376–388.

Gillot, L. (2020). Archaeologists and tourism: Symbiosis or contestation? In D.J. Timothy, & L.G. Tahan (Eds.), *Archaeology and tourism: Touring the past*, pp. 26–40. Bristol: Channel View Publications.

Giudici, E., Melis, C., Dessì, S., & Ramos, B.F.P.G. (2013). Is intangible cultural heritage able to promote sustainability in tourism? *International Journal of Quality and Service Sciences, 5*(1), 101–114.

Gonzalez, M.V. (2008). Intangible heritage tourism and identity. *Tourism Management, 29*, 807–810.

Graham, B., Ashworth, G.J., & Tunbridge, J.E. (2000). *A geography of heritage: Power, culture, and economy*. London: Arnold.

Gray, B. (2007). The process of partnership construction: Anticipating obstacles and enhancing the likelihood of successful partnerships for sustainable development. In P. Glasbergen, F. Biermann, & P.J. Mol (Eds.), *Partnerships, governance and sustainable development*, pp. 29–48. Cheltenham: Edward Elgar.

Hafstein, V. (2009). Intangible heritage as a list: From masterpieces to representation. In L. Smith, & N. Akagawa (Eds.), *Intangible heritage*, pp. 93–111. London: Routledge.

Harrison, D., & Price, M.F. (1996). Fragile environments, fragile communities? An introduction. In M.F. Price (Ed.), *People and tourism in fragile environments*, pp. 1–16. Chichester: Wiley.

Hazell, P., & Diao, X. (2005). The role of agriculture and small farms in economic development. In *The future of small farms: Proceedings of a research workshop*. Wye, UK, 26–29 June.

Hedlund, T., Marell, A., & Gärling, T. (2012). The mediating effect of value orientation on the relationship between socio-demographic factors and environmental concern in Swedish tourists' vacation choices. *Journal of Ecotourism, 11*(1), 16–33.

Herbert, D.T. (1989). Leisure trends and the heritage market. In D.T. Herbert, R.C. Prentice, & C.J. Thomas (Eds.), *Heritage sites: Strategies for marketing and development*, pp. 1–15. Aldershot: Avebury.

Hewison, R. (1988). Great expectations—hyping heritage. *Tourism Management, 9*(3), 239–240.

Hinch, T., & Butler, R. (2007). Introduction: Revisiting common ground. In R. Butler, & T. Hinch (Eds.), *Tourism and indigenous peoples: Issues and implications*, pp. 1–14. Oxford: Butterworth-Heinemann.

Ho, P.S., & McKercher, B. (2004). Managing heritage resources as tourism products. *Asia Pacific Journal of Tourism Research, 9*(3), 255–266.

IRCI (2012). *The first ICH-researchers forum: The implementation of UNESCO's 2003 Convention*. Osaka: IRCI.

Kennedy, K., & Dornan, D. (2009). An overview: Tourism non-governmental organizations and poverty reduction in developing countries. *Asia Pacific Journal of Tourism Research, 14*(2), 183–200.

Khaznadar, C. (2012). Ten years after—Pandora's Box. In *The first ICH-researchers forum: The implementation of UNESCO's 2003 Convention*. Osaka: IRCI.

Kim, S., Whitford, M., & Arcodia, C. (2019). Development of intangible cultural heritage as a sustainable tourism resource: The intangible cultural heritage practitioners' perspectives. *Journal of Heritage Tourism*, *14*(5–6), 422–435.

Kuper, A. (2003). The return of the native. *Current Anthropology*, *44*(3), 389–402.

Lai, L.W.C. (2020). Sustainable development of heritage conservation and tourism: A Hong Kong case study on colonial heritage. *Sustainable Development*, *28*(5), 1181–1188.

Lake of Stars (2012). About Lake of Stars – Malawi festival. Online at: http://www.lakeofstars.org/about.htm

Lenao, M. (2016). Rural cultural tourism development and agriculture: Evidence from residents of Mmatshumu village in Boteti region of Botswana. In H. Manwa, N. Moswete, & J. Saarinen (Eds.), *Cultural tourism in southern Africa*, pp. 132–144. Bristol: Channel View Publications.

Lepeyre, R. (2016). From hunting-gathering to hospitality? Livelihoods and tourism use of Bushman paintings in the Brandberg Mountain, Namibia. In H. Manwa, N. Moswete, & J. Saarinen (Eds.), *Cultural tourism in southern Africa*, pp. 145–164. Bristol: Channel View Publications.

Lepp, A. (2007). Residents' attitudes towards tourism in Bigodi Village, Uganda. *Tourism Management*, *28*(3), 876–885.

Lewicki, R., Gray, B., & Elliot, M. (2003). *Making sense of intractable environmental conflicts: Concepts and cases*. Washington, DC: Island Press.

Loulanski, T., & Loulanski, V. (2011). The sustainable integration of cultural heritage and tourism: A meta study. *Journal of Sustainable Tourism*, *19*(7), 837–862.

Manwa, H., Moswete, N., & Saarinen, J. (Eds.) (2016). *Cultural tourism in southern Africa*. Bristol: Channel View Publications.

Marschall, S. (2008). Zulu heritage: Between institutionalized commemoration and tourist attraction. *Visual Anthropology*, *21*(3), 245–265.

Mbaiwa, J.E. (2013). *Community-based natural resource management (CBNRM) in Botswana: CBNRM status report of 2011–2012*. Gaborone: Kalahari Conservation Society.

McElroy, J.L., & de Albuquerque, K. (2002). Problems for managing sustainable tourism in small islands. In D.J. Gayle, & Y. Apostolopoulos (Eds.), *Island tourism and sustainable development: Caribbean, Pacific, and Mediterranean experiences*, pp. 15–31. Westport, CT: Praeger.

Mitchell, J., & Ashley, C. (2010). *Tourism and poverty reduction: Pathways to prosperity*. London: Earthscan.

Monare, M. (2013). *Local people's attitudes and perceptions towards benefits and costs of cultural tourism: Case studies of Bahurutshe and Motse cultural villages, Botswana*. Unpublished thesis, University of Botswana, Gaborone.

Moswete, N., Saarinen, J., & Monare, M. (2015). Perspectives on cultural tourism: A case study of Bahurutshe Cultural Village for tourism in Botswana. *Nordic Journal of African Studies*, *24*(3&4), 279–300.

Ndoro, W., Chirikure, S., & Deacon, J. (Eds.) (2018). *Managing heritage in Africa: Who cares?* London: Routledge.

Nic Eoin, L., & King, R. (2013). How to develop intangible heritage: The case of Metolong Dam, Lesotho. *World Archaeology*, *45*(4), 653–669.

Nuryanti, W. (1996). Heritage and postmodern tourism. *Annals of Tourism Research*, *23*(2), 249–260.

Okech, R.N. (2009). Promoting cultural tourism in Kenya: Focus on funerals in Luo and Luhya communities. *Tourism Review International*, *12*, 243–258.

Park, H.Y. (2013). *Heritage tourism*. London: Routledge.

Robinson, M., & Picard, D. (2006). *Tourism, culture and sustainable development*. Paris: UNESCO.

Saarinen, J. (2007). Cultural tourism, local communities and representations of authenticity: The case of Lesedi and Swazi cultural villages in Southern Africa. In B. Wishitemi, A. Spenceley, & H. Wels (Eds.), *Culture and community: Tourism studies in Eastern and Southern Africa*, pp. 140–154. Amsterdam: Rozenberg.

Saarinen, J., Moswete, N., & Monare, M. J. (2014). Cultural tourism: New opportunities for diversifying the tourism industry in Botswana. *Bulletin of Geography: Socio-economic Series*, *26*, 7–18.

Smith, L. (2006). *Uses of heritage*. London: Routledge.

Smith, L. (2015). Intangible heritage: A challenge to the authorised heritage discourse? *Revista d'etnologia de Catalunya*, *40*, 133–142.

Smith, L., & Akagawa, N. (Eds.) (2008). *Intangible heritage*. London: Routledge.

Smith, M., & Richards, G. (Eds.) (2013). *The Routledge handbook of cultural tourism*. London: Routledge.

Stefano, M.L., & Davis, P. (Eds.) (2017). *The Routledge companion to intangible cultural heritage*. London: Routledge.

Stone, M.T. (2015). Community-based ecotourism: A collaborative partnerships perspective. *Journal of Ecotourism*, *14*(2–3), 166–184.

Stone, M.T., & Nyaupane, G. (2014). Rethinking community in community-based natural resource management. *Community Development*, *45*(1), 17–31.

Stone, M.T., & Stone, L.S. (2020). Challenges of community-based tourism in Botswana: A review of literature. *Transactions of the Royal Society of South Africa*, *75*(2), 181–193.

Timothy, D.J. (2007). Empowerment and stakeholder participation in tourism destination communities. In A. Church, & T. Coles (Eds.), *Tourism, power and space*, pp. 199–216. London: Routledge.

Timothy, D.J. (2015). Cultural heritage, tourism and socio-economic development. In R. Sharpley, & D.J. Telfer (Eds.), *Tourism and development: Concepts and issues*, 2nd Edn, pp. 237–249. Bristol: Channel View Publications.

Timothy, D.J. (2020). Heritage consumption, new tourism and the experience economy. In M. Gravari-Barbas (Ed.), *A research agenda for heritage tourism*, pp. 203–217. Cheltenham: Edward Elgar.

Timothy, D.J. (2021a). *Cultural heritage and tourism: An introduction*, 2nd Edn. Bristol: Channel View Publications.

Timothy, D.J. (2021b). Democratising the cultural past: Western values, the Global South and cross-cultural perspectives in heritage tourism. In J. Saarinen, & J.M. Rogerson (Eds.), *Tourism, change and the Global South*, pp. 33–47. London: Routledge.

Ulrich-Cloher, D. (1998). Maori communities and sustainable development. Paper presented at the *Third Biennial New Zealand Tourism and Hospitality Research Conference*, Akaroa, New Zealand, 1–4 December.

UNESCO (2003). Convention for the Safeguarding of the Intangible Cultural Heritage: List of ratifying states parties. Online at: http://www.unesco.org/eri/la/convention.asp?KO=17116&language=E&order=alpha

UNESCO (2021). Intangible cultural heritage. Online at: https://ich.unesco.org/en/home

van den Berghe, P.L. (1995). Marketing Mayas: Ethnic tourism promotion in Mexico. *Annals of Tourism Research*, *22*(3), 568–588.

van Veuren, E.J. (2001). Transforming cultural villages in the spatial development initiatives of South Africa. *South African Geographical Journal*, *83*(2), 137–148.

Wain, A. (2014). Conservation of the intangible: A continuing challenge. *AICCM Bulletin, 35*(1), 52–59.

World Tourism Organization (UNWTO) (2008). *Policy for the growth and development of tourism in Botswana.* Madrid: UNWTO.

Yang, L. (2011). Cultural tourism in an ethnic theme park: Tourists' views. *Journal of Tourism and Cultural Change, 9*(4), 320–340.

Yang, L., Wall, G., & Smith, S. (2008). Ethnic tourism development: Chinese government perspectives. *Annals of Tourism Research, 35*(3), 751–771.

Zeppel, H. (1998). Tourism and aboriginal Australia. *Tourism Management, 19*(5), 485–488.

Zeppel, H. (2002). Cultural tourism at the Cowichan Native village, British Columbia. *Journal of Travel Research 41*(1), 92–100.

Zeppel, H., & Hall, C.M. (1992). Arts and heritage tourism. In B. Weiler, & C.M. Hall (Eds.), *Special interest tourism,* pp. 47–69. London: Belhaven.

Ziff, B., & Rao, P.V. (Eds.) (1997). Introduction to cultural appropriation: A framework for analysis. In B. Ziff, & P.V. Rao (Eds.), *Borrowed power: Essays on cultural appropriation,* pp. 1–30. New Brunswick, NJ: Rutgers University Press.

3

INDIGENOUS TOURISM IN AFRICA

Kokel Melubo

Introduction

Indigenous tourism is one of the fastest expanding segments of cultural tourism, with some observers estimating that as much as 37% of international tourists incorporate visiting indigenous peoples during their tours (Novelli, 2015). Indigenous tourism has been presented as a new vehicle for spurring cultural interaction and revival, allowing indigenous peoples to retain their relationship with the land, bolster employment, and empower their communities. Despite the hundreds of indigenous communities with a wealth of diverse cultures throughout Africa, little attention has been given to indigenous tourism in the region. However, Saarinen and Rogerson (2015) observe that, in the recent past, with culture and people becoming a product of interest, indigenous tourism is becoming a lucrative niche in Sub-Saharan Africa. For example, in South Africa, over one-third of international tourists visit indigenous tourism enterprises. This chapter briefly explores key trends and issues associated with indigenous tourism in Africa.

The Indigenous People of Africa

Africa is home to numerous self-identifying indigenous people. Out of at least 370 million indigenous people worldwide, the United Nations Population Fund (2011) estimates that 50 million reside in Africa (see Table 3.1). They include various 'Pygmy' groups (e.g., Mbuit, Bageyli, and Baka) of the tropical forests of Central and Western Africa, desert dwellers of North Africa and the Sahel, various highlanders of East Africa, and the desert and grassland inhabitants of southern Africa (Mamo, 2020). Most of these people are largely hunter-gatherers and nomadic pastoralists, all of whom share a strong attachment to the land and distinctive cultural practices (Mamo, 2020). Many of the indigenous communities in Africa live in

DOI: 10.4324/9781003153955-3

TABLE 3.1 Examples of indigenous communities of Africa

Region of Africa	*Indigenous Community*	*Approximate Population*	*Countries with Largest Populations*
North Africa and the Sahel	Amazigh (Berber)	50–70 million	Algeria, Libya, Mali, Mauritania, Morocco, Niger
	Toubou	2 million	Chad, Libya, Niger, Sudan
	Tuareg	4.2 million	Algeria, Burkina Faso, Libya, Mali, Niger
West and Central Africa	Bagyeli (Gyele)	n/a	Cameroon, Gabon, Equatorial Guinea
	Baka	20,000–30,000	Cameroon, Congo, Gabon
	Fulani	45–50 million	Benin, Burkina Faso, Cameroon, Gambia, Guinea, Guinea-Bissau, Mali, Niger, Nigeria, Senegal
	Gola	304,000	Liberia, Sierra Leone
	Mbororo	100,000	Cameroon, Central African Republic, Chad, DR Congo, Niger, Nigeria
	Mbuti	35,000	DR Congo
	Ogoni	2 million	Nigeria
East Africa	Akie	5,300	Tanzania
	Barabaig	50,000	Tanzania
	Benet	8,500	Uganda
	Endorois	60,000	Kenya
	Hadzabe	1,300	Tanzania
	Karamojong	371,700	Uganda
	Maasai	2 million	Kenya, Tanzania
	Ogiek	52,500	Kenya
	Orma (Oroma)	159,000	Ethiopia, Kenya
	Pokot	913,000	Kenya, Uganda
	Rendille	96,300	Kenya
	Samburu	333,471	Kenya
	Sengwer	33,187	Kenya
	Turkana	1.02 million	Kenya
Southern Africa	Himba	51,000	Angola, Namibia
	Nama	135,000	Botswana, Namibia, South Africa
	San	105,000	Angola, Botswana, Namibia, South Africa, Zimbabwe

Source: Compiled from multiple sources.

extreme poverty, are disempowered socially and politically, and geographically isolated (Mamo, 2020). They are often last to receive basic services and infrastructure, and face multiple barriers to participating fully in the formal economy, in political processes, and in decision-making for homeland natural resources (World Bank, 2021). Other problems include land insecurity and inadequate political representation (Mamo, 2020). However, the terms ‘indigenous’ and ‘indigeneity’ pose thorny definitional problems because some scholars argue that most Africans are indigenous to the African continent in the sense that they were there before colonial invasions. Therefore, the term 'indigenous' is not applicable (African Commission on Human and People's Rights [ACHPR], 2006). Others argue that using the term indigeneity is particularly inappropriate in the context of Africa because almost all Africans have close ties to ancestral land and all Africans are culturally distinctive (Sylvain, 2002). In Africa, it is far harder to delineate indigenous tribes and communities than it is in settler societies and in other areas, such as South and Southeast Asia (Timothy, 2021).

The ACHPR (2006) lists several characteristics of indigenous people in Africa as communities who have been overlooked or repressed by the dominant groups since colonialism. Their cultures, linguistics, and phenotypes are distinct from the dominant society and are under threat of extinction. They have seen their lives and lifestyles ridiculed, negatively looked upon, destined to disappear, and considered less developed and less advanced. Their way of life includes pastoralism, hunting and gathering, and small-scale farming. They are land-dependent and seen by some outsiders as detrimental to conservation efforts, uneconomical, and lacking a place in the modern world. They often live in geographically isolated regions and suffer from various forms of political and social marginalization. This conceptualization of indigenous people in Africa is unlike other world regions where 'indigenous' usually means aboriginal (those who came first), who have been conquered, colonized, displaced, and even exterminated by the newcomers.

Although subjugated, the colonization of Africa has been less malignant than in some other regions. Indigenous people are defined in relation to a well-demarcated and precisely identified area of land to which they have a strong cultural attachment. Other defining characteristics of indigenous communities include 'unique ties and attractions to traditional habitats and ancestral territories and natural resources in these habitats and territories' (Butler & Hinch, 2007, p. 5). The distinction between indigenous and minority people is limited. However, from an analysis of human rights issues, the ACHPR (2006) states that indigenous people are formulated as having collective rights, whereas minority rights are individual rights. Rights to land, territory, and natural resources are central to indigenous people, whereas the right to enjoy their own culture, practice their own religion, use their own language, and establish their own associations are some specific rights exercised by individuals belonging to nations and linguistic minorities.

Most African governments do not recognize the 'indigenous peoples and other like minorities and ethnic groups as doing so implies a commitment to let their views, values and aspirations guide their own development' (Saugestad, 2001, p. 31). Denying native people the status of 'indigenous' often disqualifies them from claiming specific collective rights such as rights to land, territory, and natural resources. For the governments in Africa, people voicing their concerns as a united indigenous people has a negative connotation because it may lead to tribalism, ethnic conflict, and disruption of the unity of the region or the state (ACHPR, 2006). Following the disapproval of the term 'indigenous people', there is now a feeling that the term 'indigenous community' is more appropriate in Africa and should be used in indigenous-related platforms. The indigenous community is considered more appropriate, as it works towards unification, preserving a sense of solidarity through a common heritage and common destiny. This term is less political, claims no statehood but self-governance, and, therefore, it is appropriate for building up trust between communities and governments (ACHPR, 2006; Mamo, 2020).

Indigenous Communities and Tourism

Indigenous Tourism

Cultural tourism in general and indigenous tourism in particular show signs of success in most Sub-Saharan and North African countries. There is a substantial overlap between 'cultural tourism' and 'indigenous tourism' (Moswete & Lacey, 2014). The World Tourism Organization (2019) affirms that while cultural tourism is a large set of products including historical, ethnic, indigenous, and built heritage (Pereiro, 2016; Saarinen, 2016), indigenous tourism is simply a form of cultural tourism based on indigenous control and ownership of resources. Akama (2002) defines indigenous tourism as a subset of cultural tourism in which indigenous people are directly involved either through control and/or by having their culture serve as the essence of the attraction. Central to indigenous tourism are the questions of cultural identity, the locus of control of traditional ways of life and land resources, and authority over a given tourism activity (Mbaiwa, 2016). Ethnic tourism is another concept intertwined with indigenous tourism. In eliminating the overlaps between the concepts, Yang (2011) and Saarinen and Rogerson (2015) state that ethnic tourism involves tourists visiting ethnic minority groups (not necessarily native to the area) for exotic experiences including watching artistic performances. In this respect, Yang and Wall (2009) observe that all indigenous tourism is ethnic tourism, but the reverse is not necessarily true.

Indigenous people's way of life, including the living heritage of the people, is a subset of cultural heritage. The natural resource-oriented tourism products, namely wildlife, beaches, and landscapes are a mainstay of tourism in Sub-Saharan Africa. The region has been represented as a supermarket of charismatic species of wild animals and untouched wilderness without civilization (Saarinen & Niskala, 2009). Most international tourists who frequent the region come to experience the unique landscapes of the Kalahari and Namib deserts, the expanse of Victoria Falls, the herbivore migration in Serengeti National Park, and Mt. Kilimanjaro, the roof of Africa. The rich and diverse African cultures have yet to become a powerful attraction for tourists compared to wildlife-based tourism.

Mbaiwa, Thakadu, and Darkoh (2008) observe that indigenous tourism has received little attention in national tourism policy development and planning in most African countries. However, there are notable variations in the level of indigenous tourism development in the region. Countries such as Kenya, Botswana, Ghana, Mali, Mozambique, Senegal, and Cote d'Ivoire have successfully developed some form of indigenous tourism compared to countries like Nigeria, Cameroon,

Eritrea, Sierra Leone, Angola, and Zambia, which have limited development (Akama, 2002; Saarinen, 2016). In Botswana and Kenya, indigenous tourism is seen as a key focus to complement nature-based tourism. Others see it as an opportunity to spread tourist flows outside the popular and occasionally congested natural attractions (Melubo & Carr, 2019). Some believe indigenous tourism is a suitable way to involve the communities in the management of tourism and improvement of their wellbeing (Saarinen, 2016). Trends show that some countries have started recognizing the diverse and unique African indigenous heritages as a beneficial tourism asset and an important element of the African tourism industry (Mbaiwa, 2016; Lapeyre, 2016; Saarinen, 2016).

Other community-centered strategies promoted by most African governments include ecotourism, rural tourism, pro-poor tourism, and inclusive tourism. The adoption of community-centered approaches should reduce the enclaved nature of Africa's tourism industry, which is criticized by indigenous communities for the leakage of economic gains back to the tourists' home origins. Akama (2002) believes the future of African indigenous tourism is promising for two reasons. First, unlike some other indigenous tourism, which relies on monumental and material-based cultural heritage, Africa's indigenous tourism relies more heavily on intangible, living culture, such as dances and other performances, and their contemporary ways of living (Tomaselli, 1999). Second, in the minds of some Westerners, Africa is the mother source of all humanity and thus African indigenous communities are part of their own ancestral heritage. Therefore, if tourists, especially those from the African diaspora, want to understand something of themselves and have a feeling of a traditional homeland, they must visit Africa and its indigenous communities in particular.

African Indigenous Knowledge and Cultural Assets as Tourism Resources

African indigenous communities are rich in locally acquired and geographically bound bodies of knowledge. Encompassing ways of knowing, seeing, and thinking, indigenous communities use such knowledge to survive and master their environments (Buzinde, 2019). There are many examples of indigenous knowledge systems in Sub-Saharan Africa. Examples include Maasai ethnoveterinary medicine, which encompasses livestock disease diagnostics and treatment; the Fulani treatment of cattle with euphoria herbs; the Ubuntu community's values of caring for their fellow human beings; the Rwandan *gacaca* justice system for reconciliation and community empowerment; various nutritional systems (e.g., the San people's use of hoodia cacti to stave off hunger and thirst during hunting expeditions); and the Oromo *gada* system for conflict resolution, reparation, and protecting women's rights.

Tourism can benefit directly and immensely from African indigenous knowledge. The restaurant industry throughout the continent serves indigenous cuisines, such as West African *maafe* or groundnut stew, jollof rice, *klouikloui* (rings of fried peanut butter) served in Benin, and the Ntaba Central African-style lamb barbecue. These culinary traditions have been passed down over many generations and constitute an important part of Africa's indigenous knowledge and living heritage. Likewise, African architectural traditions have shaped the designs of many accommodation facilities. Indigenous games such as bao, bow and arrow, bullfighting, dodge ball, hide and seek, toss and catch (*kora*), MANCALA II—rows board game, spinning cone, stick fighting, and wrestling are manifestations of Africa's native living heritage (UNESCO, 2021). Cultural ceremonies are also a crucial part of Africa's living heritage and include, for example, *enkipaata*, *eunoto*, and *olng'esherr* rites of passage, which mark important occasions in life among the Maasai. Others include the *incwala* (first fruits festival) and Umhlanga reed dance festivals in Eswatini; cultural rituals and spiritual sites, such as *Oldoinyo Osira Lenkai* in Tanzania, Lekhubu Island (Botswana), and the 'Hand of Fatima' Mountain, in Mali. Likewise, the very unique languages of the San and Khoikhoi in South Africa and Namibia, and the Hadzabe in Tanzania, are only a few examples of the multitudes of cultural ceremonies and living cultures that are of local importance and have the potential for tourism.

Other more tangible indigenous resources from which tourism could benefit more include San rock art paintings and carvings, rural landscapes, tasteful and nutritional cuisines such as *aliya* (sun-dried meat), *mukimo* (potatoes mashed with pumpkin), and vineyards. Village life and architecture are an important component of indigenous heritage and something in which many tribal peoples take great pride. The traditional homes (*manyatta*) among the Maasai in Kenya and Tanzania, and San in southern Africa; cultural villages, or 'living museums' (e.g., Matsieng's Foot Prints, Toutswemogala Iron Age Site, Motse) in Botswana and Namibia; and UNESCO World Heritage Sites such as Ngorongoro, the sacred Mijikenda Kaya Forest, and Tsodilo Hills World Heritage Site are all important tourism resources connected to Africa's indigenous populations (Manwa et al., 2016; Tomaselli, 2012).

Natural sacred areas such as forests and dramatic landscapes, waterfalls, groves, rivers, lakes, lagoons, caves, islands, springs, and ancient trees are regarded by indigenous communities as holy places that are venerated and ritualized in everyday practice and through periodic pilgrimages (Verschuuren et al., 2021). Most of these sacred locales are beautiful and frequently rich and diverse in wildlife. According to Kideghesho (2008), some indigenous sacred sites such as the *Naimina Enkiyo*, *Mijikenda Kaya*, and *Karima* indigenous forests in Kenya, the Chewa Graveyard Forest in Malawi, and the Sheka Forest in Ethiopia contain a variety of endemic plant species. Mountains such as *Oldoinyo Osira Lenkai* and *Maasai Kinyiraga* (*Aberdare*) are considered sacred shrines by the Maasai and Kamba of Kenya. Tourism is a cornerstone at some African sacred sites, but there is significant potential for developing spiritual tourism among tourists who desire to visit sacred sites for rehabilitation, reconnection with nature or deity, making offerings, undertaking cleansing ceremonies, reawakening their souls, and utilizing local herbs for healing.

Opportunities for Indigenous Tourism

For indigenous communities, tourism is a double-edged sword. While native communities welcome tourism as a 'silver bullet' for addressing socioeconomic challenges and view tourism as an opportunity for self-determination and improved wellbeing (Buzinde, 2019), they are cognizant of the contradictions it brings.

On the one hand, tourism has significant potential to benefit the lives of indigenous communities, but it also brings problems. In Botswana, Moswete, Thapa, and Lacey (2009) discovered that with community or family-owned and operated cultural village enterprises, a significant number of African indigenous people, especially women, have benefited. Living heritage villages such as Bahurutshe Cultural Village, Motse Cultural Village, and Bangwaketse Village (Botswana); Shakaland Zulu Cultural Village (South Africa); and Damara Cultural Village (Namibia) have been praised for allowing indigenous women of the Gcwihaba, XaiXai, and Hurutshe communities to engage fully in tourism. In these 'living museums' women's creative works, such as decorations (e.g., *lolwapa* for the Tshana people), traditional music and dance, narrations of ways of life, production of handicrafts such as baskets from reeds and beads, pottery making, wearing traditional attire, and making food form a part of the tourism product. Through these activities, women earn money to sustain their families, increase self-confidence and wellbeing, acquire leadership skills, reduce their reliance on their husbands for income, and build their entrepreneurial skills (Mkhize & Cele, 2017; Tshabalala & Ezeuduji, 2016). Melubo and Carr (2019) observe that at Manyatta, Tanzania, through money-making ventures, women have become increasingly empowered in socioeconomic terms.

> Maasai women are the skillful crafters of almost all jewelry and handcrafts such as beadwork...They enact an active role in welcoming tourists with the performances of songs and dance. Most handicrafts, including beadwork, are sold to tourists at the bomas.
>
> *(Melubo & Carr, 2019, p. 226)*

Through these efforts, Maasai women have become the main beneficiaries of tourist spending. The economic contribution women make to family income has earned them respect, dignity, and leadership in a patriarchal society (Akama, 2002). In sum, employment in tourism has allowed women to work outside the domestic realm, amplify their voice in the community, and own property (Boonabaana, 2014). While Maasai women are the largest demographic group involved in the production and sale of tourist crafts, men control the business end of the enterprise. Lack of capital, business, and marketing skills, access to markets, mobility constraints, and restrictive gender roles are some of the factors that hold back indigenous women (Manwa et al., 2016).

One of the main tangible benefits of evolving indigenous-run tourism in Africa is the creation of employment opportunities. Studies in Botswana and Namibia indicate where opportunities to earn cash income are limited; most of the employees serving in cultural villages are elderly men and women of the area (Manwa et al., 2016). Terry (2001) notes that at least 7,500 indigenous San women and men in KwaZulu Natal, South Africa, were involved in basketry production alone. They also work as tour guides, traditional healers, craft and curios shopkeepers and owners, cultural performers, and poets. The money earned from these cultural initiatives is not spent frivolously but rather used to improve their standard of living by buying basic needs and paying school fees. Indeed, 'living museums' and craft-related initiatives have reduced the rural-to-urban migration trend in these communities.

In Tanzania, tourist lodges and tour operators employ Maasai youth to serve as safari guides and security personnel due to their extensive knowledge about local customs and ability to speak the language, their understanding of local plants and animals, their ability to detect and avoid predators, and their familiarity with the local landscape (Charnley, 2005). Employing local guides can contribute a great deal to community empowerment and the quality of the visitor experience. However, some tour operators feel that the indigenous Maasai people lack English language and driving skills, which limits their employability (Charnley, 2005).

Indigenous tourism also has the potential to protect and revive cultures (Ruhanen & Whitford, 2019). In Botswana, the Bahurutshe and Bangwaketse people have established a commercial enterprise called 'cultural lodges'. These traditionally designed lodges are thatched and decorated with red earth. Their purpose is twofold: to showcase, protect and preserve the cultures of the respective indigenous communities, and to earn money through tourism. Traditional activities offered at the cultural lodges include arts, crafts, traditional dancing, and educational cultural tours. Poor road networks and lack of needed entrepreneurial, marketing, professional, and managerial skills are some of the challenges precluding the community from benefiting further from tourism.

The homelands of the African indigenous communities are rich in biodiversity, including abundant wildlife, which is the prime target of most tourism products and services. Land and natural resources are the most significant tool for the community to negotiate with interested parties, including tourism investors. However, those who own the land also control and use its resources, which is a critical concern in African indigenous communities. Lack of community ownership and control over the land and resources inhibits indigenous communities from investing and benefiting from tourism opportunities (Coria & Calfucura, 2010; Fuller et al., 2005). Fletcher, Pforr, and Brueckner (2016, p. 1106) argue that ownership and control over land resources in Namibia are central to the nature and 'type of tourism activity and development, the capacity to attract investment and ultimately the achievement of the full tourism potential'. Land rights enable indigenous communities to maintain traditional systems, control the quality and authenticity of cultural products offered, and exercise local economic and sociocultural autonomy (Zeppel, 2006).

Across the region, private tourism investors have entered into joint management agreements with indigenous communities to develop new tourism enterprises, such as lodges. The agreements usually involve the community offering a rich land and authentic indigenous experience, while the private company often owns the infrastructure and services (e.g., the lodge or camp facilities), brings in clients (tourists), invests capital, and reaps most of the profits (van Veuren, 2001).

Success stories in Africa include Ololosokwan Village, Tanzania, where the Maasai people have leased 21,000 acres to &Beyond Klein's Camp Safari Lodge. Under this agreement, the community has stopped cultivation and encroachment that threaten the Serengeti ecosystem, and in return, the company pays a fixed land rent, regardless of the number of guests, and a bed levy per guest to the village (Snyder & Sulle, 2011). Other direct benefits to the community include the employment of 37 staff, plus medical staff whose salaries are paid by the company, and entrepreneurial projects such as honey production (Burgoyne & Mearns, 2017). With this arrangement, Ololosokwan Village earned approximately USD $100,000 from the lodge in 2007 (Nelson, 2012).

African indigenous communities' engagement in tourism has been enhanced with managerial, financial, and social capital skills. Indigenous communities could benefit more from tourism by supplying goods and services to tourism enterprises, which provide a local market for products. In Botswana, indigenous communities sell fruits, vegetables, fish, and crafts to tourism establishments. Crucial factors for the success of African indigenous community businesses include local leadership, legally registered and community-owned land resources with clearly defined borders, security of land tenure, a high level of community engagement, substantial capital input, tourism and business skills, and mutual recognition of community and tourism private company expectations (Ashley & Jones, 2001; Brown & Bird, 2011).

The Ugly Side of Tourism in Indigenous African Communities

Tourism has been accused of misrepresenting, marginalizing, and exploiting indigenous communities in Africa (Giraudo, 2018). For example, the colonial forms, images, and perceptions of the Maasai have been that of a backward, warlike, and belligerent people. These images have been marketed by tour operators and travel agents to prospective tourists as objects of fascination, mystique, and untouched by the forces of modernization—always without the involvement or consultation of the Maasai (Salazar, 2015). Their romanticized image has been circulated across the globe in coffee-table books, playing cards, postcards, movies, films, magazines, and on websites portraying them as authentic Africans—a priceless tourist attraction that epitomizes the 'African warrior' (Salazar, 2015). For thousands of Western tourists who travel to East Africa, the Maasai are a people who walk tall amidst deadly wild beasts and are depicted as a backward community that provides additional anecdotes to international tourists seeking a 'wild' Africa (Salazar, 2015). Salazar (2015) notes that colonial prejudices continue even among Africans. In Tanzania, the guides from the Meru tribe have donned Maasai robes and weapons to attract tourists and have staged fake Maasai culture, often representing and peddling a fantasy about the Maasai as backward and racially distinctive Tanzanians. These projections are dangerous because they not only distort and re-construct peoples and places erroneously (Bruner, 2002), but they also misappropriate culture and rob people of their dignity, making them somehow less equal as human beings.

Much tourism development in Africa's indigenous destinations has been enclavic in nature (Mbaiwa, 2005), which means that most tourist facilities, lodges, resorts, and food services are owned and controlled by foreign investors and non-indigenous people. Therefore, most of the profits from tourism in Africa's indigenous communities go to non-native actors and corporations. Foreign dominance results in the leakage of profits to areas outside the destination through outsourcing many supplies and management skills. This results in high salary disparities as locals receive less consideration for senior positions that are usually filled by outsiders. In the Okavango Delta, Mbaiwa (2005) notes that the vast majority of tourist facilities, lodges, and camps that occupy prime tourist areas are foreign-owned. The Basarwa community (comprised of Bushmen or San, such as the Nharo, the Hambukushu, and Wayei) owns no tourist facilities except the cultural performances that are gazed upon by foreign tourists (Giraudo, 2018). The dominance of non-indigenous people reduces local control over local resources and frequently results in the 'selling out' of native land resources to accommodate the requirements of 'others' and the exclusion of native people in decisions regarding the use of resources (Mbaiwa, 2005). Likewise, most of these foreign businesses outsource even locally available agricultural products ostensibly because the latter are of poorer quality, have unreliable supplies, suffer untimely delivery, and experience price instability.

In addition, the vast majority of employees within tourist facilities are non-indigenous. In Ngorongoro, the indigenous Maasai are usually employed in low-income and unskilled jobs, such as lodge guards, while better-paying positions are in the hands of non-Maasai (Charnley, 2005). In addition to a good salary, most non-Maasai employees receive gratuities, home leaves, days off, children's education allowances, and housing allowances. Mbaiwa (2005) observes that tourism has often been less sensitive to the needs and aspirations of indigenous communities and has failed to create opportunities for socioeconomic development, in part because of limited ownership and control of land resources. In situations where the land owner is an indigenous entrepreneur, there is a strong sense of ownership, commitment to cultural conservation, and education and job creation (Fletcher, Pforr & Brueckner, 2016).

Africa's indigenous communities are disadvantaged by the very tourism industry which is happening at their doorsteps (Manyara & Jones, 2007). Instead of being active participants, most indigenous communities play a passive role. The vicious cycle of poverty which has permeated daily life precludes many of them from the sea of tourism opportunities. In most African countries, indigenous people have less access to education, national health systems, and appropriate medical care. They may also suffer nutritional problems and are less likely to live in decent housing, compared to their non-indigenous counterparts. Other influential factors include climate change, which has led to food insecurity and reduced availability of water, land dispossession, and a lack of recognition of their traditional livelihood economies, particularly pastoralism and subsistence hunting. In the words of Psacharopoulos and Patrinos (1994), being indigenous is being poor. For instance, the National Bureau of Statistics (NBS) (2017) found that 50% of the Ngorongoro Conservation Area's (NCA) Maasai residents (72.2% women) are illiterate, 22.1% experience severe water scarcity, and 70% suffer from hunger. Overall, indigenous people are by and large at the bottom of the development scale in every dimension of life (Mamo, 2020). Being financially the poorest, yet among the richest in indigenous knowledge and living in regions enriched with abundant natural resources is a tragic irony.

About 80% of protected ecosystems are within the territories of indigenous communities. Corrigan et al. (2018) observe that large portions of such protected areas are in Africa, Asia, and Latin America, which are home to the majority of the world's indigenous populations. African indigenous communities have depended on these ecosystems to survive, and for spiritual, cultural, and social nourishment. These native groups are the guardians of the world's traditional knowledge and intangible heritage. Sadly, for indigenous communities, the troubles of living within tourism-oriented protected areas outweigh the benefits. Due to Western-inspired exclusionary conservation and top-down management models, many African indigenous communities have lost their land to the protection of wildlife and tourism (LeBlanc & LeBlanc, 2010). In addition, other lands are held in private concessions, state-held ranching corporations, and large agricultural plantations. Indigenous people have lost access to prime grazing areas, water sources, and spiritual sites that are fundamental to their lives. For example, in Tanzania, the Maasai have lost the Tarangire and parts of the Ngorongoro, Mkomazi, Arusha, and Serengeti because of these areas' conservation and recreational values (Brockington, 1999). In Uganda, the Batwa have lost the Bwindi, Mgahinda, and Echuya forests. In Kenya, the Endorois have lost Lake Bogoria, Lake Nakuru, and Lake Elementatia. The hunting and gathering community of the BaAka Pygmies in Central Africa have lost their traditional hunting grounds, which are now part of Lobeke National Park. Because of their exceptional diversity of wild animals and scenic landscapes, some of these areas have been designated World Heritage Sites and are becoming flagship tourist attractions. The removal of indigenous communities equals a loss of livelihoods, history, memory, sense of place, and representation.

In terms of livelihoods, Melubo and Lovelock (2019) provide an account of the denial of land rights of Maasai pastoralists and the continued threats to evict them. The community was removed from the Western Serengeti in 1959 based on the claim that water reserves and the land were too fragile for wildlife to survive and a safe environment devoid of people was necessary. This forced relocation had profound consequences on the ability of the natives to continue their pastoralist activities. In recent years, the Maasai community has lost access to Ngorongoro, Empakai, and Olmoti craters and *Entim Oolturot*, the Northern Highland Forest, which were important locations for Maasai settlements, cultural and spiritual activities, rituals, and ceremonies (Melubo & Lovelock, 2019). These were also the only sources of salt and water in the dry season. Making grazing illegal has caused many NCA Maasai residents to remain very poor inside a high-end tourist destination (NBS, 2017), which generated over 123 billion Tanzanian shillings in 2019/2020. To add salt to the wound, the NCA has a plan to resettle around 82,000 Maasai people, including 40,000 'immigrants' out of the area, expand the area from 8100 km^2 to 12,083 km^2 by annexing other Maasai areas of Loliondo and Lake Natron Game Controlled Areas (GCA), and divide the area into four zones: conservation core zone, conservation sub-zone, transition zone, and settlement and development zone. The plan presents a grave threat to the continued survival of the Maasai as it restricts community access to about 82% of the total area (Ministry of Natural Resources and Tourism, 2019). Other protected areas and community challenges include wildlife invasions. Wishitemi and Okello (2003) found that over 60% of the Maasai community in the Amboseli ecosystem lose livestock annually because of frequent incursions by wildlife. Consequently, the Maasai adjacent to protected areas experience unending wildlife conflicts, which have plunged them further into a cycle of poverty.

The establishment and management of protected areas have taken a top-down approach by government agencies, without consent or meaningful participation of indigenous communities (Melubo & Lovelock, 2019). True participation requires the recognition of indigenous people as the rightful owners and equal partners in the management of protected areas while appreciating their values, ethics, and attitudes towards the environment by incorporating these aspects into policies and agendas (Timothy, 2007). Amougou-Amougou and Woodburne (2014, p. 109) note that the creation and management of protected areas in most African countries have been done 'with minimal and sub-standard participation of indigenous peoples'. Melubo and Lovelock (2019) observe that the participation of indigenous Maasai in management and decision-making in the affairs of the NCA is a matter of grave concern. The Maasai are under-represented in the NCA Board of Directors and within top management levels (Melubo & Lovelock, 2019). Similarly, the indigenous forest Twa Pygmies claim never to have been consulted during the creation of the Kahuzi-Biega National Park in DR Congo, nor were they participants in the preparation of the park's management plans (Buhereko, 2014). Around 580 Twa families and up to 6,000 individuals were expelled brutally from the Kahuzi-Biega forest to protect the habitat of endangered gorillas. This under-representation means the concerns, interests, wishes, and views of the Twa are not adequately acknowledged in the management of the area.

As previously noted, there is a strong connection between indigenous people and their native lands. For African indigenous youth, the loss of land disconnects them further from their cultural roots. The reduced territories of Cameroon's Baka people curtail young people's ability to recognize forest animals or the plants needed for an independent life in the forest. The Bagyeli and Mbororo youth of Cameroon can no longer hunt and have lost the skills of making snares from vines. The Batwa people in Rwanda and Burundi have become wage laborers, potters, and beggars, and no longer hunt, following their removal from their ancestral lands. The loss of access to their traditional lands has deep implications for indigenous people's rights to life, health, and cultural continuity.

Cultural Incursions, Commodification, and Intellectual Property Rights

In Africa, cultural incursion is a prevalent concern. From a cultural perspective, the foreign tourist-indigenous community relationship has been an ugly one. Language is a key part of African indigenous identity, and something of which most communities are most proud. Unfortunately, tourism development has created pressure for indigenous youth to reject the local vernacular. To host international tourists, indigenous communities are incentivized to learn English, Chinese, Spanish, German, and French, as opposed to maintaining their native language. While both are possible, many young people tend to prefer speaking a foreign language over their native tongues because of the status it implies. The disuse of local languages in tourism settings may lead to the loss of native languages, which in turn can affect a community's ability to sustain indigenous practices and traditions, including the maintenance of biocultural resources (Whitney-Squire, 2016). The loss or deterioration of a language is synonymous with the loss of rich indigenous knowledge that is inherent within it.

Likewise, indigenous youth often are said to imitate tourists by changing their dress code, which involves wearing clothing and styles previously not common in their areas, reflecting the well-documented behavior of destination residents copying the behaviors, spending patterns, and trends of the tourists, known as the 'demonstration effect' (Mathieson & Wall, 1982). Mbaiwa and Stronza (2010) note, for instance, after the introduction of tourism in the Okavango Delta, some XaiXai youth have copied culturally unacceptable behaviors, including wearing clothes that expose body parts and targeting tourists from America and Europe for commercial sex. Other reported deviant activities include incidences of smoking, alcoholism, and youth idling along highways waiting for tourists to take pictures in exchange for money (Akama, 2002; Charnley, 2005). Tourists also misbehave. In Botswana, the community at Goo-Moremi noted that some tourists indulge in sexual activities during the day and swim in pools, both of which have the potential to upset the ancestors.

The commodification of African indigenous communities is prevalent. The tourism business world and governments enthusiastically commodify elements of indigenous cultures, such as lifestyles, artwork, religious rites, and cultural performances (Giraudo, 2018). Likewise, realizing the profits derived from selling cultural products, indigenous communities themselves, such as the Maasai of Kenya and Tanzania, the ǂKhomani Bushmen of Botswana, and the Ju/'Hoansi of NyaeNyae in Namibia, have joined the movement to trade their cultures for money (Koot, 2018; Melubo & Carr, 2019). For example, the ethnographic study by Avieli and Sermoneta (2020) among Maasai interlocutors finds that Maasai men have become champions of manipulation by using their culture and looks to attract tourists. They invite visitors to take pictures with them and then required payment afterward.

The production of culture for tourists' consumption has resulted in changing most of the meanings, forms, and functions of tribal heritage. For example, carved masks are now sold in Kenya and Tanzania as authentic African art, while anthropological evidence shows that these pieces of art are not original to the region (Timothy, 2021). Because of tourists' propensity to spend and their desire for cheap mementos, some cultural artworks and performances are now mass-produced in Africa the way it has been done for decades in other parts of the world (Mathieson & Wall, 1982). Melubo and Carr (2019) provide an account of the Manyatta tourist center where the cultural meaning of *Osingolio* (a dance), celebrations, and rituals of the Maasai have been altered considerably. For example, while blood-letting and blood drinking from a cow was a ritual exclusively practiced by Maasai warriors, and must not be witnessed by women, at Manyatta, slaughtering and drinking is done away from the eyes of Maasai women but freely done in front of tourists, including women tourists (Melubo & Carr, 2019). Welcoming every tourist vehicle that enters with a dance is also uncultured in the eyes of traditional Maasai because dances should be performed for special events, such as ceremonies and rituals. Some handicrafts (e.g., anklets, arrows, bracelets, chokers, cuffs, and earrings) on display, which are offered as 'authentic' depictions of original pieces, have become cheapened and modified in terms of design, colors, patterns, quality, and rate of production to suit the demands of the consumers (Bruner & Kirshenblatt-Gimblett, 1994; Melubo & Carr 2019).

The presentation of cultural products and performances accepted by tourists as original, but regarded by indigenous communities as inaccurate, is called 'staged authenticity' (Acheraïou, 2011; MacCannell, 1973). Staged authenticity creates tensions and conflicts between indigenous communities who may be reluctant to commercialize their sacred culture and tour operators and travel agents who wish to commoditize local culture for tourist consumption (Timothy, 2021). The survey by Mbaiwa (2011) at Goo-Moremi, Botswana, shows that elderly people in the community are unhappy with efforts to sell their sacred culture. To the elders, culture is a private affair that outsiders should not be allowed to see. Conversely, to tour operators, Goo-Moremi's living heritage is a major asset for tourist consumption. Due to resistance among spiritual leaders to deny tourists access to sacred sites such as the graves of *Kgosi Mapulane*, whose spirit directs the affairs of the village, some guides have shown tourists the wrong sites and disseminated incorrect information.

Cultural performances presented in tourist settings are likely to misrepresent and objectify indigenous communities. For example, Bruner (2002) observed Maasai performers dancing in tourist settings such as in hotels or at specially constructed indigenous spaces such as Manyatta, situated mostly along tourist routes. The Lesedi Cultural Village in South Africa and Ligugu Lemaswati Swazi cultural village in Eswatini are other classic examples of the same phenomenon. Traditional tribal dances include various displays, narratives, and spectacles, while the addition of demonstrating leopard skins, feathers, and spears is made up to satisfy the thirst of international tourists who stop by for a short visit (Saarinen, 2016). Disappointingly, about 96% of the revenue accrued from indigenous cultural enterprises, such as Manyatta's, is unevenly distributed, as much of the money remains in the hands of tour sellers, whereas little is shared with the local inhabitants (Melubo & Carr, 2019).

Tour operators and travel agents market, promote, and organize visits to cultural settings to prospective tourists to increase sales and profit margins (Akama, 2002). Often, tour operators misrepresent indigenous communities, sometimes equating them with wild animals. For example, the Maasai are seen as part of the landscape, living in harmony with dangerous fauna. Because of the promotional efforts of tour companies, from the tourists' perspective, the Maasai people seem to be the only inhabitants of Tanzania, even though the country is home to over 120 distinct ethnic groups (Salazar, 2015). Business operators want their clients to see the Maasai as unchanging people, who continue to practice nomadic pastoralism, regardless of the fact that modern Maasai are business people or politicians with farms and cattle, drive motor vehicles, use mobile phones, wear suits, and live in modern two-story houses (Akama, 2002). These tourism-induced obsolete representations of indigenous Africans were largely creations of their European colonizers and today are perpetuated by tourism in an effort to preserve the idealized notions of 'native' Africans, or what the visitors (neo-colonizers) assume the indigenous people *should* look like in order to satiate their own preconceived prejudices and imaginings.

For some indigenous communities, the issue is not necessarily the exposure of their culture but rather who controls and benefits from its exposure. Mbaiwa (2016) notes indigenous communities are happy not only to directly control the meaning, content, use, and trajectory of their cultural performances, but also to reap the fruits of sharing. Sadly, for the most part, they feel their participation is restricted and their performances are manipulated to accommodate the desires and tastes of international tourists.

In tourism settings, African indigenous community knowledge systems, objects, products, and expressions are often deviated, reproduced, replicated, faked, and misappropriated, and what is thought to be original is typically rather inauthentic (Johnston, 2000). The Western systems that foster commercialization of cultural heritage have turned local artifacts into tourist kitsch and non-tourist consumer goods. For instance, the colorful shawls, bright beads, and red-checkered prints with Maasai expressions and designs have been appropriated by major fashion brands, souvenir manufacturers, and the automotive industries without proper attribution. High-end fashion houses (e.g., Louis Vuitton and Masai Barefoot Technology), moccasin maker Minnetonka, and motor vehicle manufacturer Land Rover have imitated Maasai imagery or iconography to lend appeal to their brands. It is estimated that more than USD $10 million is earned each year from Maasai-inspired brands. In 2012, for instance, the Moët Hennessy Louis Vuitton fashion conglomerate sold Maasai-inspired scarves, duffle bags, beach towels, hats, and shirts at 1,000 euros each (Leleto, 2019). Unfortunately, there is no way for the Maasai who are portrayed in thousands of postcards and tourist souvenirs to receive a portion of the proceeds generated by the fashion industry, the automobile sector, and the souvenir marketplace.

Outside artisans who use indigenous resources are not bound by traditional customs of authenticity. They do not practice traditional methods of production or principles of design, which may be time-consuming. For example, the production of an authentic cultural symbol may take two weeks, but in the hands of a tourism business, it may take only a day or less. Sadly, most tourists are unable to distinguish indigenous handiwork from non-indigenous fakes. In fact, in most cases, tourists will demand brightly colored artwork that is not indigenously produced or sanctioned by the native community (Johnston, 2000).

In the area of medicine and health, there are concerns that Western companies and researchers are patenting African indigenous community treatments and cures, often without proper information, recognition, or economic returns to native people (Johnston, 2000). Herbal products are an illustrative example. In Simanjiro, Tanzania, the Maasai have accused the Swiss Lauman company of patenting *oloisuki* (*Zanthoxylum chalybeum*), a medicinal plant used to treat malaria, stomach aches, and measles, and prevent fevers (Laltaika, 2011). The company is particularly interested because *oloisuki* has been pre-tested and pre-screened for toxicity and contraindications. Similarly, in Uganda, the Kalinzu-Maranaganbo Forest community has accused a French company for patenting *Prunus africana*, a native plant used for the treatment of benign prostatic hyperplasia (Bodeker, van't Klooster, & Weisbord, 2014).

While it is evident that the indigenous communities of Africa have many natural resources of ethnobotanical medicinal value, the region lacks adequate expertise, funding, and technology to exploit its grassroots curative knowledge. African indigenous knowledge technologies, know-how, and practices about traditional medicine may be considered possessions of specialized healers, preserved within a family, or shared among and between members of a clan or community and their neighbors (Elvin-Lewis, 2007) and distributed across jurisdictions. For instance, the Oromo of Ethiopia consider coffee an integral part of their life in cultural and ritual functions. The preparation and consumption of coffee are thought to bring blessings and safeguard the household because it has spiritual powers. The women use coffee ritually as a symbol of penance for offenses committed. The preparation is done by women only (Sereke-Brhan, 2010), but Ethiopian coffee is now seen as being among the world's finest and has become, together with local coffee cultures, an important point of pride in the country's tourism schematic (Jolliffe, 2010; Woyesa & Kumar, 2021).

Intellectual property rights are one way of curbing the appropriation of indigenous cultural resources. For example, the hunter-gatherer San-speaking people signed a benefit-sharing agreement with the American pharmaceutical giant Pfizer. Through the agreement, the company could harvest the hoodia cactus, an appetite-suppressant used by the San to stave off hunger during hunts. In return, the San were to receive 14% of royalties and payments from licenses. The plant is considered a potential weight-loss drug and cure for obesity (Martin & Vermeylen, 2005). However, the adequacy and appropriateness of the Western economic and legal-styled intellectual property rights in Africa for the protection of community resources have been questioned (Johnston, 2000). Reasons for these challenges are severalfold: in the view of indigenous communities, heritage has no owner, exclusivity, or possessory rights; herbal remedies are a communal property; and the community does not advocate for commercialization or privately owned use of natural resources. The Maasai and the San view their monopoly of knowledge as being beneficial for the community to ensure that the knowledge does not die in the hands (or head) of the knowledge holder. Furthermore, it is also very difficult for poor indigenous communities to meet some of the evidence requirements for test cases involving intangible culture, and it is prohibitively expensive to hire experts in cases of litigation (Johnston, 2000).

Discussion and Conclusion

Africa is home to many indigenous communities, which have diverse cultures and many heritage resources. Besides cultural assets, indigenous lands are rich in flora and fauna. Despite their cultural wealth, most African countries have been slow in identifying and integrating indigenous cultures into their overall tourism development strategies. Encouragingly, countries such as Kenya, Botswana, Namibia, and Tanzania have started to recognize indigenous tourism as a key element of cultural tourism and as a valuable add-on product of the region's normative wildlife tourism. Indigenous tourism has proven beneficial for revitalizing cultures and helping to protect unique natural resources, including the NCA, Tanzania, and Tsodilo Hills World Heritage Site, Botswana—both UNESCO World Heritage Sites (Mbaiwa, 2016; Melubo & Lovelock, 2019). In many regards, the negative impacts of tourism on indigenous communities appear to outnumber the positives in Africa. The loss of prime native lands, loss of human dignity, the commoditization of art and culture, diminishing cultural values, and the sanitization of cultural resources for tourist consumption are some of the negative impacts brought to native communities through tourism (Mbaiwa, 2016).

There are several critical issues facing African indigenous communities. The ownership and control of cultural resources by African native communities is far from being satisfactory. While some indigenous groups may legally own traditional homelands and resources, most of them are unable to exercise control over them. Without control of land or cultural assets, they have been disempowered and have largely been precluded from profiting from the cultural tourism sector that uses their heritage as its primary resource (Butler & Hinch, 2007).

References

Acheraïou, A. (2011). *Questioning hybridity, postcolonialism and globalisation*. London: Palgrave MacMillan.

African Commission on Human and People's Rights (ACHPR) (2006). *Indigenous peoples in Africa: The forgotten peoples? The African Commission's work on indigenous peoples in Africa*. Copenhagen: International Work Group for Indigenous Affairs.

Akama, J. (2002). The creation of the Maasai image and tourism development in Kenya. In J. Akama & P. Stery (Eds.), *Cultural tourism in Africa: Strategies for the new millennium, proceedings of the ATLAS Africa international conference*, pp. 43–54. Arnhem, Netherlands: Association for Tourism and Leisure Education.

Amougou-Amougou, V., & Woodburne, O. (2014). The Sangha Trinational World Heritage Site: The experiences of indigenous peoples. In S. Disko & H. Tugendhat (Eds.), World Heritage Sites and indigenous peoples' rights, pp. 189–220. Copenhagen: International Work Group for Indigenous Affairs.

Ashley, C., & Jones, B. (2001). Joint ventures between communities and tourism investors: Experience in Southern Africa. *International Journal of Tourism Research*, *3*(5), 407–423.

Avieli, N., & Sermoneta, T. (2020). Maasai on the phone: Materiality, tourism, and the extraordinary in Zanzibar. *Humanities and Social Sciences Communications*, 7(1), 1–10.

Bodeker, G., van't klooster, C., & Weisbord, E. (2014). Prunus africana (Hook. f.) Kalkman: The overexploitation of a medicinal plant species and its legal context. *The Journal of Alternative and Complementary Medicine*, *20*(11), 810–822.

Boonabaana, B. (2014). Negotiating gender and tourism work: Women's lived experiences in Uganda. *Tourism and Hospitality Research*, *14*(1–2), 27–36.

Brockington, D. (1999). Conservation displacement and livelihoods: The effects of eviction on pastoralists moved from the Mkomazi Game Reserve, Tanzania. *Nomadic Peoples*, *3*, 74–96.

Brown, J., & Bird, N. (2011). *Sustainable natural resource management in Namibia: Successful community-based wildlife conservation*. London: Overseas Development Institute.

Bruner, E.M. (2002). The representation of African pastoralists: A commentary. *Visual Anthropology*, *15*(3), 387–392.

Bruner, E.M., & Kirshenblatt-Gimblett, B. (1994). Maasai on the lawn: Tourist realism in East Africa. *Cultural Anthropology*, *9*(4), 435–470.

Buhereko, R.M. (2014). Kahuzi-Biega National Park: World Heritage Site versus the indigenous Twa. In S. Disko & H. Tugendhat (Eds.), *World Heritage Sites and indigenous peoples' rights*, pp. 131–146. Copenhagen: International Work Group for Indigenous Affairs.

Burgoyne, C., & Mearns, K. (2017). Managing stakeholder relations, natural resources and tourism: A case study from Ololosokwan, Tanzania. *Tourism and Hospitality Research*, *17*(1), 68–78.

Butler, R., & Hinch, T. (Eds.) (2007). *Tourism and indigenous peoples: Issues and implications*. London: Routledge.

Buzinde, C.N. (2019). Tourism and Indigenous communities: Linking reterritorialisation and decolonisation in North Africa. In D.J. Timothy (Ed.), *Routledge handbook on tourism in the Middle East and North Africa*, pp. 83–94. London: Routledge.

Charnley, S. (2005). From nature tourism to ecotourism? The case of the Ngorongoro Conservation Area, Tanzania. *Human Organization*, *64*, 75–88.

Coria, J., & Calfucura, E. (2010). Ecotourism and the development of Indigenous communities: The good, the bad and the ugly. *Ecological Economics*, *73*, 47–55.

Corrigan, C., Bingham, H., Shi, Y., Lewis, E., Chauvenet, A., & Kingston, N. (2018). Quantifying the contribution to biodiversity conservation of protected areas governed by Indigenous peoples and local communities. *Biological Conservation*, *227*, 403–412.

Elvin-Lewis, M. (2007). Professional responsibilities to custodians of traditional knowledge. *African Journal of Traditional, Complementary and Alternative Medicines*, *4*(4), 380–381.

Fletcher, C., Pforr, C., & Brueckner, M. (2016). Factors influencing Indigenous engagement in tourism development: An international perspective. *Journal of Sustainable Tourism*, *24*(8–9), 1100–1120.

Fuller, D., Buultjens, J., & Cummings, E. (2005). Ecotourism and indigenous micro-enterprise formation in northern Australia: Opportunities and constraints. *Tourism Management*, *26*, 891–904.

Giraudo, R.F. (2018). San cultural tourism: Mobilizing Indigenous agency in Botswana. In A.C. Bunten & N.H.H. Graburn (Eds.), *Indigenous tourism movements*, pp. 117–139. Toronto: University of Toronto Press.

Johnston, A. (2000). Indigenous peoples and ecotourism: Bringing indigenous knowledge and rights into the sustainability equation. *Tourism Recreation Research*, *25*(2), 89–96.

Jolliffe, L. (Ed.) (2010). *Coffee culture, destinations and tourism*. Bristol: Channel View Publications.

Kideghesho, J.R. (2008). Co-existence between the traditional societies and wildlife in western Serengeti, Tanzania: Its relevancy in contemporary wildlife conservation efforts. *Biodiversity and Conservation*, *17*(8), 1861–1881.

Koot, S. (2018). The Bushman brand in southern African tourism: An Indigenous modernity in a neoliberal political economy. *Senri Ethnological Studies*, *99*, 231–250.

Laltaika, E. (2011). *Oloisuki–A tale of biopiracy in the Maasai steppe*. Bonn: Church Development Service.

Lapeyre, R. (2016). From hunting-gathering to hospitality? Livelihoods and tourism use of Bushman paintings in the Brandberg Mountain, Namibia. In H. Manwa, N. Moswete, & J. Saarinen (Eds.), *Cultural tourism in southern Africa*, pp. 145–164. Bristol: Channel View Publications.

LeBlanc, J., & LeBlanc, V. (2010). National parks and Indigenous land management: Reshaping tourism in Africa, Australia and Canada. *Ethnologies*, *32*(2), 23–57.

Leleto, N.L. (2019). Maasai resistance to cultural appropriation in tourism. *Indigenous Peoples' JL Culture & Resistance*, *5*, 21–35.

MacCannell, D. (1973). Staged authenticity: Arrangements of social space in tourist settings. *American Journal of Sociology*, *79*(3), 589–603.

Mamo, D. (Ed.) (2020). *The indigenous world 2020*. Copenhagen: International Work Group for Indigenous Affairs.

Manwa, H., Moswete, N., & Saarinen, J. (Eds.) (2016). *Cultural tourism in southern Africa*. Bristol: Channel View Publications.

Manyara, G., & Jones, E. (2007). Community-based tourism enterprises development in Kenya: An exploration of their potential as avenues of poverty reduction. *Journal of Sustainable Tourism*, *15*(6), 628–644.

Martin, G., & Vermeylen, S. (2005). Intellectual property, indigenous knowledge, and biodiversity. *Capitalism Nature Socialism*, *16*(3), 27–48.

Mathieson, A., & Wall, G. (1982). *Tourism: Economic, physical, and social impacts*. London: Longman.

Mbaiwa, J.E. (2005). Enclave tourism and its socio-economic impacts in the Okavango Delta, Botswana. *Tourism Management*, *26*(2), 157–172.

Mbaiwa, J.E. (2011). Cultural commodification and tourism: The Goo-Moremi Community, Central Botswana. *Tijdschrift voor Economische en Sociale Geografie*, *102*(3), 290–301.

Mbaiwa, J.E. (2016). The commodification of World Heritage Sites: The case study of Tsodilo Hills in Botswana. In H. Manwa, N. Moswete, & J. Saarinen (Eds.), *Cultural tourism in southern Africa*, pp. 101–120. Bristol: Channel View Publications.

Mbaiwa, J.E., & Stronza, A.L. (2010). The effects of tourism development on rural livelihoods in the Okavango Delta, Botswanad. *Journal of Sustainable Tourism*, *18*(5), 635–656.

Mbaiwa, J.E., Thakadu, O.T., & Darkoh, M.B.K. (2008). Indigenous knowledge and ecotourism-based livelihoods in the Okavango Delta in Botswana. *Botswana Notes and Records*, *39*, 62–74.

Melubo, K., & Carr, A. (2019). Developing Indigenous tourism in the bomas: Critiquing issues from within the Maasai community in Tanzania. *Journal of Heritage Tourism*, *14*(3), 219–232.

Melubo, K., & Lovelock, B. (2019). Living inside a UNESCO World Heritage Site: The perspective of the Maasai community in Tanzania. *Tourism Planning & Development*, *16*(2), 197–216.

Ministry of Natural Resources and Tourism (2019). *The multiple land use model of Ngorongoro Conservation Area: Achievements and lessons learnt, challenges and options for the future.* Dodoma, Tanzania: MNRT.

Mkhize, G., & Cele, N. (2017). The role of women in tourism in KwaZulu-Natal: Case studies from the south coast of KwaZulu-Natal. *Agenda, 31*(1), 128–139.

Moswete, N., & Lacey, G. (2014). "Women cannot lead": Empowering women through cultural tourism in Botswana. *Journal of Sustainable Tourism, 23*(4), 600–617.

Moswete, N., Thapa, B., & Lacey, G. (2009). Village-based tourism and community participation: A case study of the Matsheng villages in southwest Botswana. In J. Saarinen, F. Becker, H. Manwa, & D. Wilson (Eds.), *Sustainable tourism in Southern Africa: Local communities and natural resources in transition*, pp. 189–209. Clevedon: Channel View Publications.

National Bureau of Statistics (NBS) (2017). *Human and livestock census for Ngorongoro Conservation Area.* Dar es Salaam: National Bureau of Statistics.

Nelson, F. (2012). *Community rights, conservation and contested land: The politics of natural resource governance in Africa.* London: Earthscan.

Novelli, M. (2015). *Tourism and development in Sub-Saharan Africa.* London: Routledge.

Pereiro, X. (2016). A review of Indigenous tourism in Latin America: Reflections on an anthropological study of Guna Tourism (Panama). *Journal of Sustainable Tourism, 24*(8/9), 1121–1138.

Psacharopoulos, G., & Patrinos, H.A. (1994). *Indigenous people and poverty in Latin America: An empirical analysis.* Washington, DC: World Bank.

Ruhanen, L., & Whitford, M. (2019). Cultural heritage and Indigenous tourism. *Journal of Heritage Tourism, 14*(3), 179–191.

Saarinen, J. (2016). Cultural tourism in southern African: The role of local cultures and ethnicity in tourism development. In H. Manwa, N. Moswete, & J. Saarinen (Eds.), *Cultural tourism in southern Africa*, pp. 145–164. Bristol: Channel View Publications.

Saarinen, J., & Niskala, M. (2009). Selling places, constructing local cultures in tourism – The role of Ovahimbas in Namibian tourism promotion. In P. Hottola (Ed.), *Tourism strategies and local responses in southern Africa*, pp. 61–72. Wallingford: CABI.

Saarinen, J., & Rogerson, C.M. (2015). Setting cultural tourism in southern Africa. *Nordic Journal of African Studies, 24*(3&4), 207–220.

Salazar, N.B. (2015). The Maasai as paradoxical icons of tourism (im)mobility. In A.C. Bunten, J. Chio, & N.H.H. Graburn (Eds.), *Cultural tourism movements: New articulations of Indigenous identity*, pp. 56–72. Toronto: University of Toronto Press.

Saugestad, S. (2001). *The inconvenient indigenous: Remote area development in Botswana, donor assistance and the first people of the Kalahari.* Uppsala, Sweden: Nordic Africa Institute.

Sereke-Brhan, H. (2010). *Coffee, culture, and intellectual property: Lessons for Africa from the Ethiopian fine coffee initiative.* Boston: Boston University, Frederick S. Pardee Center for the Study of the Longer-Range Future.

Snyder, K.A., & Sulle, E.B. (2011). Tourism in Maasai communities: A chance to improve livelihoods? *Journal of Sustainable Tourism, 19*(8), 935–951.

Sylvain, R. (2002). Land, water, and truth: San identity and global indigenism. *American Anthropologist, 104*(4), 1074–1085.

Terry, M.E. (2001). An overview of the Botswana handicraft sector. *Botswana Notes & Records, 33*(1), 129–144.

Timothy, D.J. (2007). Empowerment and stakeholder participation in tourism destination communities. In A. Church & T. Coles (Eds.),Tourism, power and space, pp. 199–216. London: Routledge.

Timothy, D.J. (2021). Heritage and tourism: Alternative perspectives from South Asia. *South Asian Journal of Tourism and Hospitality, 1*(1), 35–57.

Tomaselli, K.G. (1999). Contextualising the San 'past': Dancing with development. *Visual Anthropology, 12*, 197–212.
Tomaselli, K.G. (2012). *Cultural tourism and identity: Rethinking indigeneity*. London: Brill.
Tshabalala, S.P., & Ezeuduji, I.O. (2016). Women tourism entrepreneurs in KwaZulu-Natal, South Africa: Any way forward? *Acta Universitatis Danubius: Œconomica, 12*(5), 19–32.
UNESCO (2021). UNESCO digitalizes traditional African games to preserve and expose traditional culture. Online at: http://www.unesco.org/new/en/communication-and-information/resources/news-and-in-focus-articles/allews/news/unesco_digitalizes_traditional_african_games_to_preserve_and/
United Nations Population Fund (2011). Population dynamics of Indigenous peoples: Some regional and national highlights, 2011 meeting of the UN IASGP on Indigenous Peoples Issues, New York, November, 2011. Online at: http://www.unfpa.org/webdav/site/global/shared/documents/events/2011/Heckadon%20IASG%20Presentation.pptx
van Veuren, E.J. (2001). Transforming cultural villages in the spatial development initiatives of South Africa. *South African Geographical Journal, 83*(2), 137–148.
Verschuuren, B., Mallarach, J.M., Bernbaum, E., Spoon, J., Brown, S., Borde, R., Brown, J., Calamia, M., Mitchell, N., Infield, M, & Lee, E. (2021). *Cultural and spiritual significance of nature: Guidance for protected and conserved area governance and management*. Gland, Switzerland: IUCN.
Whitney-Squire, K. (2016). Sustaining local language relationships through indigenous community-based tourism initiatives. *Journal of Sustainable Tourism, 24*(8–9), 1156–1176.
Wishitemi, B.E.L., & Okello, M.M. (2003). Application of the protected landscape model in southern Kenya. *Parks, 13*(2), 12–21.
World Bank. (2021). Indigenous Peoples. Online at: https://www.worldbank.org/en/topic/indigenouspeoples
World Tourism Organization (2019). Recommendations on sustainable development of Indigenous tourism. Madrid: UNWTO.
Woyesa, T., & Kumar, S. (2021). Potential of coffee tourism for rural development in Ethiopia: A sustainable livelihood approach. *Environment, Development and Sustainability, 23*(1), 815–832.
Yang, L. (2011). Ethnic tourism and cultural representation. *Annals of Tourism Research, 38*(2), 561–585.
Yang, L., & Wall, G. (2009). Ethnic tourism and entrepreneurship: Xishuangbanna, Yunnan, China *Tourism Geographies, 10*(4), 522–544.
Zeppel, H. (2006). *Indigenous ecotourism: Sustainable development and management*. Wallingford: CABI.

4

CULTURAL EVENTS AND TOURISM IN AFRICA

Ikechukwu O. Ezeuduji

Introduction

During the past several decades, the tourism industry has become increasingly important for governments and regional agencies seeking socioeconomic development. This is especially true in the Global South where increasing tourism demand has created direct and indirect income and employment opportunities within host regions and among previously marginalized local communities (Kavita & Saarinen, 2016). Communities in Africa possess undeniably appealing natural and cultural assets. However, these communities must develop the capabilities to manage their tourism resources and create holistic and unique tourism experiences (Ezeuduji, 2015a).

Ezeuduji (2015a) notes that tourism is driven by the traveling public and the attractions that draw them in. These attractions derive from the uniqueness and diversity of geographical and cultural features of place. The geographical and cultural resources of Africa, both tangible and intangible, combine to provide substantial tourism development potential, although much of the continent's potential remains unrealized (Ankomah & Crompton, 1990; Dieke, 2013; Novelli, Adu-Ampong & Ribero, 2021). The cultural uniqueness and diversity of Africa provide many settings for rustic and sophisticated tourism offerings in many countries and in a multitude of destinations. The range of Africa's heritage manifestations is manifold and includes arts and handicrafts, religious practices, agricultural and hunting traditions, food, music and dance, family dynamics, village celebrations, cultural performances, and many other elements of the living legacy of people's inheritances from their ancestors. The region also benefits from more recent cultural forms, such as art, film, and fusion cuisine. Africa's living heritage has paved the way for the development of cultural events-based tourism, most of which spotlights some elements of the region's cultural heritage.

DOI: 10.4324/9781003153955-4

Tourists who are motivated by knowledge-seeking are more likely to participate in art festivals and other cultural events (Timothy, 2021a; Yan & Halpenny, 2019). Festivals and events are key tourist attractions and have the ability to benefit a destination by reducing seasonal lulls and boosting tourist flows (Connell et al., 2015). Hence, promoting festivals and events has become an essential marketing strategy for many destination managers. Ezeuduji (2015b) presents a way of harnessing the natural, historical, and cultural attributes of Africa to create strategic event-based rural tourism offerings. This chapter limits its analysis and discussions to the cultural attributes of Africa that can be harnessed to create event-based tourism products. The chapter starts by exploring the background of local interest in cultural events as the foundation of tourism development. It then discusses the significance and implications of cultural events for tourism development, and the symbiotic relationships between heritage and events tourism, highlighting some examples of cultural events related to tourism in Africa, before concluding and recommending actions to optimize the use of cultural events to support tourism development.

The Significance of Cultural Events in Tourism

Tourism is globally one of the fastest growing industries and is viewed as an important tool for regional socioeconomic development, especially in developing countries (Moswete et al., 2008). Boucher, Cullen, and Calitz (2018) suggest that cultural tourism has established a market share of about 20% of the global tourism industry, although other estimates place it much higher (Timothy, 2021a). In some destinations, cultural or heritage tourism comprises the largest proportion of the tourism economy. Cultural heritage tourists are known to spend more and stay longer in the destination than other tourists; therefore, heritage tourism brings many socioeconomic benefits to the destinations and their residents. Hence, residents generally have a positive attitude toward cultural heritage events.

In common with other forms of tourism, the impacts of events-based tourism can be economic, social, cultural, political, or environmental, and the positive impacts are the principal driver of support for, and increasing popularity in, festivals and events at local, national, and international scales. It is therefore important, if events are seen as agents of development, to measure the returns on an often-significant financial investment, and justify the continuation of planned events (Tassiopoulos & Nicolaides, 2017). The cultural sector everywhere relies heavily on public funding (Baldi, 2018; Mulino, 2016) and for the most part, public funding is justified on the grounds that this sector provides public goods and services (Getzner, 2017) and has significant potential to contribute to economic development and social transformation (O'Hagan, 2016).

The economic and sociocultural impacts of cultural festivals and heritage events have been increasingly scrutinized by academics, policy-makers, investors, tourists, and planners (Addo, 2011; Blešić et al., 2013; Getz, 2008; Laing, 2018; Liu, 2014; Quinn, 2019; Richards & Wilson, 2004; Tichaawa, 2016). Boucher et al. (2018) argue that event tourism can promote economic resilience with regard to national income and increased employment and foreign direct investment. Event tourism is significant in strengthening a place's competitive advantage. It is known to be a driver of economic mobility, new services, creating a broader awareness by attracting tourists, attracting outside capital investments, encouraging entrepreneurship, and improving infrastructure and amenities that serve both locals and visitors (Addo, 2011; Akama & Kieti, 2007; Blešić et al., 2013; Getz, 2008; Tassiopoulos & Nicolaides, 2017). One of the main benefits of cultural events is the opportunities they create for social inclusion, including the involvement of Indigenous people, who might otherwise have few economic alternatives (Higgins-Desbiolles, 2016; Snowball, Collins & Tarentaal, 2017).

In addition to building the local economy, cultural events can foster social capital by encouraging residents to take pride in their community, reinforce their allegiance to the destination, and build community solidarity (Blešić et al., 2013; Colombo, 2016; Getz, 2008; Gössling et al., 2016; Okazaki, 2008). Many types of events, especially cultural and heritage festivals, are used by destination marketers and community leaders to create and enhance a sense of place, or place identity, both internally by strengthening local culture and improving a sense of identity within the local community, as well as externally by brand positioning, in a world where places are becoming very similar and homogenous. Festivals and events are a major segment of the cultural industries (Bob et al., 2019). They can be key opportunities to showcase a destination's cultural heritage attributes and interests, and they often serve as a catalyst or brand for marketing a destination (Hemmonsbey & Tichaawa, 2019).

Festivals, as a manifestation of cultural heritage practice, may also be useful tools for cultural maintenance and regeneration and culture sharing (Higgins-Desbiolles, 2016). In addition to economic development, Yolal et al. (2016) illustrate the impacts of festivals and events on residents' well-being and the cultural and educational benefits of hosting these events. Festivals and events are crucial, for example, in South Africa where colonial and apartheid structures systematically impaired and marginalized traditional cultural practices and Indigenous peoples' ability to explore and live their cultural expressions and genres (Bob et al., 2019) and build a sense of community. Many studies have shown the symbiotic relationships between cultural heritage and tourism, emphasizing the important role that heritage tourism can play in cultural protection through conservation programmes and funding (Throsby, 2016; Timothy, 2021a; Torre & Scarborough, 2017; van der Merwe, 2016). This is particularly crucial in the Global South, where heritage tourism has yet to realize its maximum potential (Rogerson & van der Merwe, 2016; Timothy, 2021b).

To create a strong sense of place, appreciation of local heritage, build community solidarity, protect vulnerable living heritage, empower communities, and provide employment opportunities in the Global South, it is crucial for destinations to engage in stakeholder-based planning and development (Kibicho, 2008; Richards et al., 2013). This community-based approach better ensures that community expectations are fulfilled and that the needs and cultures of tourists clash less with the needs and cultures of residents (Lynch et al., 2010). In this way, the community will have a stronger buy-in to any efforts to develop cultural events-based tourism (Nicolaides, 2017). Community-based collaborative networks can help ensure that the process of event tourism development is positive for residents and tourists alike (Moswete et al., 2008; Ryan et al., 2011; Stevenson, 2016).

Events play a significant role in the local community, providing important activities and spending outlets for both residents and visitors, and enhancing the image of the destination (Tassiopoulos, 2005). Events provide an avenue for visitors to sample local foods and cultural traditions, and participate in games and other entertainment activities. Many events of different sizes and scopes can keep the domestic market active and enhance the authenticity of places and their tourism offerings (Getz, 1997, 2007; Tassiopoulos, 2005). Because of these benefits, cultural festivals and events are increasingly being integrated into regional development strategies (Luonila & Johansson, 2015). Quinn (2019) argues that festivals and events are a cornerstone of the urban landscape, significantly contributing to urban development and renewal, and appearing increasingly frequently in tourism policies.

Cultural Events-Related Tourism in Africa

Cultural heritage-based festivals and events are commonplace throughout the world. In Africa, festivals and other celebrations are an essential part of ordinary cultural life (Ezenagu & Olatunji, 2014). Despite their relative prevalence in Africa, cultural festivals and heritage events are not as well-known there as in many other parts of the world, and most of them appeal to a local or regional audience (Rogerson, 2015). Many African events derive from ancient traditions that have been commonly practiced since the earliest human inhabitants roamed the continent. These include such occasions as funerals, weddings, harvest and hunting celebrations, religious ceremonies, and coming-of-age rites and rituals. Some of these draw significant attention among regional visitors (Jindra & Noret, 2011), while others have begun to draw the attention of international tourists (Okech, 2008; Ran Forte, 2009; Rogerson, 2015; Wachira, 2006). Examples include, among thousands of others, the female cult festivals in Nigeria (Akintan, 2013) and the hunting celebrations of the Maasai in Kenya and Tanzania (Goldman, De Pinho & Perry, 2013).

In addition to age-old festivities and celebrations, many new cultural events have been introduced as purposive tourism products. Among the most common recent inventions are film, music, and art festivals, as well as events that focus on specialty fares like beer, wine, and street food (Kruger & Saayman, 2016; Pretorius, 2016; Rogerson & Collins, 2015). South Africa leads the continent in the number and distribution of modern purposive festivals (Pretorius, 2016; Visser, 2017), but many other countries in North Africa and Sub-Saharan Africa are strengthening their efforts to establish a festival-oriented tourism economy (Snowball, 2016; Weber & Ali-Knight, 2012). The nearly half-century-old South African National Arts

Festival is illustrative of the purposefully planned events that aim to promote the arts and create economic growth (Snowball, 2016). Table 4.1 illustrates several examples of current African festivals and the heritage they commemorate.

Regardless of the timeframe of their development, ancient and modern festivals in Africa are an important part of the continent's cultural heritage and reflect many heritage principles that have been elucidated elsewhere: preserving cultural traditions, capitalizing on living heritage for economic growth, encouraging participatory community development, building social capital, strengthening a sense of place, and building bridges of understanding between peoples.

In addition to the examples depicted in Table 4.1, multitudes of cultural heritage-oriented festivals or other events take place throughout Africa each year. Some are one-off events, organized to celebrate certain events or milestones. Some have been held for decades or centuries, while others are now in their initial stages of development and are planned to become a regular part of the future tourismscapes

TABLE 4.1 Examples of cultural heritage-based festivals and events in Africa, 2021–2022

Festival/event name	*Event location*	*Main event theme*
Festival Culturel Porte Ouverte	Lubumbashi, DR Congo	DR Congo's traditional cultures
Tanzania International Cuisine & Food Festival	Dar Es Salaam, Tanzania	International foods and food cultures
Festival de Fès de Musique Sacrées de Monde	Fes, Morocco	Sacred music around the world
Panafest & Emancipation Celebration	Accra and Kumasi, Ghana	The African diaspora and freedom from slavery
Sweet Potatoes and Yams Festival	Accra, Ghana	Ghana's culinary heritage
Oga Bole Festival	Benin City, Nigeria	The heritage of street food
Zanzibar Reggae Festival	Stone Town, Zanzibar, Tanzania	The heritage of Reggae music
Haapi Festival	Durban, South Africa	Indigenous food, music, film, and cultures of Africa
Luju Food and Lifestyle Festival	Malkerns, Eswatini	The tastes and sounds, sights, and aromas of Eswatini and Southern Africa
Ouidah International Voodoo Festival	Ouidah, Benin	The Voodoo religion, dance, shrines, and animal sacrifices
Maulidi Festival	Lamu, Kenya	The annual celebration of the birth of the Prophet Mohammad
Kartong Festival	Kartong, The Gambia	Traditional village life in The Gambia
Mauritian Cultural Fair	Bel Air, Mauritius	The diverse cultures of Mauritius
Festival International du Sahara de Douz	Douz, Tunisia	The rich Arab–Berber heritage of North Africa

Source: Compiled from multiple sources.

of places. Cultural festivals and events in Africa face a number of benefits and challenges. Many lessons can be learned from the various events that take place on the continent, particularly as regards community-based tourism; events as a means of preserving culture and growing a sense of place and community pride; events as a branding and economic development tool; and events as mechanisms for creating satisfied customers and business success.

Community-Based, Participatory Tourism

Participatory (community-based) development better ensures that local residents will buy into the notion of tourism. When they have a voice in what elements of their culture should be presented to visitors and how tourism should best be developed, they are more inclined to support it and gain from its benefits (Timothy, 1999). Thus, community-based cultural events tourism may be a powerful tool in building support for tourism and enhancing the positive socioeconomic impacts in developing regions.

Okech (2011) recounts how sustainable cultural event-based tourism has been promoted and practiced in Lamu Old Town, a UNESCO World Heritage Site in Kenya and one of the country's top tourist destinations. Lamu Old Town is acclaimed to be the oldest and best-preserved Swahili settlement in East Africa, and it has retained many of its traditional forms and functions. Lamu has hosted major Muslim religious festivals since the nineteenth century, including the annual *Maulidi* and other cultural events. In relation to this, the city has been featured as a significant centre for the study of Islamic and Swahili cultures. Each year, Lamu holds the Lamu Cultural Festival in which at least 80% of the community participates. The three-day celebration highlights traditional dances and rivalrous competitions between neighbourhoods. This festival also encourages Kiswahili poetry competitions, donkey races, dhow races, traditional handicraft displays, henna painting, Swahili bridal ceremonies, and a Swahili food bazaar. According to Okech (2011, p. 198), 'the festival offers a modern sample of these time-honoured traditions against a rich backdrop of the beauty and splendour of the Lamu archipelago'. Approximately 80% of Lamu's residents receive some income from occasional sales of handicrafts and curios. More than 70% of the residents are employed on a casual or part-time basis in a tourism-related activity, largely stimulated by the annual event.

Another example of community-based, participatory event tourism comes from The Gambia, where staged events are used to represent elements of authentic traditional celebrations in the villages of Berefet and Njawara to entertain tourists and acquaint them with African arts and cultures, while generating income for local communities (Ezeuduji & Rid, 2011). Many rural parts of Africa rarely see the income benefits or other positive outcomes of tourism (Rid et al., 2014). In their work in The Gambia, Rid and his colleagues (2014) identified a development path that could diversify the nation's tourism sector and reach further into rural areas. The path includes event-based rural tourism initiatives centred on local cultural traditions that align with the motivations of target markets and which could reduce economic leakage. Rid et al. (2014) recommend that event-based rural tourism in The Gambia would do well to focus on the theme of the Gambia River, as the river is home to rich natural resources and is the fount of the country's living cultural heritage, especially for the communities located along its banks. It is the involvement of the community members and their willingness to share their cultural heritage that makes these events successful.

When a community realizes the social and economic value of its heritage, it is more inclined to appreciate and protect it, resulting in group solidarity and a common purpose to promote cultural well-being and heritage protection. Many festivals and events in Africa help develop and solidify a sense of place and cultural solidarity, especially when they help protect heritage and emphasize its value. For example, the *uMkhosi WoMhlanga* (Reed Dance) and Indoni cultural events in KwaZulu-Natal, South Africa, evoke a sense of national and cultural pride among the locals and have significantly contributed to tourism development and growth in that region (Nkosi, 2019). The Olojo and Ojude Oba festival in Nigeria has also been shown to promote community cultural heritage values, ethnic and cultural unity, and other events that strengthen group solidarity with a desire to protect and promote heritage (Agbabiaka et al., 2017).

Events as an Avenue for Preserving Local Culture and Developing New Heritage

Tourism is sometimes regarded as a means of protecting or preserving cultural heritage, or even resurrecting elements of lost heritage (Lussetyowati, 2015; Timothy, 2021a). Cultural events may play a similar role. The Standard Bank National Art Festival (previously known as the National Arts Festival) has proven its longevity and has grown to be one of the leading cultural events in Southern Africa (De Bruyn, 2009). Its objectives are to deliver excellence, create opportunities, and encourage innovation and the development of emerging South African arts. In the same way, the Isingqesethu Wild Coast cultural festival has put the small coastal community of Port St Johns on the tourist map. The aim of the festival is to bring back into focus the traditions and heritage of the province. Likewise, one of the main aims of the Morija Arts and Cultural Festival in Lesotho each year is to help keep the Basotho culture alive through music, dance, poetry, and handicrafts (Monaheng, 2016).

Some cultural celebrations focus more on protecting and sharing a region's heritage than on making an economic impact. This is a notable emphasis on cultural events in South Africa's North West Province, where numerous villages host festivals and celebrations that emphasize sociocultural outcomes over financial growth (Tseane-Gumbi, Manwa & Dlamini-Boemah, 2019). There, villagers share dances and living heritage with tourists, believing that imparting their traditions will help them preserve their cultural heritage as community pride and solidarity result in increased efforts to conserve living traditions. Alves, Cerro, and Martins (2010) conclude that small-scale festivals, such as in villages and rural areas, do yield economic impacts, but their sociocultural impacts are more significant than their economic value.

Heritage is not always ancient, but it does entail humankind's inheritance from the past. Newer heritage elements, or non-traditional heritage, are becoming an important part of the event tourismscapes of Africa. Wine has become an important part of South Africa's agricultural heritage since the country's early colonial days, but its viticultural industry has been highlighted a great deal in the past 30 years in a tourism context (Ferreira & Hunter, 2017). South African wine is now considered one of the world's finest, and a significant tourism sector has developed around its production, distribution, and consumption, including the development of wine routes and festivals. Wine festivals bring in substantial economic benefits to South Africa and add positively to tourists' quality of life (Kruger, Rootenberg, & Ellis, 2013). Particularly well-known wine destinations in the Western Cape Province include Stellenbosch, Franschoek, Wellington, Tulbach, Helderberg, and Paarl in the Cape Winelands, as well as Constantia Valley and Durbanville within Cape Town's city limits.

Branding and Economic Development

Many efforts are underway in Africa to improve the continent's image as a tourism destination using festivals as marketing and branding mechanisms. The 'Lake of Stars' in Malawi is a popular festival on the shores of Africa's third largest lake and a year-round project to promote Malawian arts and tourism. The festival is international in scope and generates over USD$1 million in visitor spending each year with around a third being in foreign currency (Lake of Stars, 2019). This event has a strong emphasis on arts and culture (e.g., music, talks, poetry, theatre, film, art, and wellness activities) and provides a positive story of Malawi and Africa, countering prevailing negative media messages about the region. The festival, born of the need to raise money for a developing economy, helps promote Malawi as a tourist destination and exposes Malawian and other African artists to an international audience.

Apleni and Henama (2020) studied the impact of events in boosting local economic development at Port St Johns in Eastern Cape Province, South Africa. The province has hosted a number of events that are having positive impacts on the host communities. These include the popular annual Standard Bank National Art Festival in Grahamstown, which recognizes emerging young artists, and the Isingqesethu Wild Coast cultural festival held in Port St Johns to showcase the various cultures of the region through food, dance, and music. Apleni and Henama call for a more concerted effort by local and provincial authorities and other stakeholders to plan, organize, and manage these events to optimize their socioeconomic benefits to their host communities and create a recognizable brand for this marginal coastal region.

According to Gozini and Tseane-Gumbi (2017), music festivals and the creative industry contribute about 3% of South Africa's national gross domestic product (GDP), as they have a history of being used as drawcards to boost local economies. Gozini and Tseane-Gumbi highlight several of South Africa's urban tourism destinations, such as Durban, Cape Town, and Johannesburg, which often host major cultural and music festivals, including the Cape Town International Jazz Festival, which contributes R685 million towards the national economy. Durban's International Jazz Experience, Fact Durban Rocks, Essence Festival, and Woza Durban have reported an economic impact of R30 million. The Joy of Jazz in Johannesburg contributes about R600 million towards the country's GDP. Buyel'Ekhaya Pan-African Cultural Music Festival in East London is also estimated to contribute R25.8 million to the national economy every year. All of these have become major branding events in these urban destinations.

These examples show how cultural event-based tourism can be a mechanism for raising destination profiles, generating economic impacts, and growing tourism in Africa (Ezeuduji, 2015b). This aligns with Higham and Ritchie's (2001) assertion that rural events offer huge potential for strategic development, in terms of integrating event production and tourism development interests. Close working relationships between event organizers, local authorities, and regional tourism organizations offer avenues to create an integrated regional event product that serves the interests of the region. Festivals and other events have the potential to raise the profile of a destination significantly (Bresler, 2011), and in the case of Sub-Saharan Africa, reduce the legacy of Afro-pessimism. Being an experience product, demand will likely increase after the first round of consumption—that is if the destination delivers on tourists' desired experience through good customer service and suitable amenities, a positive brand image can begin to develop (Amara, 2017; Bresler, 2011).

Customer Satisfaction and Business Success

Ezeuduji and Rid (2011) report the success of rural events in The Gambia and suggest that by offering unique experiences through rural festivals and events, rural tourism can gain an exclusive competitive advantage within the global tourism industry. This is particularly true in the Global South, where most major cultural events take place in urban settings (Dines, 2021; Quinn, 2010; Visser, 2017). When cultural events-based tourism is developed in rural Africa, host communities must realize that this development is an entrepreneurial process which requires them to audit their capabilities and knowledge base necessary for this type of tourism development. Stakeholders with local communities at the core may collaborate and manage event development proactively. Building capabilities for rural tourism in Africa requires stakeholders, including local community members, not-for-profit organizations, and local government institutions, to take responsibility, and be held accountable for the performance of planned events (Ezeuduji, 2015a, 2015b).

For cultural event organizers, it is important to continually explore the gap between consumers' expectations and actual satisfaction with events. Regular analyses of the perceived quality of cultural events, including service quality, allows public and private organizers to identify critical issues, improve event planning, and increase efficiency, profitability, and overall performance (Meleddu, Melis, Pulina & Zapata-Aguirre, 2019). After exploring a generic competitive strategy with which rural Africa can favourably position itself in the global tourism marketplace, Ezeuduji (2015b) presents an emergent event-based rural tourism strategy (experience setting through events) in which locals can stage events for tourists that may lead to establishing unique rural tourism experiences. These events, the author suggests, will be more successful if they are based on elements of traditional tangible and intangible cultural heritage, including both ancient and traditional celebrations and newly minted cultural and artistic displays. This strategy may help enhance the perception of African tourism among foreign tourists as offering a 'rural idyll' experience.

Hough and his colleagues (2011) suggest that a cultural events strategy may help a destination compete successfully against other alternatives and emphasize product differentiation by supporting resident values, empowering communities, and offering unique cultural events. Destination marketing organizations (DMOs) can use rigorous market research to identify and understand the needs of the market (Rid, Ezeuduji & Pröbstl-Haider, 2014), but the value proposition of the cultural event must be aligned to the customers' needs and/or attitudes (Boucher et al., 2018). Hough et al. (2011) and Ezeuduji (2015a, 2015b) emphasize that the unique selling proposition of a cultural event must highlight the exclusivity of the event compared to its competitors in other localities. This may be established by investigating the core resources and capabilities within the destination and within the event.

Nonetheless, there is a deep need for research focussed on the profiles, experiences, and impacts of festivals and events, especially when these events are supported by government funds (Getz, 2019; Getz & Andersson, 2016; Swart et al., 2018). Andersson and his colleagues (2017) argue event quality needs to be assessed based on participants' satisfaction levels. Such efforts will help inform future planning and management. Likewise, although many studies have assessed the socioeconomic effects of festivals and events worldwide, there is a dearth of knowledge about these impacts in Africa, where more work is sorely needed (Oyekunle & Sirayi, 2018).

Challenges

Despite all the successes and opportunities outlined above, cultural events-based tourism in Africa is not without challenges. Nicolaides (2017) admonishes that cultural events, like other forms of tourism, have major positive and negative impacts on the community and require ethical handling. Traffic congestion and overcrowding are a major issue in many popular tourist destinations (Boucher et al., 2018; Chhabra, 2010). This not only exacerbates mobility problems that already exist, but also causes host communities to become disillusioned by tourism's growth. Although such conditions might be found in some African urban areas, in most of the continent's rural areas, overcrowding has not yet become a major concern, given a generally inadequate infrastructure to support large-scale tourism. This has both positive and negative elements. It is positive in that underdeveloped infrastructure prevents overtourism or too much tourism, which can have profound social and ecological impacts. By the same token, inadequate infrastructure translates into insufficient access to places that have good potential for tourism. Therefore, many places miss out on the opportunity to adopt tourism as a meaningful economic development sector. As noted before, most cultural festivals in Africa occur in urban areas. There is much ongoing discussion about introducing event tourism to the continent's rural areas, but the inadequate infrastructure and lack of public transportation make mainstreaming this effort difficult at best (Ezeuduji & Rid, 2011; Okech, Haghiri & George, 2015).

In many parts of the Global South, there is a significant problem associated with cultural appropriation. This is less the case in Africa as it is in other places, especially among Indigenous populations, but the same could easily happen in Africa if tourism were to grow too rapidly with too many interested parties from non-local areas. Losing control of culture and tradition is something many ethnic and cultural groups fear in the modern world. Engagement with stakeholders (i.e., DMOs, residents, business owners, government agencies, and tourists) and the application of the principles of community-based planning are the foundations of successful cultural events and can be a valuable tool in helping societies protect their cultural heritage and retain their rights over it (Byrd, 2007; Ezeuduji, 2015a, 2015b; Getz et al., 2012; Hall, 2011; Moswete et al., 2008; Okazaki, 2008; Ryan et al., 2011; Vafadari, 2008). From this perspective, the destination community must decide what it wants to celebrate through a festival or other event and what elements of their culture they wish to share with outsiders. Without such input and ownership, communities may become disillusioned with tourism and resist efforts to become involved in heritage events. This has implications beyond the destination residents themselves. For example, Ryan et al. (2011) found that community resistance to tourism development, because they were not actively engaged, decreases tourist arrivals.

A commonly voiced concern about successful tourism, including events tourism, in Africa, is the general lack of education and knowledge about tourism and its management needs, and the related long-held belief that women have little to add to tourism management and entrepreneurship (Chili & Ngxongo, 2017; Durant, 2018; Moswete & Lacey, 2015). In Africa, community resistance may partly be due to a lack of formal education or understanding of tourism, which may be manifested in the non-willingness of previously disadvantaged communities to visit heritage sites (Marschall, 2012) and a general lack of awareness of the value and potential of community heritage in the Global South (Timothy, 2021b). Hence, the proactive engagement of communities may help create awareness of the social and heritage value of local culture, the knowledge of tourism's socioeconomic significance, and a realization of the need for adequate tourism management training. The training also needs to address the important role of women and ethnic minorities in event planning, management, and fruition.

Security is another major challenge facing events tourism in Africa in both urban and rural contexts. Arab Spring events in North Africa, terrorist attacks in Tunisia and Egypt in recent years, and government warnings against travel to the interior Sahara Desert dissuade tourism in general in that part of Africa and have created difficulties in planning and carrying out major cultural events (Avraham, 2019). Many parts of Africa are also imbued with tainted images, due to crime, violence, successive coups d'etats, and other political upheavals and protests (Olorunfemi & Raheem, 2008). The security images of Africa in major global markets may be difficult to overcome in planning and developing international-scale cultural festivals. In cases such as this, regional and local endeavours might be better suited until geopolitical tensions ease considerably.

Conclusion

Tourism has been a key force in promoting festival growth and expansion (Okech, 2011), both in terms of long-held traditional celebrations and recently minted festivals. Towns, villages, and cities are increasingly eager to share their living cultures, built environments, and spending opportunities with visitors by promoting festivals. As elsewhere, this is increasingly true in Africa. Regarded as a sociocultural process, festival tourism is a vehicle to bring together people of diverse backgrounds, and reinforce the reality of globalization as tourists intersect areas of cultural differences (Rogerson, 2019). The social changes resulting from the interactions between the tourists and the host community affect employment and income. The number of cultural festivals and events has increased in recent decades, mainly because of their significant potential contribution to regional economic development (Moscardo, 2007) and their potential to support economic restructuring and revitalization, place branding and marketing, community pride, business investments, and revenue generation, especially during off-peak season (Quinn, 2010).

To achieve these goals, it is critical that planning cultural festivals and events involves destination residents as a key stakeholder group (Gozini & Tseane-Gumbi, 2017), for it is their cultural heritage that is usually on display for visitors to consume and enjoy. It is also important for private businesses, non-government organizations, and government agencies to be involved to ensure that not only local businesses benefit, but also individual residents, either directly (through income) or indirectly through social benefits (e.g., better education, improved infrastructure, better health care, and overall improved quality of life). Residents' interests should be explored to maximize the benefits of cultural events and to support their sustainability. Residents must have a voice in what elements of their cultural heritage they want to share with visitors and elements they would like to keep to themselves. If community members have a voice in decision-making, they are more inclined to support a cultural celebration and help ensure its long-term success. This is especially true in the developing world, where historically outsiders have arrived to impose their own ideas on what 'locals' should or should not do in economic development through tourism (Akama & Kieti, 2007; Kibicho, 2008; Timothy, 1999).

Cultural heritage-based event tourism in Africa requires a clear and unique value position, operational excellence, and strong customer relations to gain a market leadership that offers the region a competitive advantage (Ezeuduji, 2015b; Ezeuduji & Rid, 2011). Competitive advantage is increasingly and more clearly associated with less imitable and rarer resources and capabilities. Socially complex resource bundles in Africa, such as cultural heritage, friendliness of the population, and climatic conditions, are difficult to imitate. Intangible resources on the continent such as the reputation of the 'rural idyll' experience to tourists using cultural attributes and unique rural landscape are rare in festival-based tourism. These assets may help Africa gain a competitive advantage with clear value positioning (embedded in unique event-based tourism), operational excellence (achievable through professionally staging cultural events that are in demand by tourists), and strong customer relations, evident in the hospitableness of the African people. The success of a cultural event-based tourism strategy depends on training and education, capacity building, social capital growth, and empowering local actors in the long-term in customer service, operational excellence, and product excellence (Ezeuduji & Rid, 2011), which can be made possible through stakeholder networks and collaboration. This will in turn lead to securing brand identity in the global tourism marketplace, while preserving unique natural and cultural heritages with which Africa is so well endowed.

For sustained success, it is important for festival managers to continue exploring the gap between consumers' expectations and their satisfaction with cultural heritage events (Meleddu et al., 2019). This will enable public and private organizers to identify critical issues and improve event planning, efficiency, profitability, and overall performance. Drawing from Okech's (2011) words of wisdom, cultural events for tourism development need to be carefully managed to sustain not only the events themselves but to promote sustainable approaches to tourism development in general. Heritage events have both cultural and social values, and they have become an instrument for tourism development, the expansion of tourism seasonality, improved destination image, and the growth of regional economies. Africa should therefore invest in community engagement as a high priority in developing and maintaining cultural festivals and events, and present the tourists with authentic experiences. The importance of including community residents and other local stakeholders in decision-making at all stages of development of cultural events tourism cannot be overstated. Developing a sense of place and public pride in their own cultures, creating recognizable brands, and seeing events tourism as a tool for preserving culture will go far in promoting more successful festivals and events-based tourism in Africa.

References

Addo, E. (2011). European heritage and cultural diversity: The bricks and mortar of Ghana's tourism industry. *Journal of Contemporary African Studies*, *29*(4), 405–425.

Agbabiaka, H.I., Omoike, A.A., & Omisore, E.O. (2017). Factors influencing patronage of traditional festivals in Sub-Saharan Africa. *International Journal of Tourism Cities*, *3*(4), 363–381.

Akama, J.S., & Kieti, D. (2007). Tourism and socio-economic development in developing countries: A case study of Mombasa Resort in Kenya. *Journal of Sustainable Tourism, 15*(6), 735–748.

Akintan, O.A. (2013). Traditional religious festivals and modernity: A case study of female-oriented cults annual festivals in Ijebuland of South Western Nigeria. *International Journal of Social Sciences & Education, 3*(4), 1035–1046.

Alves, H.M.B., Cerro, A.M.C., & Martins, A.V.F. (2010). Impacts of small tourism events on rural places. *Journal of Place Management and Development, 3*(1), 22–37.

Amara, D. (2017). Events tourism: A potential to build a tourist destination branding—the case of cultural events in Egypt. *International Journal of Heritage, Tourism and Hospitality, 11*(2), 1–12.

Andersson, T.D., Armbrecht, J., & Lundberg, E. (2017). Linking event quality to economic impact: A study of quality, satisfaction, use value and expenditure at a music festival. *Journal of Vacation Marketing, 23*(2), 114–132.

Ankomah, P.K., & Crompton, J.L. (1990). Unrealized tourism potential: The case of sub-Saharan Africa. *Tourism Management, 11*(1), 11–28.

Apleni, L., & Henama, U. (2020). The impact of events in boosting local economic development: A case study of Port St Johns, South Africa. *African Journal of Hospitality, Tourism and Leisure, 9*(1), 1–14.

Avraham, E. (2019). Image repair strategies adopted by MENA national tourism boards during and following crises. In D.J. Timothy (Ed.), *Routledge handbook on tourism in the Middle East and North Africa*, pp. 241–256. London: Routledge.

Baldi, C.E. (2018). Public funding in cultural sector: The arduous process for shared guidelines. *Aedon, 2*, 1–7.

Blešić, I., Pivac, T., Dordević, J., Stamenković, I., & Janićević, S. (2013). Cultural events as part of cultural tourism development. Case study: Sombor and Apatin (Serbia). *Acta Geographica Slovenica, 54*, 381–390.

Bob, U., Swart, K., Gounden, R., Gumede, A., & Nkambule, S. (2019). Socio-economic impacts of festivals and events: A case study of the Mzansi Golden Economy Programme in South Africa. *GeoJournal of Tourism and Geosites, 27*(4), 1236–1250.

Boucher, S., Cullen, M., & Calitz, A. (2018). Factors influencing cultural event tourism in Nelson Mandela Bay, South Africa. *Journal of Tourism and Cultural Change, 16*(5), 539–551.

Bresler, N.C. (2011). Tourist considerations in hosting a mega sport event: 2010 FIFA world cup in South Africa. *South African Journal for Research in Sport, Physical Education and Recreation, 33*(2), 17–33.

Byrd, E.T. (2007). Stakeholders in sustainable tourism development and their roles: Applying stakeholder theory to sustainable tourism development. *Tourism Review, 62*(2), 6–13.

Chhabra, D. (2010). *Sustainable marketing of cultural and heritage tourism*. London: Routledge.

Chili, N.S., & Ngxongo, N.A. (2017). Challenges to active community involvement in tourism development at Didima Resort–a case study of Umhlwazini community in Bergville. *African Journal of Hospitality, Tourism and Leisure, 6*(2), 1–15.

Colombo, A. (2016). How to evaluate cultural impacts of events? A model and methodology proposal. *Scandinavian Journal of Hospitality and Tourism, 16*(4), 500–511.

Connell, J., Page, S.J., & Meyer, D. (2015). Visitor attractions and events: Responding to seasonality. *Tourism Management, 46*, 283–298.

De Bruyn, P. (2009). *Frommer's South Africa*, 6th edn. Hoboken, NJ: Wiley.

Dieke, P.U.C. (2013). Tourism in sub-Saharan Africa: Production-consumption nexus. *Current Issues in Tourism, 16*(7–8), 623–626.

Dines, N. (2021). Moroccan city festivals, cultural diplomacy and urban political agency. *International Journal of Politics, Culture, and Society, 34*, 471–485.

Durant, V. (2018). *Women's empowerment and gender equality in tourism destination: Analysing men's perceptions on women's work—case study of Atakora region, Benin.* Unpublished master's thesis, European Master of Tourism Management, University of Girona, Spain.

Ezenagu, N., & Olatunji, T. (2014). Harnessing Awka traditional festival for tourism promotion. *Global Journal of Arts Humanities and Social Sciences, 2*(5), 43–56.

Ezeuduji, I.O. (2015a). Building capabilities for sub-Saharan Africa's rural tourism services performance. *African Journal for Physical, Health Education, Recreation & Dance, 21*(1), 68–75.

Ezeuduji, I.O. (2015b). Strategic event-based rural tourism development for sub-Saharan Africa. *Current Issues in Tourism, 18*(3), 212–228.

Ezeuduji, I.O., & Rid, W. (2011). Rural tourism offer and local community participation in The Gambia. *Tourismos, 6*(2), 187–211.

Ferreira, S.L., & Hunter, C.A. (2017). Wine tourism development in South Africa: A geographical analysis. *Tourism Geographies, 19*(5), 676–698.

Getz, D. (1997). *Event management and event tourism.* New York: Cognizant.

Getz, D. (2007). *Event studies: Theory, research and policy for planned events.* Oxford: Butterworth-Heinemann.

Getz, D. (2008). Event tourism: Definition, evolution, and research. *Tourism Management, 29*(3), 403–428.

Getz, D. (2019). Wine and food events: Experiences and impacts. In M. Sigala & R.N.S. Richard (Eds.), *Wine tourism destination management and marketing*, pp. 143–164. London: Palgrave Macmillan.

Getz, D., & Andersson, T. (2016). Analysing whole populations of festivals and events: An application of organisational ecology. *Journal of Policy Research in Tourism, Leisure and Events, 8*(3), 249–273.

Getz, D., Svensson, B., & Gunnervall, A. (2012). Hallmark events: Definition, goals and planning process. *International Journal of Event Management Research,* 7(1), 47–67.

Getzner, M. (2017). The role of governments in conserving and funding cultural institutions. In B. Unger, D. van der Linde, & M. Getzner (Eds.), *The public or private goods? Redefining res publica*, pp. 223–252. Cheltenham: Edward Elgar.

Goldman, M.J., De Pinho, J.R., & Perry, J. (2013). Beyond ritual and economics: Maasai lion hunting and conservation politics. *Oryx, 47*(4), 490–500.

Gössling, S., Ring, A., Dwyer, L., Andersson, A., & Hall, C.M. (2016). Optimizing or maximizing growth? A challenge for sustainable tourism. *Journal of Sustainable Tourism, 24*(4), 527–548.

Gozini, N., & Tseane-Gumbi, L.A. (2017). Economic impacts of Buyel'Ekhaya Pan-African Cultural Music Festival on East London residents, Eastern Cape Province, South Africa. *African Journal of Hospitality, Tourism and Leisure, 6*(2), 1–12.

Hall, C.M. (2011). A typology of governance and its implications for tourism policy analysis. *Journal of Sustainable Tourism, 19*(4–5), 437–457.

Hemmonsbey, H., & Tichaawa, T.M. (2019). Using non-mega events for destination branding: A stakeholder perspective. *GeoJournal of Tourism and Geosites, 24*(1), 252–266.

Higgins-Desbiolles, F. (2016) Sustaining spirit: A review and analysis of an urban Indigenous Australian cultural festival. *Journal of Sustainable Tourism, 24*(8–9), 1280–1297.

Higham, J.E.S., & Ritchie, B. (2001). The evolution of festivals and other events in rural southern New Zealand. *Event Management,* 7, 39–49.

Hough, J., Thompson, A.A., Jr., Strickland III, A.J., & Gamble, J.E. (2011). *Crafting and executing strategy: Creating sustainable high performance in Southern Africa—Text, readings and cases*, 2nd edn. Cape Town: McGraw-Hill Higher Education.

Jindra, M., & Noret, J. (2011). Funerals in Africa: An introduction. In M. Jindra & J. Noret (Eds.), *Funerals in Africa: Explorations of a social phenomenon*, pp. 1–15. Oxford: Berghahn Books.

Kavita, E., & Saarinen, J. (2016). Tourism and rural community development in Namibia: Policy issues review. *Fennia, 194*(1), 79–88.

Kibicho, W. (2008). Community-based tourism: A factor-cluster segmentation approach. *Journal of Sustainable Tourism, 16*(2), 211–231.

Kruger, M., & Saayman, M. (2016). Are you willing to pay more for the arts? *Journal of Economic and Financial Sciences, 9*(2), 392–408.

Kruger, S., Rootenberg, C., & Ellis, S. (2013). Examining the influence of the wine festival experience on tourists' quality of life. *Social Indicators Research, 111*(2), 435–452.

Laing, J. (2018). Festival and event tourism research: Current and future perspectives. *Tourism Management Perspectives, 25*, 165–168.

Lake of Stars (2019). About Lake of Stars – Malawi festival. Online: https://lakeofstars.org/about, accessed 14 July 2020.

Liu, Y. (2014). Cultural events and cultural tourism development: Lessons from the European capitals of culture. *European Capitals of Culture, 22*(3), 498–514.

Luonila, M., & Johansson, T. (2015). The role of festivals and events in the regional development of cities: Cases of two Finnish cities. *Event Management, 19*(2), 211–226.

Lussetyowati, T. (2015). Preservation and conservation through cultural heritage tourism. Case study: Musi Riverside Palembang. *Procedia-Social and Behavioral Sciences, 184*, 401–406.

Lynch, M., Duinker, L., Sheehan, L., & Chute, J. (2010). Sustainable Mi'kmaw cultural tourism development in Nova Scotia, Canada: Examining cultural tourist and Mi'kmaw perspectives. *Journal of Sustainable Tourism, 18*(4), 539–556.

Marschall, S. (2012). Sustainable heritage tourism: The Inanda Heritage Route and the 2010 FIFA World Cup. *Journal of Sustainable Tourism, 20*(5), 721–736.

Meleddu, M., Melis, G., Pulina, M., & Zapata-Aguirre, S. (2019). Expectations and service quality: Perceived performance at low-season events. *International Journal of Event & Festival Management, 10*(2), 110–137.

Monaheng, T. (2016). Integrating Indigenous knowledge in the development of cultural tourism in Lesotho. In H. Manwe, N. Moswete, & J. Saarinen (Eds.), *Cultural tourism in Southern Africa*, pp. 31–46. Bristol: Channel View Publications.

Moscardo, G. (2007). Analyzing the role of festivals and events in regional development. *Event Management, 11*(1–2), 23–32.

Moswete, N., & Lacey, G. (2015). "Women cannot lead": Empowering women through cultural tourism in Botswana. *Journal of Sustainable Tourism, 23*(4), 600–617.

Moswete, N., Thapa, B., Toteng, E., & Mbaiwa, J.E. (2008). Resident involvement and participation in urban tourism development: A comparative study in Maun and Gabarone, Botswana. *Urban Forum, 19*, 381–394.

Mulino, I.I. (2016). New challenges for cultural activities facing globalisation and digitisation. *L'industria, 2*, 225–236.

Nicolaides, A. (2017). Promoting ethical corporate social responsibility in the events industry. *African Journal of Hospitality, Tourism and Leisure, 6*(1), 1–11.

Nkosi, G.S. (2019). The role of uMkhosi WoMhlanga (Reed Dance) and Indoni cultural events through the eyes of young women in KwaZulu-Natal: A cultural tourism approach. *African Journal of Hospitality, Tourism and Leisure, 8*(2), 1–12.

Novelli, M., Adu-Ampong, E.A., & Ribero, M.A. (Eds.) (2021). *Routledge handbook of tourism in Africa*. London: Routledge.

O'Hagan, J. (2016). Objectives of arts funding agencies often do not map well on to societal benefits. *Cultural Trends*, *25*(4), 249–262.

Okazaki, E. (2008). A community-based tourism model: Its conception and use. *Journal of Sustainable Tourism*, *16*(5), 511–529.

Okech, R., Haghiri, M., & George, B.P. (2015). Rural tourism as a sustainable development alternative: An analysis with special references to Luanda, Kenya. *CULTUR-Revista de Cultura e Turismo*, *6*(3), 36–54.

Okech, R.N. (2008). Promoting cultural tourism in Kenya: Focus on funerals in Luo and Luhya communities. *Tourism Review International*, *12*(3–4), 243–258.

Okech, R.N. (2011). Promoting sustainable festival events tourism: A case study of Lamu Kenya. *Worldwide Hospitality and Tourism Themes*, *3*(3), 193–202.

Olorunfemi, F., & Raheem, U.A. (2008). Sustainable tourism development in Africa: The imperative for tourists/host communities security. *Journal of Sustainable development in Africa*, *10*(3), 201–220.

Oyekunle, O.A.A., & Sirayi, M. (2018). The role of creative industries as a driver for a sustainable economy: A case of South Africa. *Creative Industries Journal*, *11*(3), 225–244.

Pretorius, C. (2016). Cultural tourism and the arts festivals. In H. Manwe, N. Moswete, & J. Saarinen (Eds.), *Cultural tourism in Southern Africa*, pp. 76–85. Bristol: Channel View Publications.

Quinn, B. (2010). Arts festivals, urban tourism and cultural policy. *Journal of Policy Research in Tourism, Leisure and Events*, *2*, 264–279.

Quinn, B. (2019). A comment on: Arts festivals, urban tourism and cultural policy. *Journal of Policy Research in Tourism, Leisure and Events*, *11*(1), 8–12.

Ran Forte, J. (2009). Marketing Vodun: Cultural tourism and dreams of success in contemporary Benin. *Cahiers d'études africaines*, *49*(193–194), 429–451.

Richards, G., de Brito, M.P., & Wilks, L. (2013). *Exploring the social impacts of events*. London: Routledge.

Richards, G., & Wilson, J. (2004). The impact of cultural events on city image: Rotterdam, Cultural Capital of Europe 2001. *Urban Studies*, *41*(10), 1931–1951.

Rid, W., Ezeuduji, I.O., & Pröbstl-Haider, U. (2014). Segmentation by motivation for rural tourism activities in The Gambia. *Tourism Management*, *40*, 102–116.

Rogerson, C.M. (2015). Restructuring the geography of domestic tourism in South Africa. *Bulletin of Geography: Socio-economic Series*, *29*, 119–135.

Rogerson, C.M. (2019). Globalisation, place-based development, and tourism. In D.J. Timothy (Ed.), *Handbook of globalisation and tourism*, pp. 44–53. Cheltenham: Edward Elgar.

Rogerson, C.M., & Collins, K.J. (2015). Festival tourism in South Africa: Characteristics and motivations of attendees at craft beer festivals. *African Journal for Physical Health Education, Recreation and Dance*, *21*(2), 76–87.

Rogerson, C.M., & van der Merwe, C.D. (2016). Heritage tourism in the global South: Development impacts of the Cradle of Humankind World Heritage Site, South Africa. *Local Economy*, *31*(1–2), 234–248.

Ryan, C., Chaozhi, Z., & Zeng, D. (2011). The impacts of tourism at a UNESCO heritage site in China – a need for a meta-narrative? The case of the Kaiping Diaolou. *Journal of Sustainable Tourism*, *19*(6), 747–765.

Snowball, J. (2016). Valuing arts festivals: A case study of the South African National Arts Festival. In C. Newbold & J. Jordan (Eds.), *Focus on world festivals: Contemporary case studies and perspectives*, pp. 164–175. Oxford: Goodfellow.

Snowball, J., Collins, A., & Tarentaal, D. (2017). Transformation and job creation in the cultural and creative industries in South Africa. *Cultural Trends*, *26*(4), 295–309.

Stevenson, N. (2016). Local festivals, social capital and sustainable destination development: Experiences in east London. *Journal of Sustainable Tourism, 24*(7), 990–1006.

Swart, K., Bob, U., Nkambule, S., & Gumede, A. (2018). Economic impacts of the touring ventures sub-category of the Mzansi Golden Economy programme in South Africa. *EuroEconomica, 37*(2), 90–103.

Tassiopoulos, D. (2005). *Events management: A professional & developmental approach*, 2nd edn. Cape Town: Juta.

Tassiopoulos, D., & Nicolaides, A. (2017). Effective codes of ethics in the events industry. *African Journal of Hospitality, Tourism and Leisure, 6*(2), 1–15.

Throsby, D. (2016). Tourism, heritage and cultural sustainability: Three 'golden rules'. In L.G. Girard & P. Nijkamp (Eds.), *Cultural tourism and sustainable local development*, pp. 31–48. Surrey: Ashgate.

Tichaawa, T.M. (2016). The geography of festival and tourism development: The case of Cameroon. In R. Donaldson, G. Visser, J. Kemp & J. de Waal (Eds.), *Proceedings of the centenary conference of the society of South African geographers*, pp. 219–225. Bloemfontein: Society of South African Geographers.

Timothy, D.J. (1999). Participatory planning: A view of tourism in Indonesia. *Annals of Tourism Research, 26*(2), 371–391.

Timothy, D.J. (2021a). *Cultural heritage and tourism: An introduction*, 2nd edn. Bristol: Channel View Publications.

Timothy, D.J. (2021b). Democratising the cultural past: Western values, the Global South and cross-cultural perspectives in heritage tourism. In J. Saarinen & J.M. Rogerson (Eds.), *Tourism, Change and the Global South*, pp. 33–47. London: Routledge.

Torre, A., & Scarborough, H. (2017). Reconsidering the estimation of the economic impact of cultural tourism. *Tourism Management, 59*, 621–629.

Tseane-Gumbi, L., Manwa, H., & Dlamini-Boemah, D. (2019). African village-based tourism in the North West Province, South Africa. *African Journal of Hospitality, Tourism and Leisure, 8*(5), 1–11.

Vafadari, A. (2008). Visitor management, the development of sustainable cultural tourism and local community participation at Chogha Zanbil, Iran. *Conservation and Management of Archaeological Sites, 10*(3), 264–304.

van der Merwe, C.D. (2016). Tourist guides' perceptions of cultural heritage tourism in South Africa. *Bulletin of Geography: Socio-economic Series, 34*, 117–130.

Visser, G. (2017). Festival tourism in urban Africa: The case of South Africa. In C.M. Rogerson & G. Visser (Eds.), *Urban Tourism in the Developing World*, pp. 101–126. London: Routledge.

Wachira, W. (2006). *Aspects of performance in traditional Kenyan rituals and ceremonies*. Santa Barbara: University of California Press.

Weber, K., & Ali-Knight, J. (2012). Events and festivals in Asia and the Middle East/North Africa (MENA) region: Opportunities and challenges. *International Journal of Event and Festival Management, 3*(1), 4–8.

Yan, N., & Halpenny, E. (2019). The role of cultural difference and travel motivation in event participation: A cross-cultural perspective. *International Journal of Event and Festival Management, 10*(2), 155–173.

Yolal, M., Gursoy, D., Uysal, M., Kim, H.L., & Karacaoğlu, S. (2016). Impacts of festivals and events on residents' well-being. *Annals of Tourism Research, 61*, 1–18.

5

RELIGIOUS HERITAGE, TOURISM, AND PILGRIMAGE IN AFRICA

Historical, Traditional, and Contemporary Perspectives

Ermias Kifle Gedecho and Siyabulela Nyikana

Introduction

Religious tourism and pilgrimage have grown exponentially over the years and have remained one of the oldest forms of tourism despite evidence of new spiritual movements emerging in recent times and a reduction in religious adherence in Western societies (Kim et al., 2020; Luz, 2020; Nmah, 2017; Nyikana, 2017; Olsen & Wilkinson, 2016; Rojo, 2007; Ron & Timothy, 2019; Timothy & Olsen, 2006). This growth is reflected in increasing academic attention to various perspectives and manifestations of religious or spiritual tourism, including tourism associated with specific religions such as Buddhism, Christianity, Hinduism, Islam, Judaism, and many others (e.g., Bruntz & Schedneck, 2020; Collins-Kreiner, 2019; Hassib & Ibtissem, 2020; Jamal et al., 2018; Olsen & Timothy, 2022; UNWTO, 2020; Zamani-Farahani et al., 2019), spiritual tourism as distinct from religious tourism (Bowers & Cheer, 2017; Kujawa, 2017), health and wellbeing-based pilgrimages (Sharpley & Jepson, 2011; Smith, 2003; Warfield et al., 2014), pilgrimages of nostalgia and pop culture (Okamoto, 2019; Olsen, 2022; Reed, 2006), and many more travel undertakings that have either spiritual or religious connotations, however individual participants might define those experiences.

This indicates that religious tourism is not limited only to visits to sacred sites. Rather, it can be both (or either) a spiritual and secular travel activity. Ron and Timothy (2019, p. 3) note that 'religious tourism is no longer seen strictly as traditional pilgrimages, wherein suffering, penitence, and travail are important elements of the cleansing experience. Instead, faith-based tourism takes a variety of forms…'. In this sense, non-traditional pilgrimages and unconventional

DOI: 10.4324/9781003153955-5

destinations may be preferred by some faith-based or spiritual travelers; thus, contemporary faith-based travel patterns are beginning to favor lesser-known regions such as Africa.

African faith travel has not yet received sufficient attention in the academic literature, compared to other regions of the world (Nyikana, 2017). As a result, religious tourism has been examined in only a small handful of African countries, including Egypt, Ethiopia, South Africa, Uganda, and Nigeria, despite the fact that pilgrimages exist in a much wider range of African countries (Fadare & Benson, 2015; Ipara, 2013). Such academic neglect has resulted in insufficient information and a limited understanding of religious tourism on the continent (Kim et al., 2020). Therefore, this chapter unpacks African faith heritage and religious tourism development, resources, challenges, and prospects, to provide a fresh perspective on the previously neglected topic of religious travel and tourism on the continent. We argue that Africa is endowed with a rich religious heritage that ought to be understood better so that the development and management of this niche area can be enhanced. First, we will highlight the brief historical background of religion and tourism in Africa. Second, the chapter turns to religion-specific travel patterns, an examination of religious resources and trends, and the implications of related travel patterns. The third section of the chapter explores this niche market's key challenges, followed by a discussion on the prospects of religious tourism, and provides suggestions for its future development in Africa.

Religious Tourism in Africa

In common with other forms of heritage (Teye, 2009), in most African countries (primarily excluding Ethiopia), religious travel may be seen from three temporal perspectives: pre-colonial, colonial, and post-colonial. During the pre-colonial era, the vast majority of Africans believed in and practiced 'traditional religion' (Ayorekire et al., 2020), which embraced nature worship, ancestor veneration, polytheism, and various animist beliefs through folklore, music, magic, and traditional medicine. During this time, long-distance travel for religious purposes was rare because people were unaware of significant spiritual localities outside their home areas that might relate to their beliefs. By nature, traditional faiths teach that familiar objects and nearby surroundings are sacred, and divine characteristics can be found within them (Ashcraft, 2018; Mbiti, 2015). When ritual practices were necessary, these took place closer to home, although ritual 'pilgrimages' of some form are known to have taken place to sacred sites near believers' normal abodes. Local celebrations, clan and tribal activities, and certain hunting practices were outward demonstrations of spiritual beliefs, with travel largely unnecessary to accomplish their aims.

However, once Islam was introduced and spread throughout most of North Africa and the Sahel, pilgrimages to Mecca began to be part of Muslims' travel requirements beginning in the Early Middle Ages. Ancient connections between the Ethiopians and the Israelites more than two millennia ago drew Jewish connections between that part of Africa and the Holy Land, including pilgrimage-like visits, as recorded in the Bible where an Ethiopian eunuch met Philip in Jerusalem and converted to Christianity (Sisay, 2009). Likewise, ancient explorers, early missionaries, and writers were heavily involved in inbound travel to the continent and encouraged African elites to travel to holy sites in other parts of the world.

During the colonial period, more than 20% of Africa's total population was converted to Christianity (mainly Catholic and Protestant churches) following intensive missionary labors. With the spread of Islam, thousands of mosques were built, and with the growth of Christianity, thousands of churches, monasteries, religious schools, and hospitals were constructed. Traveling abroad for religious studies and pilgrimages, especially to Jerusalem, flourished in this period. Missionary service prevailed as a motivation for religious travel by international visitors and continues today.

Following the independence of most African states (i.e., post-colonization), religion continued to spur outbound and domestic travel. The desire for self-determination and freedom from the coerced embracing of Western culture brought significant religious freedom in the post-colonial era (Ayorekire et al., 2020). This has enabled African pastors, evangelists, and prophets to establish churches and send missionaries domestically and regionally to spread their respective faiths, including the growth of many African-based evangelical movements, such as Harris's prophetic movement in West Africa and Simon Kimbangu in DR Congo, which converted millions of people to Christianity (Encyclopaedia Britannica, 2018). Such movements were prevalent in Sub-Saharan Africa during the twentieth century but less so in northern and northeastern Africa, where the majority populations were already converted to Islam or Christianity. Besides the African missionary movement, Western missionaries such as Bohnke and his church have contributed substantial roles in the spread of Christianity (Encyclopaedia Britannica, 2018) and more recent international Christian churches (e.g., The Church of Jesus Christ of Latter-Day Saints and Seventh-Day Adventists) have grown in population since the late 1900s and early 2000s (Ron & Timothy, 2019). As we will discuss in more detail later, religious events for Christians and sacred sites and built heritage for Muslims and Orthodox Christians have become major products and attractions. In addition to the sacred sites of Christians and Muslims, the shrines and faith events associated with traditional religions have been studied, marketed, and experienced more as cultural tourism products than they have been treated as religious tourism places and experiences.

Figure 5.1 reveals changes in religious composition over time. More than 58% of Africa's population adhered to traditional religion before the beginning of the twentieth century, with very few Christian believers. However, with time, Christianity has grown considerably as a proportion of the population since the 1970s, which reflects the impact of Western missionaries and the African missionary movement. By contrast, Islam has experienced no significant change over time.

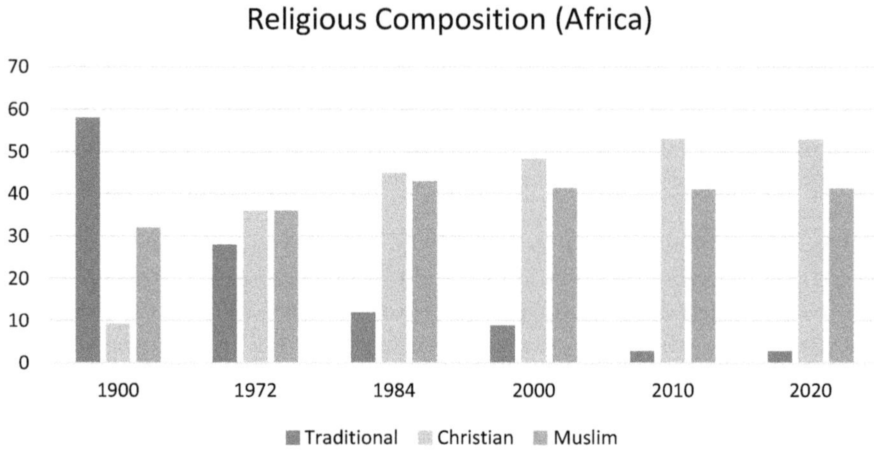

FIGURE 5.1 Trends in Africa's religious composition.
Data sources: Mbiti (2015); Pew Research Center (2015).

African Religions and Their Tourism Implications

Religion is a vital element of Africa's social and cultural life and living heritage. As noted previously, with the exception of small populations of Hindus, Buddhists, and Jews, African religious life is comprised primarily of traditional animist beliefs, Christianity, and Islam. Under this section, the nature and tourism potential of these faiths are highlighted.

Traditional Religion and Tourism

According to Mbiti (2015), traditional religions are indigenous, shared, and passed from generation to generation as a manifestation of cultural heritage. They have no founder, formal organization, missionary program, or desire to proselytize, but they shape people's culture and social life deeply. Traditional religion is comprised of five parts: beliefs, practices, values, objects, and officials. A supernatural being (or beings) is the fundamental belief of most African traditional religions. They entail performing ritual practices, celebrating festivals, and the veneration of gods and the deceased. One can find deities, or the sacred, in persons, forests, rivers, groves, shrines, rituals, and even masks. These objects and places are deemed sacred and, as such, are deeply respected by the faithful. In many ways, African traditional religion helps protect and preserve the region's natural resources and guides people's behavior, ways of living and treatment of others, and how they think about the world around them.

Table 5.1 highlights a selection of traditional religions in Africa and their distribution. West Africa is home to more traditional religions than the other sub-regions. Excluding multiple mythologies, southern Africa is represented primarily by the San religion. In contrast, the other sub-regions of Africa are home to more

TABLE 5.1 Examples of traditional religion distribution in Africa

Subregion	*Traditional religion*	*Distribution*
East and Central Africa	Dinka religion, Hausa animism	South Sudan, Chad, Gabon
	Gikuyu people's culture and beliefs	Kenya
	Dini Ya Msambwa	Kenya (Bungoma)
	Waaqeffanna and Waaqism	Ethiopia
North Africa	Ancient Egyptian religion	Egypt, Sudan
	Punic religion	Tunisia, Algeria, Libya
	Traditional Berber religion	Morocco, Algeria, Tunisia, Libya, Egypt, Mauritania, Mali, Niger, Chad, Burkina Faso
	Hausa animism	Sudan
Southern Africa	San religion	South Africa
West Africa	Akan religion	Ghana, Ivory Coast
	Asaase Yaa	Bono people of Ghana and Ivory Coast
	Hausa animism	Benin, Burkina Faso, Cameroon, Côte d'Ivoire, Ghana, Niger, Nigeria, Togo
	Edo religion	Benin, Nigeria
	Abwoi, Odinani, Ijo traditional	Nigeria
	Dogon religion	Mali
	West African Vodun (Voodoo)	Ghana, Benin, Togo, Nigeria
	Yoruba religion	Nigeria, Benin, Togo
	Serer religion (A Fat Roog)	Senegal, Gambia, Mauritania
	Dahomean religion	Benin, Togo

Source: Compiled from multiple sources.

than one traditional religion. On a country level, Nigeria embraces more traditional religions than any other state in Africa.

Despite the widespread distribution and the large number of traditional religions, which have existed for millennia, after colonization, the number of people who openly claimed to be believers in traditional religions declined dramatically, as noted earlier in Figure 5.1. Emmanuel (2014) believes that foreign religions brought to Africa have diluted the significance and relevance of traditional African faiths and have diminished the value of many spiritual customs, beliefs, and practices. Exposure to the outside world by colonizers, missionaries, and traders resulted in many wrong impressions and misconceptions outside the continent about African religious traditions, which hastened the call by Europeans and Americans to Christianize the people they encountered. Likewise, reports, news stories, and novels portrayed the region as a 'dark continent' with no religion. Many foreign researchers used incorrect and misleading terms to describe African religion, which resulted in many inaccurate portrayals of the region's indigenous faith traditions throughout the world (Awolalu, 1976; Mbiti, 2015).

Despite the success of Christian missionaries during the past two centuries, many people continue adhering to traditional faiths, sometimes even mixing animist beliefs and practices with those of Islam and Christianity into unique hybrid religious manifestations (Awolalu, 1976). Elements of Native African religion are found everywhere on the continent and exist in music, ceremonies, art, dance, symbolisms, the names of places and people, rituals, sacred spaces, festivals, and many other elements of daily life (Mbiti, 2015). Because of this, it can be argued that traditional religion lies at the roots of tourism in many localities as it offers tangible and intangible attractions and experiences. Natural resources such as forests, water bodies, and landscapes are among the most sacred spaces that are commonly consumed by tourists. For example, Osun Oshogbo's sacred grove and shrine is a UNESCO World Heritage Site (WHS) in Nigeria (Okonkwo & Eyisi, 2020; Woosnam et al., 2018). It has been well preserved for centuries owing to its veneration by Yoruba traditional beliefs. At this location, the Osun River is used in a variety of sacred ways, including a fish festival, and there are 40 shrines, two palaces, and a suspension bridge all connected to the Yoruba belief system. In addition to Osun Oshogbo, the Arochukwu shrine, the Ogbuide shrine of Oguta Lake, and the Amadioha Ozuzu shrine are monumental national heritage of religious origin in Nigeria (Iheanacho, 2015). Likewise, UNESCO has reaffirmed the heritage importance of traditional religious landscapes by inscribing the Stone Circles of Senegambia, in Senegal and The Gambia; Asante traditional buildings (shrines), Ghana; Sacred Mijikenda Kaya Forests, Kenya; and Maloti Drakensberg Park, Lesotho, as universally significant traditional religious heritage.

Given the wide variety of religious heritage sites and activities, attending traditional religious events can provide a unique experience for visitors (Teye, 2009). Traditional, faith-oriented festivals include the Argungu Fishing Festival, the Gelede Festival of Ife, Eyo Festival, Sharo/Shadi Festival, Shungo Festival, and the New Yam Festivals of the Igbo cultures in Nigeria (Iheanacho, 2015). The Osun-Oshogbo Festival is a 12-day ritual remembrance that attracts over 40,000 visitors annually. Because Osun is a god of fertility, according to Yoruba tradition, 'It is a widely-held belief among the Yoruba that Osun dwells within the Sacred Grove and the Oshogbo River; that those visiting are blessed with increased fertility' (Woosnam et al., 2018, p. 141).

Mbiti (2015) surmises that indigenous African religion is firmly attached to the people and forms an important part of their identity. As a result, it moves with migrants to other parts of the world where it is often retained by people of the African diaspora, where African traditional religions are an important part of the cultural heritage-based tourism phenomenon in places such as Brazil, Cuba, and Haiti (Hernandez-Ramdwar, 2013; Séraphin & Nolan, 2015). Many tourists from the African diaspora participate in spiritual events in these countries. With the exception of Orisha, a syncretic belief system in Trinidad which is only now beginning to be developed as a cultural tourism product, spiritual and religious traditions of African origin such as Santeria in Cuba, Candomblé in Brazil, and Vodou (Voodoo) in Haiti attract many spiritual tourists and other curiosity-seekers (Goldberg, 1983;

Hernandez-Ramdwar, 2013). Tourists attend spiritual and religious events, and many perform and participate in ceremonial elements of the religion.

The main challenge with traditional African religion and its travel practices is the general difficulty in identifying tourists who visit a destination specifically to attend traditional religious events. A visitor who is motivated to explore the exotic and deep culture of a community or local people could be considered a religious tourist or pilgrim. What people observe in traditional music, dance, and celebration largely derives from traditional religion. The exotic community is there because it has retained its traditional religion. It might have been challenging to see unique cultural practices if people were converted fully to perhaps Christianity. A second challenge is that people who seek to attend the traditional festival could also be identified as pilgrims. Social scientists might be attracted to learn about people with divergent traditions. For whatsoever reasons, the travel and experience at traditional religious sites or events need to be considered religious tourism to enhance tourists' experience and improve the income generated from tourism. Interpretations from the traditional religious perspectives improve tourists' experiences. Beyond this, acknowledging and preserving them might also initiate local and international travel. Hence, those regions with more traditional religions could capitalize on their faith heritage for better economic gains.

Christian Tourism Resources and Features

Christian pilgrimage began shortly after the death of Christ, particularly from Europe across the Mediterranean to the Holy Land and later to sites in Italy, Turkey, Spain, and Greece that were associated with the ministries of Jesus' apostles (Ron & Timothy, 2019). Christian (Roman Catholic and Orthodox) pilgrimage continued to grow as the religion spread throughout Europe from about the fourth century CE until the Protestant Reformation of the sixteenth century and the widespread bubonic plague discouraged the practice. Since that time, diverging travel patterns and destination choices have come to characterize the pilgrimage and religious tourism preferences of different Christian denominations. The Holy Land continues to be the primary destination for most Christians, while other localities have emerged as denomination-specific destinations, such as Catholics in Rome and Santiago de Compostela, Lutherans to sites associated with St. Olaf in Norway, and other Protestants who participate in tours of Reformation sites in Europe associated with the lives of Ulrich Zwingli, John Wesley, and John Calvin (Ron & Timothy, 2019). Due to canonical law or traditional practices, some religions do not allow non-members to enter their sacred spaces or visit their religious heritage. However, most faiths do allow non-members to visit. For example, the Roman Catholic Church encourages non-adherents to visit its sacred sites. The following paragraphs provide a brief overview of Catholic, Orthodox, and Protestant Christians' religious travel and tourism practices for context.

The Catholic Church is the largest Christian denomination in the world and in Africa. Due to churchwide doctrines, standardized worship, and a communal leadership system, unique practices from church to church are uncommon, although as previously noted, some societies in Africa have adapted church practices to include elements of Indigenous faith. The church offers its members spiritual nourishment. The Catholic Church operates churches, monasteries, and boarding schools, which offer different experiences. Local and regional pilgrimage, as well as festivities associated with commemorating All Saints Day and other holidays, are common religious actions that induce people to travel to strengthen their spiritual lives and seek blessings through prayer and meditation. As the Catholic footprint is relatively recent in Africa, there are few internationally recognized pilgrimage centers, festivities, and Catholic heritage sites. However, some of the better-known ones include but are not limited to the Uganda Martyrs' celebration, which attracts about two million visitors each year (Ayorekire et al., 2020). In addition, historic buildings and shrines in Uganda and Kenya have some widespread pilgrimage and cultural appeal, such as the Uganda Martyrs Catholic Shrine Namugongo, St. Mary's Rubaga Cathedral (Kampala), St. Paul Namirembe Cathedral (Kampala), and the Mai Mahiu Catholic Church in Kenya, built in 1942 by an Italian architect. These sites are seen as potentially important sacred and cultural assets that could further enhance religious tourism and cultural tourism on the continent, owing to the rich religious heritage that they exhibit from colonial times.

Orthodox Christianity is popular in Ethiopia, Egypt, and Eritrea. The Ethiopian, Egyptian, and Eritrean Orthodox churches belong to the Oriental Orthodox group and are of ancient origin. The Ethiopian Orthodox Tewahido Church (EOTC) is the second-largest Orthodox church in the world after the Russian Orthodox Church. It has more than 40 million followers, 30,000 priests, and more than 10,000 churches. The EOTC was established in the fourth century and was Ethiopia's state religion until 1966. Administratively, it had been under the Coptic Orthodox Church of Alexandria for more than 1,500 years until 1959. The church's living and built heritage has been well preserved because Ethiopia remained uncolonized by European metropoles while its neighbors fell under the colonial rule of the British, French, and Italians. This has resulted in much of the country's heritage being religious and ancient in nature and receiving thousands of pilgrimage visitors and other tourists every year.

Orthodox Christian travel to the Holy Land is a common practice by those who can afford it. Pilgrimage domestically and to the Holy Land is an important part of their liturgy because for them, pilgrimage enhances their spiritual life and reinforces cultural identity. The diplomatic and religious ties between Ethiopia and Israel are of ancient origin. Today, there are seven Ethiopian churches and monasteries in the Holy Land, and Israel is the main outbound travel destination for most African Orthodox Christians.

For international travelers, the EOTC offers diverse and unique spiritual and heritage experiences. Among UNESCO's four listed intangible heritages in Ethiopia, Timket (Ethiopian Epiphany) and Meskel (the commemoration of the finding of the True Cross) Festivals are religious celebrations in the EOTC. The church claims to have the True Cross and the Ark of the Covenant which are both internationally significant religious icons. The former is kept at Mount Gishen, the top domestic pilgrimage destination (Gedecho & Yebo, 2014), while the ark is ostensibly kept in the Zion Mary Church, Axum, also a UNESCO WHS (Hancock, 1992). Annual, monthly, and weekly religious holidays draw a steady flow of domestic pilgrims. The UNESCO-listed Rock-Hewn Churches of Lalibela and the world-famous Lake Tana monasteries and churches are the cultural pride of the country and have long drawn tourists from all over the globe (Bayih, 2018; Ndivo & Cantoni, 2016).

The Coptic Orthodox Church of Egypt is likewise a pre-colonial Christian church. It claims a membership of more than 10 million. Though the church is one of the oldest Christian churches and a founding member of the World Christian Council, it has experienced oppression at the hands of Egypt's Islamic rulers during the past several centuries. Nonetheless, the church has been able to preserve its tangible and intangible heritage. Alexandria, the Sinai Desert, and the faith's monasteries and churches are visited by tourists every day of the year. It is the largest Christian group in the Arab states, followed by various sects in Syria and Lebanon. Saint Catherine's monastery in Sinai was established in the sixth century and is widely regarded as one of the oldest functioning Christian monasteries on Earth and is a WHS. Egypt is rich in religious heritage and home to a few Coptic pilgrimage centers at tombs of martyrs in the Nile Valley and Nile Delta region (Ron & Timothy, 2019). The main religious destination in Egypt is Mount Sinai and its Saint Catherine's Monastery (Mohamed & Alseyoufi, 2018; Shackley, 1998). This site is sacred to Christians, Jews, and Muslims and was designated a WHS in 2002. Other internationally significant destinations include the monastery of Saint Abu Mena and Wadi El-Natrun (Kamil, 2000).

Unlike in Catholicism and Orthodox Christianity, buildings and places have less religious and spiritual significance in other Christian faiths (Ron & Timothy, 2019), although several Protestant churches in Africa (e.g., Moedergemeente in South Africa, Christuskirche in Namibia, and the Hanging Church in Egypt) are important tourist attractions in their respective countries. For Protestants, gathering with other believers and the social cohesion they experience is more important than any single locality or building, although Protestants too are avid travelers to sacred sites in Europe, the Middle East, and North America. Conventions, conferences, workshops, retreats, youth events, and festivals are more of a focus among protestant groups and draw large crowds of believers from all over the world. An increasing number of protestant denominations has resulted in the growth of religious mega-events (Dowson, 2017, 2022; Schedneck, 2022). Reinhard Bonnke, a German evangelist, spent years organizing religious mega-events in Africa (Christ for all Nations, 2020). His largest event was attended by 1.6 million people in 2000 in Lagos, Nigeria (Dowson, 2017). Even though Bonnke passed away in 2019, his church plans to continue holding these events in Africa.

In some countries, Pentecostal churches organize mega-events that attract both domestic and international pilgrims. The Zion Christ Church (ZCC) organizes an annual religious festival where more than a million people (most of whom are domestic) gather to celebrate Easter in Moria, Limpopo, South Africa. This event is well developed and organized, playing an important socioeconomic role in Moria (Apleni & Henama, 2020; Kruger & Saayman, 2016; Nyikana, 2017; Saayman et al., 2014). The United African Apostolic Church's annual week-long gathering is the second-largest gathering in South Africa, after the one in Limpopo (Dafuleya et al., 2017). This event is estimated to contribute some ZAR 2.14 million to the local economy and provides opportunities to improve morality, social solidarity, and a sense of community pride (Dafuleya et al., 2017).

Christian crusades, festivals, and seminars are held regularly in Nigeria and other countries. The Shiloh Festival held by The Living Faith Chapel (aka Winners Chapel), Holy Ghost Service (a monthly festival), the annual Holy Ghost congress sponsored by The Redeemed Christian Church, and 'The Experience' Festival by House on the Rock church are a few examples of Pentecostal events that draw both domestic and foreign believers to Nigeria (Iheanacho, 2015). Iheanacho (2015) argues that visits made by distinguished African politicians and personnel to Prophet Joshua, head of the Synagogue Church of All Nations, are a form of religious tourism. He has been visited by heads of states, government officials, and celebrities from South Africa, Ghana, and other countries. Other religious leaders in Nigeria also organize special religious events. The growing popularity of these and other events plays an important role in the development of religious tourism in Africa.

There is growing interest among international tourists for spiritual and religious experiences in Africa. Okonkwo and Nzeh (2009) noted that some churches in the United States have travel programs that provide outreach and solidarity visits to various places in Africa. Various pilgrimages and short-term volunteer missions to natural disaster zones and poverty-stricken areas are especially popular among evangelical Christians. Increasing numbers of conventions and conferences are also being held in African countries, attracting people from throughout the continent and overseas (Ron & Timothy, 2019).

Protestant travel is favorable for Africa. Instead of traveling only to the Holy Land, many Christians now travel to places not mentioned in holy scripture or which might not be connected to sacred events or people. In these cases, visits are usually of a volunteer/service nature or fellowshipping with co-religionists. Protestant Christians regularly seek to satisfy biblical exhortations to serve the needy in their travel plans. Attending a conference or dramatic presentation, participating in volunteer activities (e.g., building homes for the poor, digging wells for remote villages, and providing medical and dental services), and visiting local Christian communities in a show of solidarity (Ron & Timothy, 2019).

Islamic Religious Tourism

Africa is rich in Islamic heritage, both living rituals and practices and built faithscapes, many of which have become important tourism assets in countries such as Morocco, Tunisia, Mali, Sudan, and Egypt (Table 5.2). Traditionally, Muslim travel has focused on the annual hajj, and most scholars have emphasized this as the singular act of Islamic pilgrimage and religious devotion. However, in the past

TABLE 5.2 Islamic tourism attractions and their distribution

Country	*Islamic heritages*
Egypt	Historic mosques (Mosque of Ibn Tulun/Al Hakim Mosque/Al-Azhar Mosque/Sultan Hassan Mosque/Alabaster 'Mohammad Ali' Mosque/El Mu'ayyad Mosque 'Red Mosque'/Abdel Monaem Read Mosque/The Citadel of Saladin)
Morocco	Mosques such as Hassan II Mosque Casablanca/Kutubiyya Mosque in Marrakesh, Gnawa
Tunisia	Qairawan Great Mosque 'Sidi Oqba'/Mosque of the Barber 'Zaouia Sidi Sahab' Kairouan/Kairouan Mosque of three doors 'Djeema Tleta Bibane'/Kairouan Mosque of the Sabres 'Zaouia Sidi Amor Abbada'/ Sousse Great Mosque of Sousse/Tunis Al Zaytuna Mosque
Mali	Old town of Djene, Timbuktu, and Askia's tomb
Algeria	Mosques
East Africa	Kibuli Mosque (Kampala city), Old Kampala National Mosque (Kampala city), Kigungu landing site (Wakiso district)

decade, scholars have focused increasingly on the broader notions of other Islamic pilgrimages, halal tourism, mosque tourism, and hajj tourism to examine Muslims' travel patterns in different contexts. Muslims undertake other pilgrimages beyond the hajj, including several in Morocco and other North African states, as well as in West Africa and Ethiopia (Dagnachew, 2020) that focus on the tombs of famous imams and sites of religious importance (Berriane, 2015; Bhardwaj, 1998). Muslim-majority countries in the region include all North African states, as well as Mali, Niger, Chad, Sudan, Somalia, Mauritania, Senegal, and Guinea (Kessler, 2015). Among these countries, Egypt is one of the top traditional Islamic destinations. According to Kessler (2013), Muslim travels can be considered Islamic tourism regardless of travel motivations. Conceptually, therefore, Muslims traveling only for spiritual or only for secular purposes either to Islamic or non-Islamic destinations are viewed as a form of Islamic tourism.

Mosques that exemplify outstanding and unique architecture and heritage features are among the most important attractions for believers and non-believers in Muslim-majority countries. In fact, several mosques have been inscribed on UNESCO's World Heritage List. Historical, architectural, religious, and touristic values are criteria that render these buildings important heritage and religious attractions. Cairo is home to the historic Al Azar Mosque and Marrakesh, Morocco, is home to the famous Kutubiyya Mosque. Kairouan, Tunisia, is known as the city of 50 mosques and the spiritual capital of the country. Likewise, the adobe mosques of Timbuktu (and various other sacred sites in the vicinity) and Djenné in Mali are WHSs and an important part of that country's heritage which, until its recent security problems, provided an important attraction base for Mali.

Intangible religious heritage is an important identity marker in much of Africa and a point of pride for many communities on the continent. The Ojude Oba and Durbar Festivals are important Islamic heritage festivals in Nigeria (Iheanacho, 2015), and many other Islamic festivals and sacred ceremonies regularly take place in Algeria, Morocco, and Egypt, as well as further south in places such as Zanzibar, Kenya, and Nigeria, many of which attract international and domestic tourists.

Jewish Heritage in Africa

As previously noted, there is a centuries-long strong connection between Ethiopia and Israel, and before many of the Ethiopian Jews (Beta Israel) migrated to Israel in the 1980s–2000s, there was a relatively large Jewish population in Ethiopia. Today, approximately 12,000 remain, and there is a significant travel connection in both directions between the two countries that hinges on Jewish religion and culture.

Likewise, prior to the foundation of Israel in 1948, there was a large Jewish population in North Africa, particularly in Morocco and Tunisia. After 1948, most of the Maghrebi Jews migrated to Israel, but their built heritage remained largely intact in North Africa. There are many famous synagogues in Marrakesh, Fes, and Casablanca, Morocco, that attract Israelis, Jews, and other visitors. Likewise, the El Ghriba Synagogue on Djerba Island, Tunisia, continues to serve the island's remaining Jewish community and is a significant Jewish pilgrimage destination surrounding the annual Ghriba pilgrimage (Carpenter-Latiri, 2010). Morocco's synagogues are maintained as part of the country's multicultural heritage milieu and serve as important heritage attractions and pilgrimage sites for Jews from Israel and the diaspora (Timothy, 2021). The synagogue of Djerba is maintained by the 2,000-strong Jewish community that remains on the Island and is a major tourist attraction for Jews and non-Jews alike (Boussetta, 2018).

Having provided a background to religious heritage-based tourism in Africa and provided the necessary context regarding the different religious movements on the continent, the following sections highlight some of the main challenges facing the development of this niche form of tourism.

Challenges to African Religious Tourism Development

Although more than 95% of Africans claim their belief in religion, inbound travel to Africa for religious purposes is very limited, although outbound religious travel is common to important Muslim and Christian holy sites elsewhere, and there is a significant level of intra-regional and domestic pilgrimage/religious tourism on the continent. The absence of internationally recognized sacred sites and the small religious tourism market share in Africa is a notable tourism lacuna and can be explained at least partially by several factors.

First, policy, planning, and regulatory issues are less supportive of developing and marketing religious tourism. In most Sub-Saharan African countries, religious tourism development guidelines and regulations are absent, and marketing and promotion strategies are weak, as this is not generally recognized as a lucrative product or consumer market (Ayorekire et al., 2020; Bayih, 2018; Gedecho, 2014). Perhaps the biggest issue in this regard is a general misunderstanding within the industry and among government agencies about what religious tourism entails. Likewise, unnecessary competition and the concomitant lack of cooperation between the dominant faith groups might have drawn a reluctance to focus on this element of tourism.

Second, religious extremism in some countries is a threat to both domestic and international travel. Terrorist groups such as Al-Shabab in Somalia, Boko Haram in Nigeria, and ISIS in North Africa are significant security threats to the continent's religious travel marketplace and have thus painted a negative image of Africa and some of its faith communities (Mohamed & Alseyoufi, 2018). Additionally, these terror groups are accountable for damaging religious shrines, mosques, and other heritage sites. These groups hinder the development of pilgrimage and most other forms of tourism in certain parts of North, East, and West Africa.

The third variable is certain destinations' inability to manage tourism's negative physical impacts, which deteriorate religious heritage, physically and symbolically, and adversely affect tourists' experiences. Some studies have found that overcrowding, pollution, and poor hygiene affect pilgrims' experience at religious events (Mwebaza et al., 2018; Nyikana, 2017; Okonkwo & Eyisi, 2020).

The fourth challenge is that religious heritage often gets a lower priority when it comes to preservation and conservation compared to nature or other forms of heritage. Making money, providing positive visitor experiences, and safety are the main concerns of most stakeholders. Studies on religious tourism in the Sinai Desert have recommended that managers set limits on visitor numbers entering religious sites while simultaneously continuing to realize the sites' commercial value (Hobbs, 1992; Shackley, 1998). Hobbs (1992) revealed that stakeholders were concerned about tourism development around Mount Sinai, while Shackley (1998) argues that the increased number of visitors to Saint Catherine's Monastery threatens its long-term viability. Due to inadequate heritage protection, many religious icons and antiquities have been illegally trafficked out of Africa, especially from Ethiopia, to museums and collectors abroad.

The fifth challenge is the fact that, comparatively, the quality of tourist services tends to be lower in many of Africa's faith destinations than in those of other well-developed destinations. As a result, most visitors are unlikely to visit again, spread positive word of mouth, and recommend the place to others. Ndivo and Cantoni (2016) note that perceived exploitation of visitors by the local guides and church workers, harassment, guides' dishonesty when dealing with clients, language barriers, and lack of signage and maps negatively affected visitors' heritage experiences at Lalibela, Ethiopia.

Finally, early travel writers, explorers, and colonizers, as well as recently the media and social media have done damage to Africa's image as a potential religious destination. Negative perceptions about African religion hinder some religious people from visiting. One early criticism of African traditional religion (Awolalu, 1976) indicates that the public was commonly misled by pejorative terms for African traditional religions such as primitive, savage, fetishism, juju, heathenism, paganism, idolatry (false god worship), and polytheism. The use of these terms might have led many non-Africans to perceive the continent's religious landscape and its heritage negatively. Indeed, such claims require further investigation (Mbiti, 2015). Compared to other regions, Africa's image is often depicted in unfavorable terms and portrayed as 'the region with nothing to offer', which creates serious challenges to marketing and promotion. Changing negative perceptions and images requires huge marketing investments and diplomatic efforts.

The weakness of Africa's religious tourism fundamentally emanates from two sources. The first is from the religious institutions that are leading and teaching their followers. This group is responsible for organizing events, preserving religious heritage, motivating people to undertake spiritual travel or pilgrimage, and networking to attract others of the faith from other countries. Except for a few institutions in a few countries, most faith organizations are unwilling to make such efforts. The rules and regulations of some religions do not allow non-members to stay long or to enter certain sacrosanct spaces. Second, government organizations and agencies have devoted very limited attention to religious tourism development in terms of policy, planning, infrastructure development, and marketing (Bayih, 2018; Gedecho, 2014). As scant research findings indicate, governments in Africa pay little regard to tourism's spiritual or religious dimensions. Instead, they choose to focus on other cultural and natural assets. In other words, there are no policies or support strategies to encourage religious institutions to develop tourism. In some countries, religious heritage is played down and is treated as being secondary to national and colonial heritage and the cultural tourism they support.

Prospects for Religious Heritage-Based Tourism in Africa

Since the fourteenth and fifteenth centuries, missionary efforts have been a major travel motive to Africa. Missionaries were a by-product of the colonial-political mission of European metropoles. Most mission work has resulted in positive outcomes; boarding schools, hospitals, and social service providers were built by the church. In modern times, mission service continues to be a major part of every denomination's relationship with Africa.

Sub-Saharan Africa is the world's most impoverished region requiring financial and development aid from many donor agencies and foreign governments. People's standards of living compared to other regions are generally very low. Many churches in Europe, North America, Latin America, and Australia arrange solidarity visits to meet their co-religionists in Africa, particularly those who are poor and needy. Many faiths provide aid to Africa's poor as part of their foundational religious beliefs and out of a desire to serve humankind. As a result, thousands of people travel to Africa each year for that purpose. From Africa's perspective, however, most of that travel has been counted as business or leisure travel, not religious or spiritual travel. Many religious people are keen to engage in voluntary services and activities, and such deeds have spiritual rewards. Some activities include helping the elderly, visiting patients, providing medical services, Bible study, agricultural assistance, and childcare. Lacey, Weiler, and Peel (2014) examine volunteer tourism experiences and motivations at a school in Kenya. They found that Australian volunteers received opportunities to meet and work with farmers in the local area. Pervasive poverty and the desire to give of themselves, as well as the school's Christian foundations, are the main attractions that pull many volunteer tourists to the location in Kenya. Some of the visitors interviewed by Lacey and his colleagues were sponsors of students at the school and desired to check in on the recipients of their sponsorship and how well their donations were being used. The school is church-based and supported by a church congregation in Australia. The volunteers provided education and farming assistance, and played sports with the students.

Volunteer tourism is now widespread throughout the world, including Africa, with much of it having a religious and spiritual connotation, but it is not commonly recognized as 'volunteer tourism', nor are its religious dimensions well understood as a manifestation of spiritual or religious tourism. Compared to many other destinations, Africa has a better competitive advantage in terms of the availability of individuals in need with a great deal of room for expanding faith-based volunteer tourism. Nonetheless, it requires substantial efforts from concerned stakeholders such as religious institutions (local and overseas), governments, NGOs, and others. Religious institutions, in collaboration with governments and non-governmental organizations, could work to create more religious tourism products.

Conclusion

This chapter discusses historical perspectives, resources, patterns, challenges, and prospects of African religious tourism and other cultural tourism based on religious heritage assets (see Figure 5.2). In 2015, the Pew Research Center projected the religious composition of Africa in 2020 to be 53% Christians, 41.4% Muslims, 2.7% traditional or folk religions, 0.2% Hindus, less than 0.1% each of Buddhist and Jews, and 3% unaffiliated. The continent has diverse spiritual, religious, and secular-religious resources, with most needing to be developed into tourism products. Given the growth in religious and spiritual tourism, and the canonical principles of some religions that encourage travel for various reasons, as Africa continues as a burgeoning tourism region, it is likely that religious tourism can also grow, especially with

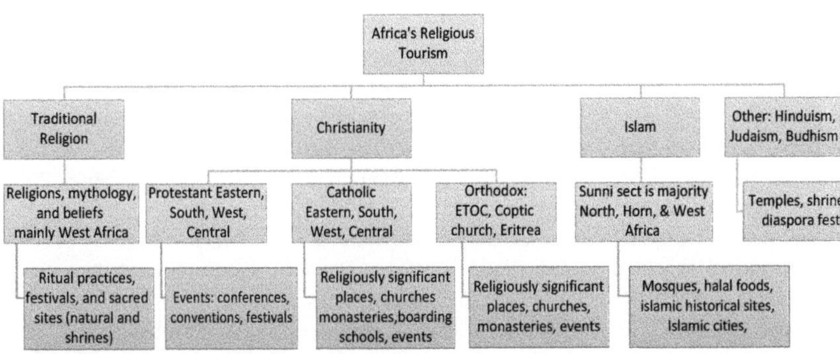

FIGURE 5.2 Religious groups and their tourism products.

regard to specific niche markets such as those associated with mosque visits, halal tourism, events and conference tourism, solidarity tourism, and volunteer tourism. However, for this to be effectively achieved, much has to be done to consolidate the management and development initiatives around this niche sector. In essence, there is plenty of potential for religious tourism development in Africa; however, this potential has not yet been well explored, resulting in a relatively poor market share and disappointing academic focus on this area. This chapter thus makes a modest contribution to enhancing our understanding of heritage and religious tourism in Africa. It should therefore ignite some interest in this niche, with more academic focus especially as dictated by the differing religious movements across the different geographical sub-regions within the continent projected to increase.

References

Apleni, L., & Henama, U.S. (2020). Visitors' perceptions of religious tourism in South Africa: The case of Moria in the Limpopo Province. *African Journal of Hospitality, Tourism and Leisure, 9*(2), 1–19.

Ashcraft, W.M. (2018). *A historical introduction to the study of new religious movements*. London: Routledge.

Awolalu, J.O. (1976). What is African traditional religion? *Studies in Comparative Religion, 10*(2), 1–10.

Ayorekire, J., Obua, J., Mugizi, F., & Byaruhanga, B.M. (2020). Opportunities and challenges of religious tourism development in Uganda: Policy, planning and institutional perspectives. *International Journal of Religious Tourism and Pilgrimage, 8*(3), 144–158.

Bayih, B.E. (2018). Potentials and challenges of religious tourism development in Lalibela, Ethiopia. *African Journal of Hospitality, Tourism and Leisure,* 7(4), 1–17.

Berriane, J. (2015). Pilgrimage, spiritual tourism and the shaping of transnational 'imagined communities': The case of the Tidjani Ziyara to Fez. *International Journal of Religious Tourism and Pilgrimage, 3*(2), 1–10.

Bhardwaj, S.M. (1998). Non-hajj pilgrimage in Islam: A neglected deminsion of religious circulation. *Journal of Cultural Geography, 17*(2), 69–87.

Boussetta, M. (2018). Reducing barriers: How the Jews of Djerba are using tourism to assert their place in the modern nation state of Tunisia. *The Journal of North African Studies, 23*(1–2), 311–331.

Bowers, H., & Cheer, J.M. (2017). Yoga tourism: Commodification and western embracement of eastern spiritual practice. *Tourism Management Perspectives, 24*, 208–216.

Bruntz, C., & Schedneck, B. (Eds.) (2020). *Buddhist tourism in Asia*. Honolulu: University of Hawaii Press.

Carpenter-Latiri, D. (2010). The Ghriba pilgrimage in the island of Jerba: The semantics of otherness. *Scripta Instituti Donneriani Aboensis, 22*, 38–55.

Christ for all Nations (2020). Reinhard Bonnke: Biography. Retrieved December 8, 2020, from https://cfan.org/reinhard-bonnke

Collins-Kreiner, N. (2019). Contemporary Jewish tourism: Pilgrimage, religious heritage and educational tourism. In D.J. Timothy (Ed.), *Routledge handbook on tourism in the Middle East and North Africa*, pp. 137–146. London: Routledge.

Dafuleya, G., Gyekye, A.B., Oseifuah, E.K., Nethengwe, T., & Sumbana, F. (2017). Religiously motivated travel and rural tourism in Vhembe District of South Africa. *Tourism Economics, 23*(4), 911–918.

Dagnachew, S. (2020). Interfaith tourism in Ethiopia: An opportunity for socio-economic development and peace-building? In I.S. McIntosh, N.F. Haddad, & D. Munro (Eds.), *Peace journeys: A new direction in religious tourism and pilgrimage research*, pp. 80–89. Newcastle upon Tyne: Cambridge Scholars Publishing.

Dowson, R. (2017). Towards a definition of Christian mega-events in the 21st century. *International Journal of Religious Tourism and Pilgrimage, 5*, 1–18.

Dowson, R. (2022). Managing complex issues in religious and spiritual events, festivals and celebrations. In D.H. Olsen & D.J. Timothy (Eds.), *The Routledge handbook of religious and spiritual tourism*, pp. 359–371. London: Routledge.

Emmanuel, O.K. (2014). Religious tourism and sustainable development: A study of Eyo Festival in Lagos, Nigeria. *International Journal of Social Sciences and Education, 4*(2), 524–534.

Encyclopaedia Britannica (2018). African religions Britannica. Retrieved December 10, 2020, from Encyclopaedia Britannica website: https://www.britannica.com/topic/African-religions

Fadare, O., & Benson, E.I. (2015). The consumption and management of religious tourism sties in Africa. In R. Raj & K. Griffin (Eds.), *Religious tourism and pilgrimage management: An international perspective*, 2nd edn, pp. 218–233. Wallingford: CABI.

Gedecho, E.K. (2014). Challenges of religious tourism development: The case of Gishen Mariam, Ethiopia. *American Journal of Tourism Research, 3*(2), 42–57.

Gedecho, E.K., & Yebo, M.F. (2014). Religious tourism potential of Gishen Derbe Kerbe Mariam, Ethiopia. *African Journal of Hospitality, Tourism and Leisure, 3*(1), 1–19.

Goldberg, A. (1983). Identity and experience in Haitian voodoo shows. *Annals of Tourism Research, 10*(4), 479–495.

Hancock, G. (1992). *The sign and the seal: The quest for the lost Ark of the Covenant*. New York: Simon & Schuster.

Hassib, R., & Ibtissem, S. (2020). Islamic tourism and sustainable development: From utopia to reality. *International Journal of Religious Tourism and Pilgrimage, 8*(3), 159–167.

Hernandez-Ramdwar, C. (2013). African traditional religions in the Caribbean and Brazil: Models of religious tourism and impacts of commodification. *Journal of Heritage Tourism, 8*(1), 81–88.

Hobbs, J.J. (1992). Sacred space and touristic development at Jebel Musa (Mt Sinai), Egypt. *Journal of Cultural Geography, 12*(2), 99–113.

Iheanacho, N. (2015). Nigerian praxis of religious tourism and pilgrimage motivations in the globalizing world. *Revista Ciências Da Religião-História e Sociedade, 13*(1), 259–284.

Ipara, H. (2013). Promoting heritage tourism through religious pilgrimages. Paper presented at the *Atlas Africa Conference*, Kigalia, Rwanda, June 4, 2013.

Jamal, A., Griffin, K., & Raj, R. (2018). *Islamic tourism: Management of travel destinations*. Wallingford: CABI.

Kamil, J. (2000). Al-Ahram weekly religious tourism as big business. *Al-Ahram Weekly*. Retrieved from http://www.ahram.org.eg/weekly/2000/469/tr2.htm

Kessler, K. (2015). Conceptualizing mosque tourism: A central feature of Islamic and religious tourism. *The International Journal of Religious Tourism and Pilgrimage, 3*(2), 11–32.

Kim, B., Kim, S., & King, B. (2020). Religious tourism studies: Evolution, progress, and future prospects. *Tourism Recreation Research, 45*(2), 185–203.

Kruger, M., & Saayman, M. (2016). Understanding the Zion Christian Church (ZCC) pilgrims. *International Journal of Tourism Research, 18*(1), 27–38.

Kujawa, J. (2017). Spiritual tourism as a quest. *Tourism Management Perspectives, 24*, 193–200.

Lacey, G.T., Weiler, B., & Peel, V. (2014). Revealing hidden attractions in a rural Kenyan periphery. *Tourism Geographies, 16*(2), 174–189.

Luz, N. (2020). Pilgrimage and religious tourism in Islam. *Annals of Tourism Research, 8*, 102915.

Mbiti, J.S. (2015). *Introduction to African religion*. Long Grove, ILL: Waveland Press.

Mohamed, T.Z.F., & Alseyoufi, T.S. (2018). Evolution and impact of terrorism in the Middle East: Implications for Egyptian travel and tourism. *International Journal of Religious Tourism and Pilgrimage, 6*(3), 86–106.

Mwebaza, S., Jjuuko, J., & Kesande, P. (2018). Religious tourism and pilgrimage: Conflicts and challenges, A case of the Namugongo Martyrs Shrines in Uganda. *Turystyka Kulturowa*, 7, 117–131.

Ndivo, R.M., & Cantoni, L. (2016). The efficacy of heritage interpretation at the Lalibela Rock-Hewn Churches in Ethiopia: Exploring the need for integrating ICT-mediation. *International Journal of Religious Tourism and Pilgrimage, 4*(3), 16–28.

Nmah, P.E. (2017). What is new in "new" religious movements in Africa? An analytical approach. *IGWEBUIKE: An African Journal of Arts and Humanities, 3*(2), 47–63.

Nyikana, S. (2017). Religious tourism in South Africa: Preliminary analysis of a major festival in Limpopo. *African Journal of Hospitality, Tourism and Leisure, 6*(1), 1–8.

Okamoto, R. (2019). *Pilgrimage in the secular age: From El Camino to anime*. Tokyo: Japan Foundation for Culture.

Okonkwo, E., & Eyisi, A. (2020). Pilgrimage circuit of Osun Osogbo sacred grove and shrine, Osun State, Nigeria. *International Journal of Religious Tourism and Pilgrimage, 8*(4), 80–88.

Okonkwo, E., & Nzeh, C.A. (2009). Faith-based cctivities and their tourism potentials in Nigeria. *International Journal of Research in Arts and Social Sciences, 1*, 286–298.

Olsen, D.H. (2022). Fan pilgrimage, religion, and spirituality. In D.H. Olsen & D.J. Timothy (Eds.), *The Routledge handbook of religious and spiritual tourism*, pp. 90–110. London: Routledge.

Olsen, D.H., & Timothy, D.J. (Eds.) (2022). *The Routledge handbook of religious and spiritual tourism*. London: Routledge.

Olsen, D.H., & Wilkinson, G. (2016). Are fast pilgrims true pilgrims? The Shikoku pilgrimage. *Annals of Tourism Research, 61*, 228–230.

Pew Research Center (2015). Religious composition by country, 2010–2050. Retrieved November 23, 2020, from Pew Research Center website: https://www.pewforum.org/2015/04/02/religious-projection-table/2020/percent/all/

Reed, A. (2006). *Gateway to Africa: The pilgrimage tourism of diaspora Africans to Ghana*. Upublished doctoral dissertation, Indiana University.

Rojo, D.M. (2007). *Religious tourism: The way to Santiago*. Unpublished masters thesis, Bournemouth University.

Ron, A.S., & Timothy, D.J. (2019). *Contemporary Christian travel: Pilgrimage, practice, and place*. Bristol: Channel View Publications.

Saayman, A., Saayman, M., & Gyekye, A. (2014). Perspectives on the regional economic value of a pilgrimage. *International Journal of Tourism Research, 16*(4), 407–414.

Schedneck, B. (2022). Religious and spiritual retreats. In D.H. Olsen & D.J. Timothy (Eds.), *The Routledge handbook of religious and spiritual tourism*, pp. 191–203. London: Routledge.

Séraphin, H., & Nolan, E. (2015). Voodoo in Haiti: A religious ceremony at the service of the 'Houngan' called 'tourism'. In J. Laing & W. Frost (Eds.), *Rituals and traditional events in the modern world*, pp. 231–241. London: Routledge.

Shackley, M. (1998). A golden calf in sacred space? The future of St Katherine's monastery, Mount Sinai (Egypt). *International Journal of Heritage Studies, 4*(3–4), 124–134.

Sharpley, R., & Jepson, D. (2011). Rural tourism: A spiritual experience? *Annals of Tourism Research, 38*(1), 52–71.

Sisay, A. (2009). *Historical development of travel and tourism in Ethiopia*. Addis Ababa: Commercial Printing Enterprise.

Smith, M.K. (2003). Holistic holidays: Tourism and the reconciliation of body, mind and spirit. *Tourism Recreation Research*, *28*(1), 103–108.

Teye, V.B. (2009). Tourism and Africa's tripartite cultural past. In D.J. Timothy & G.P. Nyaupane (Eds.), *Cultural heritage and tourism in the developing world: A regional perspective*, pp. 165–185. London: Routledge.

Timothy, D.J. (2021). Tourism and the multi-faith heritage of the Middle East and North Africa: A resource perspective. In C.M. Hall & S. Seyfi (Eds.),*Cultural and heritage tourism in the Middle East and North Africa: Complexities, management and practices*, pp. 34–53. London: Routledge.

Timothy, D.J., & Olsen, D.H. (Eds.) (2006). *Tourism, religion and spiritual journeys*. London: Routledge.

UNWTO (2020). *Buddhist tourism in Asia: Towards sustainable development*. Madrid: World Tourism Organization.

Warfield, H.A., Baker, S.B., & Foxx, S.B.P. (2014). The therapeutic value of pilgrimage: A grounded theory study. *Mental Health, Religion and Culture*, *17*(8), 860–875.

Woosnam, K.M., Aleshinloye, K.D., Ribeiro, M.A., Stylidis, D., Jiang, J., & Erul, E. (2018). Social determinants of place attachment at a World Heritage Site. *Tourism Management*, *67*, 139–146.

Zamani-Farahani, H., Carboni, M., Perilli, C., & Torabi Farsani, N. (2019). Islamic tourism in the Middle East. In D.J. Timothy (Ed.), *Routledge handbook on tourism in the Middle East and North Africa*, pp. 125–136. London: Routledge.

6

THE SPATIAL MISMATCH BETWEEN CITIES AND HERITAGE TOURISM SITES IN AFRICA

Gustav Visser

Introduction

Heritage tourism is an increasingly popular form of tourism across many destinations in the Global South and is often presented as a major growth market in both natural (Hottola, 2009; van Beek & Schmidt, 2012) and urban settings (Rogerson & Rogerson, 2021a; Rogerson & van der Merwe, 2016). Heritage is generally framed as 'our legacy inheritance from the past, what we pass on to future generations, including the natural and cultural, living and built elements of culture, tangible as well as intangible cultural aspects with cultural significance of value' (Lwoga & Adu-Ampong, 2021, p. 69). This tourism niche is considered a tool with the potential to promote local development opportunities in the Global South, including in Africa (Hottola, 2009; Rogerson & Rogerson, 2021a). At the outset, it must be appreciated that across Africa, as expressed by Rogerson and Rogerson (2021b) and described in detail by Lwoga and Adu-Ampong (2021), the phenomenon of travel for various reasons to cities, in particular, is not new. These scholars note that from the birth of urban settlements, people with the compulsion, discretionary means, and inclination to travel became tourists in order to experience and trade in these spaces resulting in melting pots supporting and influencing culture, art, music, architecture, and religions. This trend has continued for centuries and we see it even now in the present day.

Relative to the Global North, the impacts of heritage tourism in the Global South have seen modest scholarly examination but do, albeit limited, consider a range of locations (Gravari-Barbas & Jacquot, 2014; Rogerson & Rogerson, 2021a). The extant scholarship on heritage tourism in Africa, and Sub-Saharan Africa specifically, is sparse, very uneven in spatial focus and type and is dominated by a handful of interconnected investigators and their particular research interests. The chapter will first consider the notion of heritage tourism products in

DOI: 10.4324/9781003153955-6

the Sub-Saharan African region and its general geography. Secondly, the lack of a strong link between heritage tourism and urban tourism is examined. Finally, the investigation considers a selection of extant literature and discusses their focus areas as well their main conclusions. The argument presented is that heritage tourism in Sub-Saharan Africa requires considerable refinement in order to play a meaningful role in the region's urban development.

Where Is Urban Africa's Heritage?

As a concept, heritage is a vexing issue. What constitutes it? Gravari-Barbas and Jacquot (2014), along with Lwoga and Adu-Ampong (2021), concur that there is no universal consensus on this matter. For the purposes of this investigation, we draw on the rather generic, and admittedly problematic, source of the United Nations Educational, Scientific and Cultural Organization's (UNESCO) understanding thereof. Knowing that heritage is rooted in a locale, context and time, for UNESCO, heritage sites are designated as having 'outstanding universal value' under the convention concerning the protection of the World Cultural and Natural Heritage. UNESCO has designated well over one hundred World Heritage Sites (WHSs) in Africa. These sites are located across 35 African countries. The current constitution of heritage tourism sites in Africa is represented in many ways. It is important to note that these sites mainly relate to those of 'international significance'. There has been considerable debate concerning the 'Western lens' through which this heritage is conceptualised, curated, managed, and for what purpose (Gravari-Barbas & Jacquot, 2014; Lwoga & Adu-Ampong, 2021; Novelli, Adu-Ampong, & Ribeiro, 2021).

Typically, these sites/regions fall into one of two broad categories: cultural or natural heritage. Each year, UNESCO's World Heritage Committee may inscribe new sites onto the list, or delist sites that no longer meet the criteria. Selection is based on ten criteria: six for cultural heritage (i–vi) and four for natural heritage (vii–x). Some sites are designated 'mixed sites and represent both cultural and natural heritage. At the time of writing, in Africa, there are 55 cultural, 44 natural, and 5 mixed sites. As a continent, Africa has the fewest number of WHSs (UNESCO, 2020). Over time, there has been a steady increase in the number of new sites registered on the continent; since the 1970s, there have generally been two to four new sites registered per annum. The first recognised sites from the continent were the Island of Gorée of Senegal and the Rock-Hewn Churches of Ethiopia, which were chosen during the list's conception and were inscribed in 1978. New sites have been slowly added to the list over the ensuing decades (Figure 6.1).

For Africa as a whole, it can be observed that UNESCO WHSs are very unevenly distributed. This is not to say that there are not countless other nationally declared heritage sites (there are in fact thousands), rather it is the case that not all of these sites are 'interpreted' as sites of 'international significance'. There is also the issue that vast parts of the continent, such as the Sahara, other extremely arid regions, as well as rain forests that are sparely settled (if at all), are not represented.

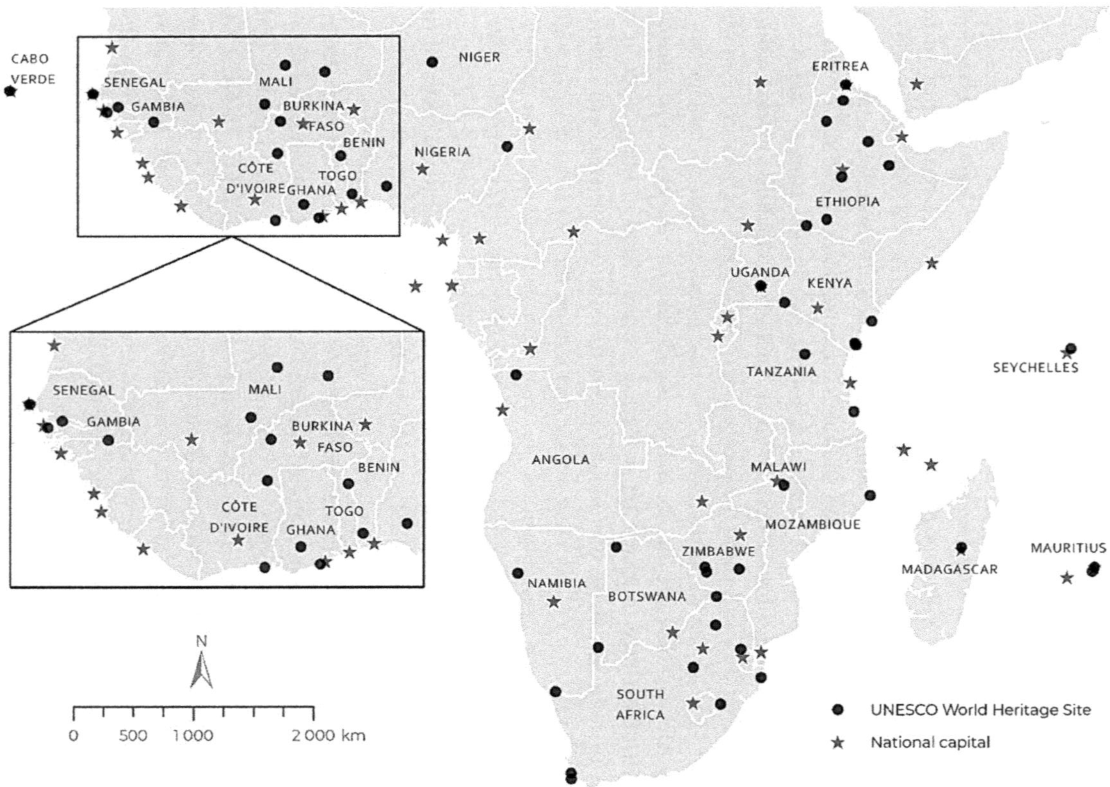

FIGURE 6.1 Distribution of UNESCO Heritage Sites relative to national capitals.

Source: Own Survey.

A reason for this is because the heritage of these regions is often located or represented in non-material formats, being in the form of past legends and/or narratives, or is alternatively due to an inability to argue their value and maintenance (Gravari-Barbas & Jacquot, 2014). The greatest density of UNESCO WHSs in Africa is located in countries on the northern and southern reaches of the continent and to a lesser extent, its eastern and western extremities.

For those in the north, nearly all UNESCO Heritage Sites relate to Ancient Egypt (third century BCE onwards) and the Greco-Roman eras (second century BCE to fifth century CE) (Lwoga & Adu-Ampong, 2021). There are also a number of sites concerning the area's previous Arab influence and Berber nations in the northwestern parts of Africa following those eras along with the expansion of Islam and associated scholars and related institutions. Moving to Sub-Saharan Africa, generally speaking, it is South Africa that is the location of the most heritage sites—ten in total (UNESCO, 2021a). These sites are mostly natural heritage and mark the early development of humans and their environments. In East Africa, Ethiopia is home to nine sites generally concerning the development of early Christianity and natural landscapes (UNESCO, 2021b), whilst further south, the role of the slave trade and Arabic architecture, along with the majestic landscape of the Serengeti, are showcased (UNESCO, 2021c). In West Africa, in countries such as Ghana, the slave trade of the seventeenth and eighteenth centuries looms large in terms of heritage declarations and there are numerous sites in the region mainly concerned with cultural heritage ranging from pre-colonial kingdoms and natural areas (UNESCO, 2021d). There are, however, nearly no UNESCO Heritage Sites in the central parts of Africa (UNESCO, 2021d). This is in large part because tangible heritage is often not permanent but is rather constantly evolving along with the registration of such sites by the relevant national governments.

This gestures towards the problematic Western notions of permanence in heritage, as argued by Lwoga and Adu-Ampong (2021). Following their argument, by not accepting the impermanent nature of much of African traditional built legacy *as* heritage, the broader international community is blinded from a very real, rich, and layered heritage. They provide an example of this in which they highlight that the treatment of traditional heritage in many places in Africa conventionally involves the consistent repairing of mud walls and thatched grass roofs. This additionally involves the regular repairing of decorations; however, there is a limited desire to imitate and restore the existing designs to what they were, rather there is a trend to build new designs that reflect contemporary social issues. According to Joffroy (2009), many would often apply entirely new decorations during repairs and in so doing, they treat their traditional houses and other built heritage as inspired by expressions of their mood and the trend of the moment. For many Africans, the inspiration and expressions that accompany the process of repairing and rebuilding their inherited built environment become engulfed with notions of authenticity.

In the general consideration of the broad spatial distribution of heritage sites, a critical observation needs to be made. Sub-Saharan African cities are home to many layers of a mainly recent human past and although some cities also contain pre-colonial archaeological heritage sites and structures, remnants of colonial architecture, and post-colonial forms and functions or urban design heritage sites of perceived international importance, they are generally not located in large urban areas. This is in stark contrast to many highly rated heritage sites in Europe. In addition, something else that needs to be highlighted is a lack of meaningful spatial clustering of key UNESCO WHSs in most of Africa. It is key to note that great care needs to be taken in the analysis of these kinds of sites but perhaps more importantly, it is necessary to remember the scale of the region under review. For example, Gall-Peters, as opposed to the more generally used Mercator map projection, provides some insight into the sheer size of the African continent relative to Europe and illustrates that many of its constituent nation states are very large in comparison, spatially distanced and so too are the many forms of UNESCO-recognised heritage. It could be argued that this spatial distribution of these sites is one of the great challenges for the deployment of heritage tourism as a developmental driver in this region. The fact that heritage sites are often also not located in cities or in relative close proximity to particularly large cities with relatively easy access to international tourist markets is a significant additional challenge. That is not to say that, for example, cities which function as large ports of entry do not have any cultural heritage, rather it is the fact that sites which have gained recognition in accordance with the UNESCO definition are, for the most part, not located at ports of entry.

Who Is Writing about African Urban Heritage Tourism and Their Findings?

When considering heritage tourism scholarship in Africa, it is important to reflect on the framers thereof in the region generally. An extensive review of the post-2000 period, using the contents of leading English-language serials of tourism and hospitality research, as well as other significant outlets for African tourism research in the Southern African region, is provided by Rogerson and Rogerson (2021a). They make several key observations: that the general publication outlets are not tourism-specific journals, that within this scholarship the largest volume of writings is produced by geographers, and that the greatest share of this literature relates to South Africa. As Binns and Nel (2002, p. 236) noted two decades ago, 'South Africa's natural and cultural heritage … is clearly one of the most appealing and rapidly growing tourism attractions in the world, (which) makes this sector, an area to which government, entrepreneurs and communities are currently giving serious consideration', and this is evident in the volume of published work on the area. In the extant scholarship, a critically important observation relates to the main thematic foci showing several research clusters (Rogerson & Rogerson, 2021a). The six clusters related to the following: (1) African cities as business tourist destinations, (2) the emergence of cities as tourism destinations (which to some degree includes heritage-related references), (3) tourism and urban economic restructuring associated with the establishment of new leisure tourism products (again sometimes related to heritage), (4) slum tourism (some heritage investigations in South Africa), (5) the growth of accommodation services, and (6) informal sector tourism. In reviewing the studies from which these categorisations were derived, only a handful speak to the notion of heritage tourism in the region linked to ports of entry or major cities.

By casting the line of investigation further, in terms of outlets (grey literature) and language (including French-written scholarship, for example, a perennial issue in scholarship on tourism), in order to include studies from across Africa, Gravari-Barbas and Jacquot (2014) concluded that, when compared on the international stage, heritage tourism scholarship in Africa is generally not extensive. That said, they did observe that the current scholarship on Africa does reflect on those places that have the greatest number of UNESCO sites (bar South African scholarship). Current writings mainly follow the density of the registered heritage sites. Consequently, we have some coverage of North Africa, a few sites in East and West Africa, and then the bulk of the work, as highlighted earlier, from Southern Africa. Given the spatial location of both heritage sites, but not necessarily author base, it must be noted that it is very difficult to speak in general terms about tourism in Africa and its relationship with heritage and development because of the diverse contextual differences across the continent (Boswell & Kane, 2011, cited in Gravari-Barbas & Jacquot, 2014). Moreover, the gaps between the most highly developed countries for tourism and those that are less developed are much more pronounced than those observed between different European countries, which results in a difficulty to conduct truly comparative research which would lead to theory building. The question then arises, what was written about heritage tourism in Sub-Saharan Africa and its relationship to cities?

The march of urbanisation is transforming the world and one of the most visible features of the contemporary age is the claim that society has entered into 'the century of the city' (Rockefeller Foundation, 2009). According to United Nations data, by 2015, more than half of the world's population resided in urban settlements—large metropolitan areas, medium-sized or secondary centres, and small towns, as opposed to rural areas. By 2050, this urban share is projected to rise to 75% with an annual growth of 65 million new urban residents (Saghir & Santoro, 2018). As a whole, the 'Global South' is viewed as the 'new epicentre of urbanism' (Oldfied & Parnell, 2014, p. 1) with the greatest projected growth for expansion being for Sub-Saharan Africa. Aall and Koens (2019, p. 1) speculate that if the next 100 years is in fact the century of the city, then 'perhaps the same can be said about urban tourism, which also is increasing rapidly compared to other forms of tourism – maybe society is entering the century of urban tourism'. Despite the fallout from the COVID-19 pandemic representing a major threat to such a prediction, in the shifting international landscape of urban tourism, destinations in the Global South are likely to become ever more significant 'growth poles'.

It needs to be made clear that in Sub-Saharan Africa, many larger cites and capital cities (often the main ports of entry) are colonial or post-colonial creations and some are both. For the most part, these settlements served as administrative and commercial locations for colonisers and later, for post-independence government administrations, but there are a number of newly created capitals to consider. The key observation here is that where heritage in the western conception (understood as per supranational institutions like UNESCO) is accepted, the focus of travel agents and marketing bodies is nearly always geographically out of sync with where the material or symbolic heritage of many Sub-Saharan nations are actually located. Of course, this mismatch can be taken back to the European division of Africa into randomly demarcated colonies with no regard for what was lived heritage across the continent. So unlike European cities in particular, there are few historic linkages between historical heritage and the location of post-independence capitals, large cities, or key international ports of entry. This issue aside, there is also a need to address the question, what are some of the key debates surrounding heritage tourism beyond the theoretical concerns put forward by scholars such as Gravari-Barbas and Jacquot (2014) and Lwoga and Adu-Ampong (2021)?

Van der Merwe (2013, p. 574) argues that for the most part, several African countries recognise the importance of pursuing heritage tourism as a vehicle for promoting national and regional tourism development, resulting in the attainment of broader developmental goals such as economic growth and employment creation. What is important to note is that heritage is often framed by its utility for tourism development and economic growth and not simply for its intrinsic value (Rogerson & Visser, 2020). Not dissimilar to the trend in nature conservation where parks are required to 'pay their way' (Wolf, Croft, & Green, 2019), the same argument can be levelled at the more physical cultural elements of heritage in cities in Sub-Saharan Africa. As nearly all countries in the region are resource-poor, heritage for the most part is conserved, preserved, or established with very limited resources on hand and many of them are also often geographically not in close proximity either to each other nationally or to ports of entry. Since we are either 'telling or selling … heritage tourism products', as Ivanovic and Saayman (2013) so presciently phrase it, it begs the question to what end: for conservation, preservation, or some form of economic and 'development' gain? The latter is often the intent across the region; however, this is seldom the outcome (Butler & Ivanovic, 2016).

There has been interest in the deployment of heritage as a tool for promoting tourism and, in particular, for developing urban tourism as a complementary tourism product to natural attractions in order to broaden economic development (Hottola, 2009; Manwa, Moswete, & Saarinen, 2016; Rogerson & van der Merwe, 2016; van der Merwe, 2014, 2016; van der Merwe & Rogerson, 2018). Local economic development (LED) underlies this ambition (Arthur & Mensah, 2006) however, many studies gesture to the potential for enhancing LED rather than attaining it, certainly in an inclusive manner (Rogerson, 2020). The recent silence of heritage and LED research is deafening; those investigations do not take place, for they, with the exception of some natural heritage sites, are not seen as vectors for economic (or any other) development. Indeed, it is important to consider the underlying motivation for investigating heritage tourism in the first place.

Most scholars view tourism through the lens of various aspects of development. This idea has, however, been questioned by Gravari-Barbas and Jacquot (2014, p. 2) who observed that 'it [is] necessary to call into question the relationship between tourism and development and, heritage and tourism and the supposedly more significant role that World Heritage can play in this tourism-based development'.

In a few instances, heritage tourism has been connected in some way to forms of urban renewal (Hoyle, 2002). Prime examples of urban renewal in South Africa certainly include Cape Town's V&A Waterfront (de Villiers & Ferreira, 2014; Ferreira & Visser, 2007) as well as development in Durban and Johannesburg (van der Merwe, 2016, 2019) and in Zanzibar's Stone Town (Hoyle, 2002; Khalfan & Nobuyuki, 2012). This trend is, however, not strongly reflected across the continent but is rather limited to a number of locations. Within the urban context, and even in major gateway cities, van der Merwe (2016) argued that one of 'the significant challenges to promoting heritage tourism in South Africa is that many heritage sites are located within ailing inner-city environments and linked to vandalism and crime which impact negatively upon visitor numbers'. In terms of some sort of urban heritage in the developing South, Frenzel (2016) asks if 'slums have become cool?' In his argument, an increasing number of tourists from across the globe seem to think so as they discover favelas, ghettos, townships, and barrios on leisurely visits to cities. But while slum tourism often evokes moral outrage, critics rarely ask about what motivates this tourism, or what wider consequences and effects it initiates. In the end, this is not a key motivator for heritage tourism in African cities, but perhaps only an 'add-on' for a business traveller, or an occasional tourist seeking out natural heritage with an urban excursion in a port of entry. This brings us to the question of who heritage visitors might be.

The deployment of heritage tourism in African cities is delimited by the constitution of the visitor cohort. Some research poses the question, who are the heritage tourists to urban heritage? In some research, it is argued that there is little to go by. Conducting a comparative investigation in Johannesburg at two heritage sites not listed by UNESCO, Khumalo, Sebatlelo, and van der Merwe (2014) found a mixed bag of responses—there was no 'generic visitor'.

Both on the east and west coasts of Africa, urban heritage has been explored. It has been argued by Teye and Timothy (2004, p. 145) some time ago that African Americans, in particular, have become the primary market for Africa's expanding heritage tourism sector demonstrating that several slave heritage sites in Africa embody the heritage of the African countries, various European countries, and Africans in the diaspora. The development of slave heritage tourism sites in Africa, they observe, involves the concept of contested heritage with emotional and painful undercurrents. Rather than really getting to know the tourists, we are reminded of the complexity of the representation of heritage for multiple tourist audiences. This circles back to the notion of the curation of heritage in urban Africa generally and Sub-Saharan Africa in particular, as noted by Gravari-Barbas and Jacquot (2014) and Lwoga and Adu-Ampong (2021).

Perceptions of, and attitudes to, heritage tourism in urban areas have also come into view (Arthur & Mensah, 2006; Teye, Turk, & Sönmez, 2011) in terms of dealing with the slave trade (Yankholmes & Akyeampong, 2010), notions of authenticity and the disconnect between attitudes, intentions and actual engagement in the conservation of built heritage (Lwoga, 2018; Masilo & van der Merwe, 2016; Viljoen & Menamo, 2017). Questions are often raised concerning the emergence of a 'public memory' of slavery and slave tourism in West Africa, and the differences between American (slave descendants) and European (slave owners) people and their representation and reception. Indeed, this opens up the issue of 'perceived authenticity' in iconic heritage sites in urban tourism generally and Sub-Saharan Africa in particular (Ivanovic, 2014). In addition, it is often unclear whether a tourist will ever revisit a particular heritage destination, in contrast to some legacy art museums in Europe and the USA, for example, this appears to be less likely in Sub-Saharan urban centres, large or small (Chami, 2018).

Investigations into the satisfaction with existing heritage sites, though not UNESCO sites, have come from a vast stream of work conducted by the Research in Economics, Environment and Society (TREES) research programme based at North-West University in South Africa. This work, however, mainly focuses on that specific region and is therefore not directly related to the continent as a whole or Sub-Saharan Africa in particular. That said, it does provide a rich source of information concerning cultural heritage events such as numerous arts festivals, visitor profiles, and their experiences. What is perhaps most important about their work is that it deals with contemporary, non-tangible heritage mostly located in areas removed from gateway or port cities which have little noteworthy built heritage.

Part of the challenge of the deployment of heritage in Sub-Saharan urban Africa is the general lack of quality built heritage in the form of curated physical experiences in museums or as a static physical built environment. This observation does pit different understandings of heritage from different cultural frameworks against each other: 'fixed' (European, for example) versus more 'fluid' heritage. Some time ago, Taylor and Lennon (2010) suggested that Africa follows a less tangible heritage tradition, one that is dominated by artefacts such as dance, music, singing, and narratives. Gravari-Barbas and Jacquot (2014) cite a UNESCO report 'CRATerre-ENSAG, Convention France' that rather bluntly and, in my view, very problematically stated,

> Most African cities do not have museums worthy of the name. Few have established a proper inventory of their natural and cultural heritage. Very few are aware of the World Heritage Convention. We can legitimately talk here of a 'spirit' which, as is the case for other sectors of the economy and society, puts Africa at risk of being marginalised.

The point is that conceptually, heritage tourism in much of Africa is not akin to that of Europe or North America, rather it is largely experiential in a non-materialist manner, or taking the form of natural landscapes in unique biomes.

Final Thoughts

In contrast to heritage tourism in UNESCO hotspots such as those in Europe, Sub-Saharan Africa and scholarships reflecting on those heritage sites do not register in any force. Generally, the main impression from the literature is that urban heritage tourism is not reaching its full potential. The central argument of the chapter is that heritage tourism in Sub-Saharan Africa, through the lens of UNESCO WHSs, is for the most part not linked to major cities or city gateways. There is spatial dissonance between notable heritage sites and ports of entry. This in turn limits the deployment of such sites as 'products' that could drive tourism-led development. For those heritage sites that are recognised, one of the challenges of heritage tourism in urban Africa is that many sites are located in ailing or poorly managed areas, or are of limited interest to a broader tourist base. Those sites that are mainly curated for a global tourism market are 'islands' in a larger urban context for international tourists, not locals, who mainly travel to Africa for business or owing to an interest in natural heritage located far away from cities and often not in gateway cities. The notion of heritage will need to be made more inclusive of a more current interpretation thereof in order to encompass not only the historic or 'monumental'. There is a need for holistic precinct planning combining various types of heritage that might rather be of local interest by providing a heritage experience that extends beyond either the deeply rooted 'local non-tangible' or 'the monumental'.

References

Aall, C., & Koens, K. (2019). The discourse on sustainable urban tourism: The need for discussing more than overtourism'. *Sustainability, 11*(5), 4228.

Arthur, N., & Mensah, J. (2006). Urban management and heritage tourism for sustainable development: The case of Elmina Cultural Heritage and Management Programme in Ghana. *Management of Environmental Quality: An International Journal,* 17(3), 299–312.

Binns, T., & Nel, E. (2002). Devolving development: Integrated development planning and developmental local government in post-apartheid South Africa. *Regional Studies, 36*(8), 921–932.

Boswell, R.R., & Kane, D. (2011). Introduction: Heritage management and tourism in Africa. *Journal of Contemporary African Studies, 29*(4), 361–369.

Butler, G., & Ivanovic, M. (2016). Cultural heritage development in post-apartheid South Africa: Critical issues and challenges. In H. Manwa, N. Moswete, & J. Saarinen (Eds.), *Cultural tourism in Southern Africa*, pp. 58–75. Bristol: Channel View Publications.

Chami, M. (2018). Assessment of tourist willingness to re-visiting cultural heritage sites in Tanzania: A case of Zanzibar Stone Town, UNESCO World Heritage Site. *Journal of Advance Research in Social Science and Humanities, 4*(60), 19–29.

de Villiers, R., & Ferreira, S. (2014). The Victoria and Alfred Waterfront as playground for Capetonians. *Urbani Izziv, 25*, 63–80.

Ferreira, S., & Visser, G. (2007). Creating an African riviera: Revisiting the impact of the Victoria and Alfred Waterfront development in Cape Town. *Urban Forum, 18*(3), 227–246.

Frenzel, F. (2016). *Slumming it: The tourist valorization of urban poverty*. London: Zed.

Gravari-Barbas, M., & Jacquot, S. (2014). World Heritage, tourism and sustainable development in Africa: Discourses, approaches and challenges. *Via Tourism Review, 2014*, 4–5.
Hottola, P. (Ed.) (2009). *Tourism strategies and local responses in Southern Africa*. Wallingford: CABI.
Hoyle, B. (2002). Urban waterfront revitalization in developing countries: The example of Zanzibar's Stone Town. *The Geographical Journal, 162*(2), 141–162.
Ivanovic, M. (2014). The perceived authenticity of iconic heritage sites in urban tourism: The case of Constitutional Hill, Johannesburg, South Africa. *Urban Forum, 25*, 501–515.
Ivanovic, M., & Saayman, M. (2013). South Africa calling cultural tourists. *African Journal of Physical, Health Education, Recreation and Dance, 19*(3), 138–154.
Joffroy, T. (2009). *Traditional conservation practices in Africa*. Rome: International Centre for the Study of the Preservation and Restoration of Cultural Property.
Khalfan, A., & Nobuyuki, O. (2012). Sustainable architectural conservation according to traditions of Islamic *waqf*: The World Heritage–listed Stone Town of Zanzibar. *International Journal of Heritage Studies, 18*(6), 588–604.
Khumalo, T., Sebatlelo, P., & van der Merwe, C. (2014). Who is a heritage tourist? A comparative study of Constitution Hill and the Hector Pieterson Memorial and Museum, Johannesburg, South Africa. *Africa Journal of Hospitality, Tourism and Leisure, 3*(1), 1–13.
Lwoga, N. (2018). Dilemma of local socio-economic perspectives in management of historic ruins in Kilwa Kisiwani World Heritage Site, Tanzania. *International Journal of Heritage Studies, 24*(10), 1019–1037.
Lwoga, N., & Adu-Ampong, E. (2021). Heritage tourism in Africa: Historical overview and contemporary issues. In M. Novelli, E. Adu-Ampong, & M. Ribeiro (Eds.), *Routledge handbook of tourism in Africa*, pp. 69–84. London: Routledge.
Manwa, H., Moswete, N., & Saarinen, J. (Eds.) (2016). *Cultural tourism in Southern Africa*. Bristol: Channel View Publications.
Masilo, H., & van der Merwe, C. (2016). Heritage tourists' experiences of 'Struggle Heritage' at Liliesleaf Museum and the Hector Pieterson Memorial and Museum South Africa. *African Journal of Hospitality, Tourism and Leisure, 5*(3), 1–20.
Novelli, M., Adu-Ampong, E., & Ribeiro, M. (Eds.) (2021). *Routledge handbook of tourism in Africa*. London: Routledge.
Oldfied, S., & Parnell, S. (Eds.) (2014). *The Routledge handbook on cities in the Global South*. London: Routledge.
Rockefeller Foundation (2009). Century of the city. Available at: https://whc.unesco.org/en/statesparties/tz [Accessed 26/01/2021].
Rogerson, C.M. (2020). In search of inclusive tourism in South Africa: Some lessons from the international experience. In J.M. Rogerson & G. Visser (Eds.), *New directions in South African tourism geographies*, pp. 147–165. Cham: Springer.
Rogerson, C.M., & Rogerson, J.M. (Eds.) (2021a). *Urban tourism in the Global South: South African Perspectives*. Cham: Springer.
Rogerson, C.M., & Rogerson, J.M. (2021b). City tourism in Southern Africa: Progress and issues. In M. Novelli, E. Adu-Ampong, & M. Ribeiro (Eds.), *Routledge handbook of tourism in Africa*, pp. 447–458. London: Routledge.
Rogerson, C.M., & van der Merwe, C. (2016). Heritage tourism in the Global South: Development impacts of the Cradle of Humankind World Heritage Site, South Africa. *Local Economy, 31*(1–2), 234–248.
Rogerson, J.M., & Visser, G. (2020). Recent trends in South African tourism geographies. In J.M. Rogerson & G. Visser (Eds.), *New directions in South African tourism geographies*, pp. 1–14. Cham: Springer.

Saghir, J., & Santoro, J. (2018). *Urbanization in Sub-Saharan Africa: Meeting challenges by bridging stakeholders*. Washington, DC: Centre for Strategic International Studies.
Taylor, K., & Lennon, J. (2010). Cultural landscapes: A bridge between culture and nature? *International Journal of Heritage Studies*, *17*(6), 537–554.
Teye, V., & Timothy, D.J. (2004). The varied colors of slave heritage in West Africa: White American stakeholders. *Space and Culture*, 7(2), 145–155.
Teye, V., Turk, E., & Sönmez, S. (2011). Heritage tourism in Africa: Residents' perception of African-American and white tourists. *Tourism Analysis*, *16*, 169–185.
UNESCO (United Nations Educational, Scientific and Cultural Organisation) (2020). Available at: https://en.unesco.org/ [Accessed 26/01/2021].
UNESCO (2021a). *South Africa*. Available at https://whc.unesco.org/en/statesparties/za [Accessed 26/01/2021].
UNESCO (2021b). *Ethiopia*. Available at: https://whc.unesco.org/en/statesparties/et [Accessed 26/01/2021].
UNESCO (2021c). *Tanzania*. Available at: https://whc.unesco.org/en/statesparties/tz [Accessed 26/01/2021].
UNESCO (2021d). *Africa*. Available at: https://whc.unesco.org/en/list/ [Accessed 26/01/2021].
van Beek, W., & Schmidt, A. (Eds.) (2012). *African hosts and their guests: Cultural dynamics of tourism*. Woodbridge: James Currey.
van der Merwe, C. (2013). The limits of urban heritage tourism in South Africa: The case of Constitution Hill, Johannesburg. *Urban Forum*, *24*, 573–588.
van der Merwe, C. (2014). Battlefields tourism: The status of heritage tourism in Dundee, South Africa. *Bulletin: Socio-Economic Series*, *26*, 121–139.
van der Merwe, C. (2016). Tourist guides' perceptions of cultural heritage tourism in South Africa. *Bulletin of Geography: Socio-Economic Series*, *34*, 117–130.
van der Merwe, C. (2019). The geography of heritage in South Africa. In J. Knight & C.M. Rogerson (Eds.), *The geography of South Africa: Contemporary challenges and new directions*, pp. 133–144. Cham: Springer.
van der Merwe, C., & Rogerson, C.M. (2018). The local development challenges of industrial heritage in the developing world: Evidence for Cullinan, South Africa. *GeoJournal of Tourism and Geosites*, *21*(1), 186–199.
Viljoen, J., & Menamo, U. (2017). Growing heritage tourism and social cohesion in South Africa. *African Journal of Hospitality, Tourism and Leisure*, *5*(3), 1–20.
Wolf, I., Croft, D., & Green, R. (2019). Nature conservation and nature-based tourism: A paradox? *Environments*, *6*(9), 104.
Yankholmes, A., & Akyeampong, O. (2010). Tourists' perceptions of heritage tourism development in Danish-Osu, Ghana. *International Journal of Tourism Research*, *12*(5), 603–616.

7

RURAL HERITAGE AND TOURISM IN AFRICA

Joseph E. Mbaiwa and Gladys B. Siphambe

Introduction

With the exception of the 2020–2022 COVID-19 pandemic, tourism has experienced sustained global growth and development over the last few decades (UNWTO, 2014). Tourism is regarded as the fastest-growing industry worldwide (Lenao, 2016; UNWTO, 2003). In Botswana, tourism is regarded as the largest potential engine of economic growth (Government of Botswana, 1990; Lenao, 2016; WTTC, 2007). This indicates that tourism is assumed to diversify the country's economy at different levels (Lenao, 2016). For example, at the national level in Botswana, tourism holds the promise of injecting more income into the economy and in the process, reducing poverty and dependence on minerals, especially diamonds (Government of Botswana, 2013; Lenao, 2016). The number of international tourist arrivals grew by 5% a year on average between 2010 and 2014, a trend that translated into more economic growth, more exports, and more jobs (UNWTO, 2014). International tourist arrivals were expected to increase by 3.8% per year for the period 2010 to 2020 (UNWTO, 2014). However, the global tourism industry suffered its worst year on record in 2020, with international arrivals dropping by 74% according to the latest data from the World Tourism Organization (UNWTO, 2021). Destinations worldwide welcomed one billion fewer international arrivals in 2020 than in the previous year, due to an unprecedented fall in demand and widespread travel restrictions caused by the COVID-19 pandemic.

Cultural tourism also experienced a significant growth before the pandemic, leading to the commodification and globalization of many cultural landscapes. Cultural tourism accounted for 37% of global tourism and was expected to grow at a rate of 15% per year (UNWTO, 2015). Novelli (2016) argues that cultural tourism is one of the fastest expanding sectors of the global tourism economy with as much as 40% of international leisure tourism involving a cultural component,

DOI: 10.4324/9781003153955-7

although in many countries it is a much higher percentage (Timothy, 2021). In tourism destinations, cultural tourism presents an ideal vehicle for local and regional development where local people can experience direct economic benefits, such as employment (Smith, 2003). Cultural heritage-based tourism is thus a major global industry that brings income and promotes national identity and the preservation of heritage (Richards, 2007). However, the pandemic has caused this growth trend in cultural tourism, like in other tourisms, to decline. Travel restrictions and lockdowns brought the tourism industry to the brink of collapse.

Prior to, and despite, the pandemic, rural heritage-based tourism played a significant role in sustaining rural livelihoods in some parts of Africa. This chapter examines the nature of rural heritage and tourism in Africa. It focuses on rural livelihoods, cultural landscapes, villages, and other elements of the realm's rural heritage. It describes forms of heritage tourism in the main subregions of the continent, highlights some of the multitudinous challenges facing heritage tourism in rural Africa, and offers some suggestions for better growth and management.

Rural Heritage

The concept of rural heritage refers to the landscapes carved out over centuries by people who live off the land and, more generally, through the exploitation of natural resources (Chiva, 1994). The buildings that make up rural architecture, whether clustered together (villages, hamlets, isolated houses, and buildings) or individual structures, are part of cultural heritage. Rural heritage also includes the local products that have adapted to local conditions and the needs of those who developed them. Agricultural landscapes, vernacular landscapes, and living culture are rural heritage products that can be presented to tourists in various formats, including villages, agricultural areas, and open-air museums where local people demonstrate elements of living culture to visitors (Davis & Corsane, 2014). Cultural heritage reflects one's own self-identity in relation to place; rural village lifestyles are the result of long traditions reflecting communities' tangible and intangible heritage characteristics and values (Khamung, 2015). Rural heritage is a mix of nature and culture, having been possessed and modified by people for many generations and includes the landscape, agricultural systems and patterns, ecosystems as natural heritage, harvest and religious festivals, languages, dress, handicrafts, religious beliefs, and lifestyles (Hall et al., 2011; Khamung, 2015). Heritage denotes a legacy or inheritance from the past, what we live with today, and what we pass on to future generations, including natural and cultural, tangible and intangible assets with significance and value (ICOMOS, 1999; Timothy, 2021).

Tangible heritage includes artefacts, natural or built landscapes, buildings, museums, and monuments (Khamung, 2015). Intangible heritage includes languages, artistic performances, music, beliefs, festivals, knowledge, folklore, values, rituals, traditions, and ways of life (Timothy, 2021). Heritage is from the past and is used and valued by present society for various purposes (Catibog-Sinha & Wechutunyagul, 2011; Khamung, 2015). Therefore, cultural heritage strongly reflects the society in which it is created, valued, and perpetuated (Khamung, 2015; Salazar & Porter, 2004).

Rural heritage has many things in common with urban heritage (e.g., architecture, cultural performances, and culinary traditions), but it is specifically and uniquely comprised of agricultural practices, harvest festivals, certain handicrafts (e.g., weaving), farmhouse styles, village characteristics, trails and pathways, hunting and gathering practices, certain pilgrimages, rural food systems and culinary specialties, certain animist belief systems and practices, ethnic and Indigenous cultures, battlefields and battle memorials, and sites associated with films and television shows (Almeida et al., 2020; Collins-Kreiner, 2019; Guzel & Apaydin, 2016; Kim et al., 2019; Laing, 2018; Reijnders, 2016; Ramírez, 2020; Sambuu, 2020).

Heritage is key in connecting people to their cultural origins (Khamung, 2015). When valued by the descendant communities whose cultural inheritance is considered, rural heritage can help improve the quality of life in remote communities through the conservation of local culture. The quality of life in rural areas is directly related to the survival and enhancement of local culture, rootedness, intergenerational knowledge transfer, built heritage, and landscapes (Eppich, 2014). The conservation of rural places creates a spirit of home that interlinks generations, including diasporic peoples, to their past and reinforces a sense of belonging and social solidarity (Eppich, 2014; Timothy, 2015). As a result, rural tourism development should foremost seek to protect local cultures and aim to improve the quality of life of rural residents by helping people set their own priorities through effective and inclusive involvement, improving local capacities, social services, and safeguarding culture (Eppich, 2014; Timothy, 2015). Rural development that does not address cultural heritage in all its forms risks endangering the community, its values, and sustainability (Eppich, 2014). One way in which rural communities have been able to safeguard their cultural heritage and rural landscapes is through tourism. Therefore, cultural heritage is an asset and well-suited for enriching socioeconomic possibilities in marginal rural areas in need of additional income and sources of livelihood, of which tourism is one of the most flourishing (Swensen, 2016).

Rural Heritage and Tourism in Africa

Destination culture is one of the main pull factors that influence people's travel decisions. Cultural resources are often considered icons of important streams of global culture (Akama, 2002; Richards, 2001), and many destinations utilize their iconic cultural heritage as branding mechanisms and marketing features (Timothy, 2021). At local and national levels, culture plays an important role in establishing and reinforcing people's unique identities and a sense of belonging to a particular locale (Akama, 2002). Heritage tourism is regarded as one of Africa's strongest development opportunities (Nkwanyana et al., 2016), as many African countries are celebrated for their rich and diverse cultural heritage (Saarinen, 2016), even though many of them are best known for their natural environments and wildlife safaris. Although heritage assets are increasingly being utilized in local development strategies (Chok et al., 2007) and are becoming a significant tourist attraction in many African countries, it has yet to generate considerable interest among tourism promoters and government overseers in many countries of the region.

Heritage tourism has thus become an increasingly attractive economic diversification option for many developing countries. In Africa, heritage tourism has untapped potential for poverty alleviation through tourism expansion and inclusive growth (Chok et al., 2007; Christie et al., 2013; Novelli, 2016), especially in rural areas. Rich traditions of music, art, and dance have long been cultural products in Africa, but they, along with other elements of tangible and intangible heritage, have the potential to play a greater role in tourism development (Christie et al., 2013). Several African countries, including but not limited to Botswana, Kenya, South Africa, Tanzania, Mozambique, Ghana, Eswatini, and Mali, have used living cultural heritage niche products to diversify their tourism economies for many years (Rogerson, 2012). Likewise, several countries' great heritage sites have been inscribed on UNESCO's World Heritage List, including South Africa's Cradle of Humankind and Mapungubwe Cultural Landscape (National Park); Tanzania's Kilwa Kiswani; and Zimbabwe's Great Zimbabwe, and all function as important rural tourist attractions in their respective countries.

In African communities, ancestral spirits are believed to be alive in forests, special or specific trees, caves, and water bodies, and several archaeological sites are also part of the continent's spiritual landscapes and hold important sacred values (Macamo, 2006; Nyamweru & Sheridan, 2008; Pwiti & Mvenge, 1996). Therefore, it is not surprising that archaeological areas such as Great Zimbabwe and the abandoned settlements of the sacred Mijikenda Kaya Forest in Kenya are revered as sacrosanct shrines by their respective descendant communities (Ndoro, 2001). Traditional management systems for these sacred landscapes are key in promoting the sustainable use of these cultural and natural resources, and to safeguarding their spiritual and cultural qualities (Mumma, 2003).

Notwithstanding all these arguments for traditional custodianship systems over Africa's cultural heritage, including belief systems, social customs, ethical values, religious ceremonies, traditional knowledge, sacred places, and archaeological sites, there is no discussion of how the commodification of African heritage for tourism has impacted the custodianship of these assets (Ndoro, 2001). In the majority of African countries, tourism has become one of the top foreign income generators (Jopela, 2011; Mkono, 2019). Therefore, diversification of the tourism product from an over-emphasis on natural assets, such as wildlife and wilderness, to include more rural culture and cultural tourism is the answer for many countries (Ayorekire et al., 2019b; Hall & Seyfi, 2021; Jopela, 2011; Lwoga & Adu-Ampong, 2021). Despite a lack of formal management systems, the role of local communities in the active custodianship of cultural landscapes and heritage tourism through living traditions has been observed in various parts of Africa. The remainder of this chapter describes the nature of cultural landscapes in Africa, providing examples from throughout the continent and focusing on rural heritage and livelihoods. It should be noted here that natural and cultural heritage are inseparable in rural Africa, and both are considered in the broader context of rural heritage tourism.

Rural Heritage and Tourism in Southern Africa

For a long period of time, tourism in most parts of Africa has emphasized wilderness and wildlife, with places like the Serengeti in East Africa, the Okavango Delta in Botswana, and Kafue National Park in Zambia being among the busiest tourist destinations on the continent (Mulale et al., 2013). Perhaps one of the most outstanding natural features in the world today is Mosi-oa-Tunya ('the smoke that thunders'—known worldwide as Victoria Falls), one of the world's greatest waterfalls straddling the border of Zimbabwe and Zambia. This spectacular treasure is one of the Seven Natural Wonders of the World and attracts millions of tourists both from within and outside of Africa. The communities that live around the falls have a different way of looking at the site. Instead of seeing it only as a natural phenomenon, they ascribe personality traits to it; to them, it is 'the physical manifestation of the activities of spirit forces' (Tasie, 2014, p. 104).

Even though questions arise about the sustainability of mass tourism, of which heritage tourism is a large part, in small communities, rural livelihoods appear to be affected positively. This is evident in the phenomenon of overseas Africans of the diaspora arranging to stay with local families for the culturally immersive experience. This has turned out to be a thriving business in southern Africa (Tasie, 2014). Similarly, in Zimbabwe, members of the Shangana ethnic group constructed a model village, where small numbers of visitors can stay overnight and participate in village activities. Botswana also has a similar setup where different ethnic groups, especially in the south, have built cultural villages that draw many tourists in search of a different type of tourism product from the wilderness and wildlife orientation found in northern Botswana. There are growing numbers of places in southern Africa that are seeking new connections between cultural heritage and tourism, which will help strengthen rather than water down local culture. These activities can raise the socioeconomic standards of local communities and improve the links between local creativity, innovation, and tourism, thereby improving community livelihoods while simultaneously protecting local heritage (Jopela, 2011; Mulale et al., 2013; Richards, 2007).

In colonial Zimbabwe (Rhodesia), the ruins of Great Zimbabwe were valued purely for their physical attributes. Great Zimbabwe's intrinsic spiritual and religious values in the local community were ignored (Ndoro, 2001). Nonetheless, scholars from Zimbabwe, Kenya, and Mozambique (e.g., Chirikure & Pwiti, 2008; Jopela, 2011; Ndoro, 2001; Ndoro & Kiriama, 2008; Taruvinga & Ndoro, 2003) observe that communities' attachment to cultural landscapes does not always imply that they embrace custodianship of these heritagescapes. For example, in Malawi, although communities living in the Chongoni landscape recognize ancient rock art as part of their heritage, 'this recognition does not permeate to conservation' (Jopela, 2011, p. 14). In most southern African countries, rich and diverse Indigenous cultures have not been developed for tourism (Munjeri, 2003; Ndoro, 2001). Limited funding, which normally translates into understaffing of state institutions in charge of heritage, means that interaction between communities and said agencies is limited (Mahachi & Kamuhangire, 2008), making it difficult for such institutions to respond to the needs of community stakeholders who live in these cultural landscapes.

In Botswana, the focus of tourism development and scholarship has been on wildlife-based tourism, ignoring the country's bountiful cultural tourism assets in rural regions (Bolaane, 2004; Lenao, 2014; Mbaiwa, 2007, 2009, 2011; Moswete, 2009). Although a small number of studies have examined access and the rights of minority groups to use natural resources in wilderness areas (e.g., Bolaane, 2004, 2005, 2013; Campbell, 2010; Hitchcock, 2000, 2004; Molebatsi & Masilo-Rakgoasi, 2005; Moswete et al., 2009), they have failed to show how minority and rural communities can derive socioeconomic benefits from cultural landscapes such as Tsodilo Hills, Gcwiihaba Caves, and Goo-Moremi Gorge. Likewise, whereas Campbell (2010) and Hitchcook (2000) provide insights into elements of cultural tourism at Tsodilo Hills, they fail to demonstrate cultural changes and how the commodification of the area's heritage has affected the social life and socioeconomic characteristics of the local communities.

The steady development of Botswana's tourism industry from a 2% contribution to the GDP in 1966 at independence (Botswana Review, 2017) to a 9.7% GDP contribution by 2007 (WTTC, 2012), has seen tourism overtake the agriculture sector to become the second largest contributor to the country's GDP after diamond mining (Dichaba, 2009; Moswete, 2009). From its earliest stages of development, Botswana's tourism industry has been overly reliant on wildlife and wilderness resources that mostly abound in the northwestern parts of the country (Lenao & Saarinen, 2015; Mbaiwa & Darkoh, 2006; Mbaiwa & Sakuze, 2009; Mbaiwa & Stronza, 2010; Moswete, 2009; Moswete & Mavondo, 2003). Nonetheless, at the Goo-Moremi Cultural landscape, Dichaba (2009), Geoflux (2009), Mbaiwa (2011), and White (2001) describe the cultural tourism product, in addition to local flora and fauna. Their research provides insights into how cultural landscapes-based tourism can benefit community development. However, the work of these authors does not provide information about the tourism impacts of these cultural landscapes on local livelihoods and conservation.

The literature on archaeological sites in Botswana only suggests how they can be utilized as tourist attractions, and the work of Denbow and Thebe (2006) and Segadika (2006, 2008) describes how communities living next to Tsodilo Hills are important in the preservation of the area's rock art. The research does not address the impact of commodifying Tsodilo Hills on the livelihoods of the local communities. Therefore, although cultural tourism is being promoted in Botswana, very little research has so far analysed the 'effects of tourism and commodification of cultural landscapes on the socio-cultural, socio-economic livelihoods of local communities' (Mbaiwa, 2011, p. 291). Mbaiwa (2011) notes the reluctance by the elders of Goo-Moremi village to develop tourism for fear of losing their cultural values, while contrarily, the youth support tourism development, considering the socioeconomic benefits to be derived from visitors, such as employment and business opportunities.

South Africa is heir to a legacy of autochthonous livelihoods (most famously those of the Khoi and the San), as well as Bantu immigration, slavery, colonization, settler economies, and liberation movements. These histories have all had a drastic effect on the make-up of South Africa's population. South Africa has been famously referred to as the 'rainbow nation' because it is comprised of so many diverse cultures and religions. South Africa is home to Zulu, Xhosa, Pedi, Tswana, Ndebele, Khoisan, Hindu, Muslim, and Afrikaner people, to name but a few. All of these people are united by their South African citizenship, and therefore their lives all contribute to forming a part of the country's heritage, identity, and culture. South Africa has nine sites inscribed on the World Heritage List, including iSimangaliso Wetland Park (1999), Robben Island (1999), Fossil Hominid Sites of South Africa (1999) (i.e., Cradle of Humankind, Taung Skull Fossil Site, and Makapan Valley), uKhahlamba Drakensberg Park (2000), Mapungubwe Cultural Landscape (2003), Cape Flora Region Protected Areas (2004), Vredefort Dome (2005), the Richtersveld Cultural Landscape (2007), and ǂKhomani Cultural Landscape (2017). Once viewed as a symbol of oppression and hardship, Robben Island has been transformed by recent history into a symbol of peace, hope, and reconciliation that has resulted in being inscribed on the World Heritage List. Robben Island is famous for the incarceration of political prisoners, including Nelson Mandela, the first democratically elected President of South Africa, who was held there for 18 of his 27 years of imprisonment.

In South Africa, iSimangaliso Wetland Park, formerly called the Greater St Lucia Wetland Park, has both one of the largest estuary systems in Africa and the continent's southernmost coral reefs. The Cradle of Humankind includes the hominid fossil sites at Swartkrans, Sterkfontein, and Kromdraai. Ukhahlamba-Drakensberg Park has outstanding natural beauty, is Africa's highest mountain range south of Kilimanjaro, and is the largest and most concentrated series of rock art paintings in Africa. Mapungubwe Heritage Site, known as a 'place of the stone of wisdom', was South Africa's first kingdom and developed into the subcontinent's largest realm, lasting for 400 years before it was abandoned in the fourteenth century. Its highly sophisticated people traded gold and ivory with China, India, and Egypt. The Cape Floral Kingdom makes up only 0.04% of the world's land area, yet it contains an astonishing 3% of its plant species, making it one of the richest areas for plants in the world and one of the globe's 18 biodiversity hotspots. The rural heritage of southern Africa is well used and recognized as an important asset for rural tourism, but much work still needs to be done to understand its opportunities and management and planning challenges.

Rural Heritage and Tourism in East and Central Africa

Tourism in East Africa tends to take place in cities and rural nature preserves, but there are many elements of rural heritage that appeal to tourists and have the potential to be developed into important tourist products. For example, the submerged and coastal marine cultural heritage of Kenya, Tanzania, Mozambique, and Madagascar is part of East Africa's rural environment and has the potential to stimulate ethical, inclusive, and sustainable growth in the region (Deisser & Njuguna, 2016). Other rural heritage landmarks in Kenya include mountains, forests, rivers, caves, beaches on the Indian Ocean, and Lake Victoria (Deisser & Njuguna, 2016). Rural Kenyans are increasingly aware of the social value of conserving nature through good environmental management, education and appreciation of culture, and involvement of ordinary citizens in state conservation policies (Deisser & Njuguna, 2016). Archaeological remains are also an important rural resource, for example, as evidence confirms the origins and continuous in situ evolution of the bao game at the Hyrax Hill Pre-Historic Sites, Kenya, where archaeological research has led to the accumulation of critical data and documentation on bao (Mwanzia, 2016). Unfortunately, much of Kenya's heritage legislation remains outmoded and vague (Mwanzia, 2016).

There are numerous Early and Late Stone Age archaeological sites in eastern Africa, such as Olorgesaillie, Kariandusi, and Hyrax Hill (Kenya) that have the potential to become heritage tourism sites (Irandu & Shah, 2016), but for the most part, they remain undeveloped, undermanaged, and under-visited. Music festivals, carnivals, and sporting events also abound and could be developed better into cultural attractions for domestic and international tourists (Irandu & Shah, 2016).

Perhaps the best-known rural heritage of East Africa is associated with the Maasai culture. The Maasai of Kenya and Tanzania have a rich cultural heritage that has over the years attracted many tourists. The Maasai are semi-nomadic Indigenous people who raise cattle and hunt wildlife and inhabit northern, central, and southern Kenya, as well as northern Tanzania. The Maasai have longstanding links to the national parks and reserves in both countries. For decades, the Maasai have sold traditional handicrafts and souvenirs and performed traditional dances for tourists. However, the Maasai handicrafts and cultural performances have been largely removed from their authentic social and cultural contexts and are now commercialized to suit the whims of mass tourists. According to Akama (2002), these commoditized handicrafts and contrived cultural performances reinforce existing stereotypes and negative images of the Maasai as a 'backward and primitive people' who form an additional anecdote to international tourists keen for exoticism and adventure.

Greenwood (1977) writes about the commodified Maasai culture and argues that tourism tends to lead to cultural changes through commercialization. For example, colourful local costumes and customs, rituals and feasts, and folk and ethnic arts become touristic services or commodities as they come to be performed or produced for touristic consumption. The issue here is that commoditization is likely to change the meaning of cultural products, rendering them less meaningful. As far back as the 1970s, worldwide experiences showed that native cultures are altered and often destroyed when staged for tourist consumption and when managed unwisely (Greenwood, 1977, p. 131). Likewise, culture is frequently expropriated by others without the consent of its inheritors, leading to the exploitation of local people and their heritage. This is a considerable challenge in Africa, as authenticity decreases when certain people connive to sell inauthentic cultural products such as spears and shields to tourists (Irandu & Shah, 2016). The commercialization of Maasai heritage for the tourism marketplace is mired in controversy and even though much of Africa's tourism takes place in cities, rural regions such as Maasai Mara, have the potential to become over-commercialized and over-touristified.

In Tanzania, cultural heritage sites include the famous Olduvai Gorge in Arusha, Isimila in Iringa, Kilwa Kisiwani and Songo Mnara World Heritage Site, Zanzibar Stone Town World Heritage Site, Bagamoyo Historic Town, and many others (Lwoga, 2016). Cultural heritage has become an increasing attraction for visitors from around the world and visits to rural tribal villages, especially Maasai, are often a highlight of safari itineraries.

The Swahili Coast, as the region is called, is a predominantly Islamic region with old mosques and coral palaces found throughout the area. Swahili culture centres on the dhow, a wooden sailing boat powered by seasonal winds. Historically, the boats connected the Swahili Coast with Arabia and India and allowed trade between the regions to flourish. Fishing remains a mainstay of coastal income in small villages throughout the area, and coconut and spice plantations continue to form an important source of export. Fishing and agriculture are a critical part of East Africa's coastal heritage and are ripe for development as an attraction.

Tanzania is endowed with unique and diverse natural and cultural tourist attractions (Lwoga, 2016). These include 16 national parks, natural and cultural world heritage sites such as the Zanzibar Stone Town, Kilwa Kisiwani and Songo Mnara, Ngorongoro Conservation Area (NCA), the Serengeti National Park, the Kilimanjaro Mountain National Park, and many others. Cultural heritage tourism practices are manifested in areas and villages surrounding the national parks and the protected cultural heritage sites such as Kaole Ruins in Bagamoyo. These involve the living culture of people living in villages surrounding the national parks as well as other cultural heritage sites within the villages' environs. The national parks in Tanzania are normally managed in partnership between villagers and the national parks authorities (TANAPA) in the form of Wildlife Management Areas (WMA) and other arrangements which involve partnership between community and the tour operators (Lwoga, 2016). The Kaole Ruins Site allows, in a special arrangement, community members to sell their cultural materials to tourists visiting the sites (Lwoga, 2016). Mountains play a significant role in landscapes as the subject of prayer and as a seat of the gods, for example, Mount Kilimanjaro in Tanzania (Eborieme, 2000).

Uganda has diverse and unique cultures and traditions across different ethnic communities that form a significant pool of potential tourism resources (Ayorekire et al., 2019a). Despite the country's high rural-based cultural tourism potential, Uganda's resources have not been mapped, documented, packaged, and marketed for tourism. In addition, there has been limited awareness of the need to protect, conserve, and promote its heritage resources (Ayorekire et al., 2019a). Although there is limited research about the use of culture for tourism in Uganda, Uganda has ten kingdoms and chiefdoms that have attempted to harness cultural heritage resources such as songs, dances, festivals, rituals, totems, shrines, kings' tombs, and palaces for tourism (Ahebwa et al., 2016; Ayorekire et al., 2019a). These cultural heritage resources are rich in Indigenous knowledge that has been used over the years to promote nature conservation and social cohesion (Ayorekire et al., 2019a; Nuwagaba & Kiwere, 2014). Unfortunately, however, cultural norms and practices have been gradually eroded and a number of cultural heritage sites neglected, degraded, and damaged (Ayorekire et al., 2019b).

Of all the countries in Central and East Africa, Ethiopia, Kenya, and Tanzania have the most developed heritage tourism sector. Ethiopia boasts of historic cities, varied ethnic cultures, and the world-famous Rock-Hewn Churches on the rural outskirts of Lalibela, a UNESCO World Heritage Site. Agricultural heritage is also a salient part of Ethiopia's rural tourism, especially as regards coffee production. Ethiopia is one of the world's leading and original coffee-producing regions, and the country has successfully capitalized on this unique form of rural tourism (Degarege & Lovelock, 2021; Woyesa & Kumar, 2021).

Kenya and Tanzania are best known for safari tourism which, although taking place largely in natural areas and game preserves, is an important element of the region's rural cultural heritage. In fact, safari tourism itself is part of East Africa's heritage and has been popular since early colonial times. Despite the success of safari tourism in rural Kenya and Tanzania, it has not been without significant controversy. For example, some host communities have been pitted against governments as they have been displaced from their traditional lands to make way for wildlife parks, and many rural communities face the chronic threats of crop destruction and attacks on humans by wild animals (Ezenagu, 2020). Governments could potentially stimulate more earnings and minimize friction with local citizenry if they would encourage other interesting aspects of safari tourism that have remained largely unexploited, including the role of totemic animals and the services of wildlife charmers (Tasie, 2014). As part of native peoples' own heritage lore, such actions might help pacify their concerns. This is particularly true in Kenya which, according to Irandu (2004, p. 139), 'no longer has a competitive edge over her rivals in Africa such as South Africa Botswana, Namibia, Zimbabwe and Tanzania in terms of wildlife tourism'. Irandu suggests expanding the wildlife-based heritage product in these ways would help diversify Kenya's tourism product to enhance the cultural heritage element and local participation in the industry to help preserve residents' cultural identity and sense of place, and improve their livelihoods. During the colonial period, art forms and other attributes of the Kenyan people were nearly eradicated by the European colonizers who were determined to convert the colonized into Christians and 'civilize' them (Irandu, 2004).

However, with the development of international tourism in Kenya since the 1970s, there has been a revival of traditional arts and crafts. International tourism in Kenya has stimulated the restoration and preservation of historical sites and monuments in cities and in the countryside.

Rural Heritage and Tourism in West Africa

West Africa is especially well known for its slave castles and other slavery-related heritage sites that are part of the region's rural heritage tourism. Tourists of all nationalities visit, but the most notable are those of the African diaspora who visit the region as a form of roots-seeking and self-identity reaffirmation. For many visitors, these sites educe deep emotional trauma. Their experiences are typically a mix of bitterness, anguish, and sadness, as well as catharsis, forgiveness, and hope for a better future (Timothy & Teye, 2004). The annual Roots Homecoming festival in The Gambia highlights the cultural ties between Africa and her diasporic peoples (Tasie, 2014).

This roots tourism has been crucial in the restoration of a number of centuries-old European-built forts and castles in both urban and rural areas of Ghana (Teye et al., 2011). Beyond the coastal forts, Ghana also has several vital cultural resource sites that are critical components of the 'Slave Route Project' (Teye et al., 2011; Yankholmes & Timothy, 2017). It is worth mentioning that ever since the late 1970s release of the television miniseries based on Alex Haley's 1976 book, *Roots*, the number of African American tourists has increased exponentially as they come to learn more about, and come to terms with, the slave trade, and visit various localities associated with their ancestors (Tasie, 2014). These 'roots tourists' also visit rural areas of other countries, including Nigeria, Benin, Togo, The Gambia, and Senegal.

> Their target destinations are mainly Goree Island, a major slave port off the coast of Senegal as well as sites in Benin, Gambia and Ghana. Haley's book, in particular, has turned the village of Jufurre in The Gambia into a Mecca of a kind, for thousands of African American tourists.
>
> *(Tasie, 2014, p. 108)*

West Africa is endowed with a rich and diverse cornucopia of cultures with meanings and values firmly maintained by contemporary ethnic groups (Twumasi-Ampofo et al., 2020). These diverse cultures are expressed in both tangible and intangible ways, including natural and built landscapes, vernacular architecture, relics and artefacts, monuments, festivals, social values, ways of life, and an abundance of agricultural, spiritual, and family-related ceremonies (Fredholm, 2016; Ndoro et al., 2018; Twumasi-Ampofo et al., 2020). Ghana's heritage on the world stage is significantly highlighted by historic battles of conquest and slavery along its coastal areas with the resulting list of forts, castles, and other monuments that represent early interactions with the Europeans. The several remaining forts and castles built between 1482 and 1786 span some 500 km along the Ghanaian coast (Anquandah, 1999; Bruku, 2015; Twumasi-Ampofo et al., 2020; Van Dantzig, 1980), and there are many other important heritage sites associated with the slave trade in rural areas of Senegal, The Gambia and Benin.

Although much of West Africa's heritage tourism product is overshadowed by the slave trade and its historic sites, traditional village life, and rural (and urban) architecture are important secondary attractions throughout the region. The latter includes the Asante Traditional Buildings, a UNESCO World Heritage Site in Ghana, and the Stone Circles of Senegambia a World Heritage Site in both Senegal and The Gambia. Likewise, with some degree of success and failure, several countries have attempted to develop rural village-based tourism that highlights farm life, rural living, and traditional celebrations (Ezeuduji & Rid, 2011; Neba, 2010), somewhat akin to what is taking place in Botswana and South Africa. Given the region's large variety of ethnic groups, physical environments, faiths, and agricultural systems, there is a lot of potential for developing rural tourism further.

The different ethnic groups of West Africa have rich living heritage that has a lot of potential for tourism (Ezenagu, 2020). The Sukur cultural landscape is an ancient settlement in Nigeria with a recorded history of iron smelting technology, flourishing trade, and strong political institutions dating back to the sixteenth century (Eborieme, 2000). Likewise, many spiritual and religious traditions in West Africa have enabled the growth of certain types of tourism, such as pilgrimages, themselves a form of heritage tourism, to Nigeria's Osun Osogbo sacred grove and other rural sites (Ezenagu, 2020). Likewise, the wood and bronze handicrafts in much of the region are one of the most important rural products that tourists consume, and they have much potential for additional tourism development in situ in rural regions (Ankomah & Crompton, 1990).

Rural Heritage and Tourism in North Africa

Heritage tourism in North Africa is complex, just as it is in other parts of the world, influenced by factors such as climate change, colonialism, war and conflict, overtourism, political instability, and a lack of funding for heritage protection (Hall & Seyfi, 2021). In this harsh climate dominated by the enormous Sahara Desert, most cultural heritage assets are in cities and towns, but nevertheless, the region has a significant rural heritage landscape. The Middle East and North Africa together (MENA) is regarded as the cradle of Western and Arab civilization and is home to several of the world's major religions. Because of this, the region is rich in heritage sites that serve as major tourist attractions and as icons of national, cultural, and religious identity (Hall & Seyfi, 2021). North Africa is rich in cultural heritage, which has become a key tourism attraction over the years. According to the World Bank (2001), the countries of North Africa are blessed with an extraordinary cultural patrimony, secular and religious, of huge importance for each country and for humankind at large.

The vast historic movements of North Africa's populations, its vast and expansive trade routes that linked parts of Africa to the Middle East and Europe in ancient days, wars and political conquests, the rise and collapse of empires and states, and the juxtaposition of civilizations, explain much about the diversity and pluralism of the region's heritage (World Bank, 2001). Living traditional arts and crafts are a socially and economically important part of North Africa's heritage and are a large industry in which millions of people are employed in medium and small enterprises. In Morocco, in the early 2000s, for instance, about two million people were working in the artisanal sector, representing some 12% of Morocco's active population and contributing close to 10% of its gross domestic product (World Bank, 2001). The region is also home to cultural assets that buttress its heritage tourism sector, including religious buildings (e.g., mosques, churches, synagogues, libraries, and monuments) and secular buildings, such as historic palaces, traditional caravansaries, and residential housing (Hall & Seyfi, 2021).

Egypt has some of the continent's most ancient and spectacular monuments and artefacts. Egypt has long been a major African destination, having received a steady stream of tourists since the late 1800s, most of whom visit the Valley of the Kings and the Pyramids of Giza, both of which are part of the country's rural heritage (even though Giza has now been subsumed by the city of Cairo) (Akama & Sterry, 2002; Mustafa, 2021). Egypt is home to one of the world's oldest civilizations whose millennia of recorded history have had a profound influence on the cultures of Europe, the Middle East, and Africa. Archaeological tourism is the traditional kind of tourism in Egypt and is, for the most part, a rural activity.

Although most of North Africa's rural heritage has been archaeology focused (e.g., Abu Simbel (Egypt), Carthage (Tunisia), Volubilis (Morocco), Sabratha (Libya), and Djémila (Algeria)), new forms of rural heritage tourism are becoming increasingly popular, especially in conjunction with desert safaris. Rural Egypt is home to several oases and nomadic Bedouin communities, and their poetry, music, dance, and other elements of intangible heritage have played an important role in rural tourism, including the increasingly popular Bedouin village homestays (Mustafa, 2021). In addition to the Bedouins of Egypt, the Indigenous Berber tribes have become increasingly involved in tourism in the Maghreb region and Libya. Among these two nomadic societies, their lifestyles, rituals, folklore, and camel markets have become important cultural assets for tourism.

Although cultural tourism is thriving in many parts of North Africa, Mustafa (2021) argues that human actions and climate change, globalization processes, political instability, improper tourism development, and social change all contribute to the erosion of the region's culture and social identity. Ikram (2013) notes how civil unrest and armed conflict in North Africa have resulted in cultural property loss, as police and security forces vanished from their posts, including from archaeological sites, which left them open to pillaging.

Conclusion

Africa is behind most other world regions with regard to rural heritage tourism development. Most countries in the realm have yet to develop robust and sustained rural tourism sectors, which may largely be a result of image and security problems, but also a lack of access and infrastructure, lack of information, and the low prioritization of most countries compared to nature-based and urban tourism. Nevertheless, the region has a great deal of potential, for 'African heritage has to be considered in its totality, including the intangible elements such as the spiritual and sacred values that are held within cultural landscapes' (Ndoro, 2001 p. 336). Africa has a rich cultural heritage and historic landscapes which need to be managed and developed for tourism purposes and simultaneously protected. In many cases, the management arrangements for the protection of cultural landscapes imbued with sacred values are ineffective due to the centralized state-based administrative systems for heritage management that dominate most African countries (Chirikure & Pwiti, 2008; Jopela, 2011).

The development of rural heritage tourism in Africa is overall at an incipient stage (Akama & Sterry, 2002; Ezeuduji, 2015; Rid et al., 2014). However, there is a great variation in the level of tourism development throughout the region (Dieke, 2000; Munjeri, 2003; Novelli, 2016). Countries such as Kenya, Mauritius, Seychelles, Morocco, Tunisia, South Africa, Zimbabwe, Cote d'Ivoire, and Senegal have well-established tourism industries, with prominent rural tourism being an important sector in some of these. These countries are sometimes referred to as so-called African 'success stories' (Akama & Sterry, 2002; Dieke, 2000), whereas other countries have limited tourism development but possess considerable potential for future development (Dieke, 2000; Mkono, 2019).

The rural heritage assets of Africa are diverse and varied. Most colonial heritage is urban, whereas most Indigenous heritage is rural. Given the colonial past, historical association with slavery, and geopolitical conflict, much of the rural heritage product is dark, but many countries are beginning to refocus their efforts on brighter heritages that celebrate the ordinary lives of ordinary people in rural landscapes.

In eastern and southern Africa, the development of tourism is narrowly focused on wildlife safaris and beach tourism (Dichaba & Thebe, 2015), with the exception of the world-renowned Maasai culture, which has long been involved in cultural tourism. The wildlife and nature-based tourism that pervades the region has become a form of heritage tourism there. Senegal, The Gambia, Cote d'Ivoire, and Ghana have developed elements of cultural tourism, the primary product being 'roots tourism', which primarily targets a narrow market segment: the African diaspora (Dichaba & Thebe, 2015; Dieke, 2000; Yankholmes & Timothy, 2017). Other rural cultural products in Africa include agricultural heritage, traditional villages and lifestyles, cultural customs and traditions, and various celebrations. Culturally immersive forms of tourism are currently being developed in West Africa, Southern Africa, and North Africa as people desire to experience more authentic folklife, particularly tourists from the Global North and diasporic Africans.

Despite its vast potential for growth in rural heritage tourism and a general lack of planning attention to its development, there are signs that cultural tourism has been over-commercialized in some areas, such as among the Maasai of Kenya and Tanzania, in some South African tribal areas, and in various nature preserves. Likewise, conflict has occurred between Indigenous people and government agencies when locals are excluded from tourism decision-making, especially with regard to restricting native people's access to land that governments wish to use for wildlife protection and tourism. Thus, careful planning and management must ensue to ensure that the people whose heritage is selected to package and sell to tourists remain in control of their own cultural inheritance and are empowered to decide what to expose to the outside world and what to keep away from the tourist gaze. This is an essential principle of sustainable heritage tourism that can help local people simultaneously capitalize on rural tourism and protect their cultural assets. The livelihood potential of heritage tourism in rural Africa is immense but needs a concerted effort to make it a reality.

References

Ahebwa, W.M., Aporu, J.P., & Nyakaana, J.B. (2016). Bridging community livelihoods and cultural conservation through tourism: Case study of Kabaka heritage trail in Uganda. *Tourism and Hospitality Research*, *16*(2), 103–115.

Akama, J. (2002). The creation of the Maasai image and tourism development in Kenya. Paper presented at the ATLAS African International Conference, Mombasa, Kenya.

Akama, J., & Sterry, P. (2002). Cultural tourism in Africa: Strategies for the new millennium. Paper presented at the ATLAS Africa International Conference, Mombasa, Kenya.

Almeida-García, F., Cortes-Macías, R., Balbuena-Vázquez, A., & Carmen-Hidalgo, M. (2020). New perspectives of residents' perceptions in a mature seaside destination. *Sustainability*, *12*(10), 4183.

Ankomah, P.K., & Crompton, J.L. (1990). Unrealized tourism potential: The case of sub-Saharan Africa. *Tourism Management*, *11*(1), 11–28.

Anquandah, K.J. (1999). *Castles and forts of Ghana*. Accra: Ghana Museums and Monuments Board.

Ayorekire, J., Obua, J., & Byaruhanga, M.B. (2019a). Broadening Uganda's tourism product base through indigenous cultural resource utilisation. In M. Mkono (Ed.), *Positive tourism in Arica*, pp. 79–90. London: Routledge.

Ayorekire, J., Obua, J., & Manyara, G. (2019b). Regional tourism in inter-governmental authority on development: A comparative policy and institutional best practice approach. *International Journal of Tourism Policy*, *9*(1), 50–70.

Bolaane, M. (2004). The impact of game reserve policy on the River BaSarwa/Bushmen of Botswana. *Social Policy & Administration*, *38*(4), 399–417.

Bolaane, M. (2005). Chiefs, hunters and adventurers: The foundation of the Okavango/Moremi National Park, Botswana. *Journal of Historical Geography*, *31*(2), 241–259.

Bolaane, M. (2013). Chiefs, hunters and San in the creation of Moremi Game Reserve, Okavango Delta: Multiracial interactions and initiatives 1956–1978. *African Study Monographs*, *35*(1), 41–64.

Botswana Review (2017). *Botswana review of commerce and industry*, 36th edn. Gaborone: Ministry of Investment, Trade and Industry.

Bruku, S. (2015). Community engagement in historical site protection: Lessons from the Elmina Castle Project in Ghana. *Conservation and Management of Archaeological Sites*, *17*(1), 67–76.

Campbell, A.C. (2010). *Tsodilo Hills: Copper bracelet of the Kalahari*. East Lansing: Michigan State University Press.

Catibog-Sinha, C., & Wechutunyagul, P. (2011). Natural and cultural heritage. Perceived importance for tourism: A case study in Thailand. *Journal of Tourism and Sustainability, 1*(1), 29–45.

Chirikure, S., & Pwiti, G. (2008). Community involvement in archaeology and cultural heritage management: An assessment from case studies in Southern Africa and elsewhere. *Current Anthropology, 49*(3), 467–485.

Chiva, I. (1994). Une politique pour le patrimoine culturel rural. Rapport présenté à M. Jacques TOUBON Ministre de la Culture et de la Francophonie. Available at: http://www.culture.gouv.fr/mpe/dossiers/rapport_chiva.htm. Accessed 12.02.2020.

Chok, S., Macbeth, J., & Warren, C. (2007). Tourism as a tool for poverty alleviation: A critical analysis of 'pro-poor tourism' and implications for sustainability. *Current Issues in Tourism, 10*(2–3), 144–165.

Christie, I., Fernandes, E., Messerli, H., & Twining-Ward, L. (2013). *Tourism in Africa: Harnessing tourism for growth and improved livelihoods*. Washington, DC: World Bank.

Collins-Kreiner, N. (2019). Pilgrimage tourism—past, present and future rejuvenation: A perspective article. *Tourism Review, 75*(1), 145–148.

Davis, P., & Corsane, G. (2014). Communities, heritage and new cultural landscapes: Ecomuseological approaches. In M. Roe & K. Taylor (Eds.), *New cultural landscapes*, pp. 118–135. London: Routledge.

Degarege, G.A., & Lovelock, B. (2021). Institutional barriers to coffee tourism development: Insights from Ethiopia–the birthplace of coffee. *International Journal of Culture, Tourism and Hospitality Research, 15*(3), 428–442.

Deisser, A.M., & Njuguna, M. (Eds.) (2016). *Conservation of natural and cultural heritagein Kenya*. London: UCL Press.

Denbow, J., & Thebe, P.C. (2006). *Culture and customs of Botswana*. London: Greenwood Press.

Dichaba, T.S. (2009). From monuments to cultural landscapes: Rethinking heritage management in Botswana. Unpublished master's thesis, Rice University, USA.

Dichaba, T.S., & Thebe, P.C. (2015). In search of identity and ownership of the MmaMotshwane Gorge: Heritage tourism and management of Mogonye Cultural Landscape. *Botswana Notes and Records, 47*, 56–68.

Dieke, P.U.C. (2000). *Tourism in Africa's economic development: Policy implications*. Dubai: The Emirates Academy of Hospitality Management.

Eborieme, J. (2000). The Sukur cultural landscape. In M. Rossler & G. Saouma (Eds.), *The World Heritage Convention and cultural landscape in Africa*, pp. 64–69. Paris: UNESCO, World Heritage Centre.

Eppich, R. (2014). *Cultural heritage, landscape & rural development: Good practice, methodology, policy recommendations & guidelines for rural communities*. Brussels: European Union & HISTCAPE.

Ezenagu, N. (2020). Heritage resources as a driver for cultural tourism in Nigeria. *Cogent Arts & Humanities*, 7(1), 1–14.

Ezeuduji, I.O. (2015). Strategic event-based rural tourism development for sub-Saharan Africa. *Current Issues in Tourism, 18*(3), 212–228.

Ezeuduji, I.O., & Rid, W. (2011). Rural tourism offer and local community participation in The Gambia. *Tourismos, 6*(2), 187–211.

Fredholm, S. (2016). Assets in the age of tourism: The development of heritage planning in Ghanaian policy. *Journal of Contemporary African Studies, 34*(4), 498–518.

Geoflux (2009). *Consultancy services for environmental impact assessment for the proposed tourism development of Moremi Manonnye Conservation Area* (Final EIA report). Gaborone: Botswana Tourism Board.

Government of Botswana (1990). *Tourism policy: Government paper no.2 of 1990*. Gaborone: Government Printer.

Government of Botswana (2013). *Best of Botswana* (Vol. 4). Gaborone: Red Pepper PR & Communications.

Greenwood, D.J. (1977). Culture by the pound: An anthropological perspective on tourism as cultural commoditization. In V.L. Smith (Ed.), *Hosts and guests: The anthropology of tourism*, pp. 129–139. Philadelphia: University of Pennsylvania Press.

Guzel, B., & Apaydin, M. (2016). Gastronomy tourism: Motivations and destinations. In C. Avcikurt, M.S. Dinu, N. Hacioğlu, R. Efe, A. Soykan, & N. Tetik (Eds.), *Global issues and trends in tourism*, pp. 394–404. Sofia, Bulgaria: St Kliment Ohridski University Press.

Hall, C.M., James, M., & Baird, T. (2011). Forests and trees as charismatic mega-flora: Implications for heritage tourism and conservation. *Journal of Heritage Tourism*, *6*(4), 309–323.

Hall, C.M., & Seyfi, S. (Eds.) (2021). *Cultural and heritage tourism in the Middle East and North Africa: Complexities, management and practices*. London: Routledge.

Hitchcock, R.K. (2000). Traditional African wildlife utilization: Subsistence hunting, poaching, and sustainable use. In H.T. Prins, J.G. Grootenhuis, & T.T. Dolan (Eds.), *Wildlife conservation by sustainable use*, pp. 389–415. New York: Springer.

Hitchcock, R.K. (2004). The roles of applied and development anthropology and archaeology among the San of Botswana. *Botswana Notes and Records 36*(1), 125–135.

ICOMOS (1999). International cultural tourism charter, managing tourism at places of heritage significance. Online at: http://www.intemationa1.icomos.org/charters/tourism_e.pd~1999–10

Ikram, S. (2013). Cultural heritage in times of crisis. *Journal of Eastern Mediterranean Archaeology Heritage Studies*, *1*, 366–371.

Irandu, E., & Shah, P. (2016). Development of cultural heritage tourism in Kenya: A strategy for diversification of tourism products. In A.-M. Deisser & M. Njuguna (Eds.), *Conservation of natural and cultural heritage in Kenya: A cross-disciplinary approach*, pp. 154–171. London: UCL Press.

Irandu, E.M. (2004). The role of tourism in the conservation of cultural heritage in Kenya. *Asia Pacific Journal of Tourism Research*, *9*(2), 133–150.

Jopela, A.P. (2011). Traditional custodianship: A useful framework for heritage management in Southern Africa. *Conservation and Management of Archaeological Sites*, *13*(2–3), 103–122.

Khamung, R. (2015). Analysis of aptitudes, aspirations, capacities, and resources for acommunity-based agritourism establishment. *Silpakorn University Journal of Social Sciences, Humanities and Arts*, *5*(2), 25–57.

Kim, S., Whitford, M., & Arcodia, C. (2019). Development of intangible cultural heritage as a sustainable tourism resource: The intangible cultural heritage practitioners' perspectives. *Journal of Heritage Tourism*, *14*(5–6), 422–435.

Laing, J. (2018). Festival and event tourism research: Current and future perspectives. *Tourism Management Perspectives*, *25*, 165–168.

Lenao, M. (2014). Rural tourism development and economic diversification for local communities in Botswana: The case of Lekhubu Island. *Nordia Geographical Publications*, *43*(2), 53–53.

Lenao, M. (2016). Rural cultural tourism development and agriculture: Evidence from residents of Mmatshumu Village in the Boteti Region of Botswana. In H. Manwa, N. Moswete, & J. Saarinen (Eds.), *Cultural tourism in Southern Africa*, pp. 132–144. Bristol: Channel View Publications.

Lenao, M., & Saarinen, J. (2015). Integrated rural tourism as a tool for community tourism development: Exploring culture and heritage projects in the North-East District of Botswana. *South African Geographical Journal, 97*(2), 203–216.

Lwoga, N.B. (2016). Tourism employment and local residents' engagement in the conservation of the built heritage in Zanzibar Stone Town in Tanzania. *WIT Transactions on Ecology and the Environment, 201*, 43–55.

Lwoga, N.B., & Adu-Ampong, E.A. (2021). Heritage and heritage tourism in Africa: Historical overview and contemporary issues. In M. Novelli, E.K. Adu-Ampong, & M.A. Ribeiro (Eds.), *Routledge handbook of tourism in Africa*, pp. 69–84. London: Routledge.

Macamo, S. (2006). *Privileged places in south central Mozambique: The archaeology of Manyikene, Niamara Songo and Degue-Mufa*. Uppsala, Sweden: University of Uppsala, Department of Archaeology and Ancient History.

Mahachi, G., & Kamuhangire, E. (2008). Administrative arrangements for heritage resources management in sub-Saharan Africa. In W. Ndoro, A. Mumma, & G. Abungu (Eds.), *Cultural heritage and the law: Protecting immovable heritage in English speaking countries of Southern Africa*, pp. 43–51. Rome: ICCROM.

Mbaiwa, J.E. (2007). Local communities' attitudes towards wildlife conservation and community based natural resources management in Ngamiland District, Botswana. In B. Schuster & O.T. Thakadu (Eds.), *Natural resources management and people*, pp. 19–26. Gaborone: IUCN CBNRM Support Programme.

Mbaiwa, J.E. (2009). Tourism development, rural livelihoods and biodiversity conservationin the Okavango Delta, Botswana. In P. Hottola (Ed.), *Tourism strategies and local responses in Southern Africa*, pp. 90–104. Wallingford: CABI.

Mbaiwa, J.E. (2011). Cultural commodification and tourism: The Goo-Moremi Community, Central Botswana. *Tijdschrift voor Economische en Sociale Geografie, 102*(3), 290–301.

Mbaiwa, J.E., & Darkoh, M.B.K. (2006). *Tourism and the environment in the Okavango Delta, Botswana*. Gaborone: Pula Press.

Mbaiwa, J.E., & Sakuze, L.K. (2009). Cultural tourism and livelihood diversification: The case of Gcwihaba Caves and XaiXai village in the Okavango Delta, Botswana. *Journal of Tourism and Cultural Change*, 7(1), 61–75.

Mbaiwa, J.E., & Stronza, A.L. (2010). The effects of tourism development on rural livelihoods in the Okavango Delta, Botswana. *Journal of Sustainable Tourism, 18*(5), 635–656.

Mkono, M. (Ed.) (2019). *Positive tourism in Africa*. London: Routledge.

Molebatsi, C., & Masilo-Rakgoasi, R. (2005). Local communities and the implementation of environmental conventions in Botswana: A case study of the United Nations Convention to Combat Desertification and the Convention on Biological Diversity. *Pula: Botswana Journal of African Studies, 19*(1), 56–65.

Moswete, N. (2009). *Stakeholder perspectives on the potential for community-based ecotourism development and support for the Kgalagadi Transfronteir Park in Botswana*. Unpublished doctoral dissertation, University of Florida, USA.

Moswete, N., & Mavondo, F.T. (2003). Problems facing the tourism industry of Botswana. *Botswana Notes and Records, 35*, 69–77.

Moswete, N., Thapa, B., & Lacey, G. (2009). Village-based tourism and community participation: A case study of the Matsheng villages in southwest Botswana. In J. Saarinen, F. Becker, H. Manwa, & D. Wilson (Eds.), *Sustainable tourism in Southern Africa: Local communities and natural resources in transition*, pp. 189–209. Clevedon: Channel View Publications.

Mulale, K., Matema, C., Funda, X., Slater-Jones, S., Njovu, D., Kangueehi, G., Hay, D., & Crookes, V. (2013). *Community-based natural resource management in Southern Africa: An introduction*. Bloomington, IN: AuthorHouse.

Mumma, A. (2003). Community-based legal systems and the management of World Heritage Sites. In E. de Merode, R. Smeets, & C. Westrik (Eds.), *Linking universal and local values: Managing a sustainable future for World Heritage*, pp. 43–48. Paris: UNESCO World Heritage Centre.

Munjeri, D. (2003). Anchoring African cultural and natural heritage: The significance of local community awareness in the context of capacity-building. *World Heritage Papers, 13*, 76–81.

Mustafa, M.H. (2021). Cultural heritage: A tourism product of Egypt under risk. *Journal of Environmental Management and Tourism, 12*(1), 243–257.

Mwanzia, J.M. (2016). Climate change and environmental sustainability action plans in Africa: A case study of Kenya, (1980–2014). Unpublished doctoral dissertation, University of Nairobi, Kenya.

Ndoro, W. (2001). *Your monument our shrine: The preservation of Great Zimbabwe*. Uppsala, Sweden: Uppsala University, Department of Archaeology and Ancient History.

Ndoro, W., Chirikure, S., & Deacon, J. (Eds.) (2018). *Managing heritage in Africa: Who cares?* London: Routledge.

Ndoro, W., & Kiriama, H. (2008). Management mechanism in heritage legislation. In W. Ndoro, A. Mumma, & G. Abungu (Eds.), *Cultural heritage and the law: Protecting immovable heritage in English speaking countries of Southern Africa*, pp. 53–62. Rome: ICCROM.

Neba, N.E. (2010). Developing rural tourism as an alternative strategy for poverty alleviation in protected areas: Example of Oku, Cameroon. *International NGO Journal, 5*(2), 50–58.

Nkwanyana, M.S., Ezeuduji, I.O., & Nzama, A.T. (2016). Cultural heritage tourism in South Africa: Perceived a panacea for rural development? *Acta Universitatis Danubius, Œconomica, 12*(6), 160–167.

Novelli, M. (2016). *Tourism and development in Sub-Saharan Africa: Current issues and local realities*. London: Routledge.

Nuwagaba, T.F., & Kiwere, N. (2014). *Totems of Uganda—Buganda Edition: Culture embracing nature*. Kampala: Taga Nuwagaba and Nathan Kiwere Publishing.

Nyamweru, C., & Sheridan, M. (2008). Introduction: African sacred ecologies. *Journal for the Study of Religion, Nature, and Culture, 2*(3), 285–291.

Pwiti, G., & Mvenge, G. (1996). Archaeologists, tourists and rainmakers: Problems in the management of rock art sites in Zimbabwe, a case study of Domboshava National Monument. In G. Pwiti & R. Soper (Eds.), *Aspects of African Archaeology: Papers from the 10th congress of the Pan African Association for Prehistory and Related Studies*, pp. 817–823. Harare: University of Zimbabwe.

Ramírez, P.E.M. (2020). What can we weave? Authority, reconstructing, and negotiating heritages through archaeological open-air museums. *Archaeologies, 16*(1), 72–98.

Reijnders, S. (2016). Stories that move: Fiction, imagination, tourism. *European Journal of Cultural Studies, 19*(6), 672–689.

Richards, G. (2001). *The development of cultural tourism in Europe*. In G. Richards (Ed.), *Cultural attractions and European tourism*, pp. 3–29. Wallingford: CABI.

Richards, G. (2007). *Cultural tourism: Global and local perspectives*. New York: Haworth Press.

Rid, W., Ezeuduji, I.O., & Pröbstl-Haider, U. (2014). Segmentation by motivation for rural tourism activities in The Gambia. *Tourism Management, 40*, 102–116.

Rogerson, C.M. (2012). The tourism-development nexus in sub-Saharan Africa: Progress and prospects. *Africa Insight, 42*(2), 28–45.

Saarinen, J. (2016). Cultural tourism in Southern Africa: The role of local cultures and ethnicity in tourism development. In H. Manwa, N. Moswete, & J. Saarinen (Eds.), *Cultural tourism in Southern Africa*, pp. 17–30. Bristol: Channel View Publications.

Salazar, N.B., & Porter, B. (2004). Cultural heritage and tourism: A public interest approach, Introduction. *Anthropology in Action, 11*(2), 2–8.

Sambuu, A. (2020). Development of ethnographic tourism in the Republic of Tuva. *Journal of Environmental Management and Tourism, 11*(2), 341–345.

Segadika, P. (2006). Managing intangible heritage at Tsodilo. *Museum International, 58*(1–2), 31–40.

Segadika, P. (2008). The Botswana National Museum and Tsdilo World Heritage Site Museum. In P. Voogt (Ed.), *Can we make a difference? Museums, Society and Development in North and South*, pp. 109–118. Amsterdam: KIT Publishers.

Smith, S.L. (2003). A vision for the Canadian tourism industry. *Tourism Management, 24*(2), 123–133.

Swensen, G. (2016). Through the lens of affect—Interpreting prison buildings of the past as heritage in the present. In T.S. Guttormsen & G. Swensen (Eds.), *Heritage, democracy and the public: Nordic approaches*, pp. 201-214. London: Routledge.

Taruvinga, P., & Ndoro, W. (2003). The vandalism of the Domboshava rock painting site, Zimbabwe: Some reflections on approaches to heritage management. *Conservation and Management of Archaeological Sites, 6*(1), 3–10.

Tasie, G.I. (2014). Promoting cultural tourism industry in Africa: The place of African Indigenous religion. *African Research Review, 8*(4), 96–110.

Teye, V.B., Turk, E., & Sönmez, S. (2011). Heritage tourism in Africa: Residents' perceptions of African-American and White tourists. *Tourism Analysis, 16*(2), 169–185.

Timothy, D.J. (2015). Cultural heritage, tourism and socio-economic development. In R. Sharpley & D.J. Telfer (Eds.), *Tourism and development: Concepts and issues*, 2nd Edn, pp. 237–249. Bristol: Channel View Publications.

Timothy, D.J. (2021). *Cultural heritage and tourism: An introduction*, 2nd Edn. Bristol: Channel View Publications.

Timothy, D.J., & Teye, V.B. (2004). American children of the African diaspora: Journeys to the motherland. In T. Coles & D.J. Timothy (Eds.), *Tourism, diasporas and space*, pp. 111–123. London: Routledge.

Twumasi-Ampofo, K., Oppong, R.A., & Quagraine, V.K. (2020). The state of architectural heritage preservation in Ghana: A review. *Cogent Arts & Humanities*, 7(1), 1812183.

UNWTO (2003). Tourism highlights, 2003 edition. Available online: http://ebookbrowse.com/unwto-tourism-highlights-2003-pdf-d302355161. Accessed 01.05.2013. UNWTO.

UNWTO (2014). *Tourism barometer: 2013 international tourism results and prospects for 2014*. Madrid: World Tourism Organization.

UNWTO (2015). Tourism highlights, 2015—UNWTO tourism highlights, 2015 Edition. Available online: http://people.unica.it/carlamassidda/files/2012/04/, Accessed 01.05.2016.

UNWTO (2021). *Glossary of tourism terms*. Madrid: World Tourism Organization.

Van Dantzig, A. (1980). *Forts and castles of Ghana*. Accra: Sedco.

White, R. (2001). *Integrated development and management plan for Moremi Gorge*. Gaborone: Moremi Manonnye Conservation Trust.

World Bank (2001). *Tourism in Africa*. Washington, DC: World Bank.

Woyesa, T., & Kumar, S. (2021). Potential of coffee tourism for rural development in Ethiopia: A sustainable livelihood approach. *Environment, Development and Sustainability, 23*(1), 815–832.

WTTC (2007). *Botswana: The impact of travel and tourism on jobs and the economy*. Online at: http://www.wttc.org

WTTC (2012). *Travel and tourism economic impact*. London: World Travel and Tourism Council.

Yankholmes, A., & Timothy, D.J. (2017). Social distance between local residents and African-American expatriates in the context of Ghana's slavery-based heritage tourism. *International Journal of Tourism Research*, *19*(5), 486–495.

8

GROWING AGRI-HERITAGE TOURISM IN AFRICA

Challenges and Opportunities

Brent Lovelock, Gebeyaw Degarege, and David Adeloye

Introduction

The term agritourism (sometimes referred to as agrotourism) is often used interchangeably with rural tourism and farm-based tourism (Barbieri & Mshenga, 2008). A useful definition is that agritourism entails any type of recreational or educational activity offered in any type of working agricultural setting, including farms (Arroyo, Barbieri, & Rich, 2013). In this chapter, we focus upon *agri-heritage* tourism, a term used in an academic sense as early as 2013 by Vafadari, but which has more recently been employed by Timothy (2021). Some uses of the term (e.g., Vafadari, 2013) specifically link agri-heritage tourism to tourism that is related to formally recognised agricultural heritage systems, such as those recognised by the FAO's Globally Important Agricultural Heritage Systems (GIAHS) (note that this is also sometimes referred to as agricultural heritage systems tourism (Tian et al., 2016)). However, for the purposes of this chapter, we adopt a broader view that agri-heritage tourism is simply agritourism that incorporates heritage elements. These heritage elements may be tangible (e.g., built structures, landscape elements) or intangible (e.g., customs, beliefs, traditions, practices, knowledge, and language) (see Torabi Farsani, Ghotbabadi, & Altafi, 2019), who propose a framework for agricultural heritage as agritourism attractions). In this chapter, we also consider specialty agri-heritage products, which may relate to crops or livestock and their related production processes and food products available on-farm, along with the broader landscape elements and activities associated with their production.

A discussion of the opportunities for agri-heritage tourism in Africa is relevant for a number of reasons. First, the relationship between tourism and agricultural heritage can be symbiotic, with farm heritage serving as a tourist attraction and tourism providing an incentive for the preservation of farm heritage (Sun et al., 2011; Tew & Barbieri, 2012). Any development of agri-tourism does need to

DOI: 10.4324/9781003153955-8

consider, however, the potential harmful impacts of tourism on agricultural heritage (Sun et al., 2019; Timothy, 2015). Perhaps more important, however, is that agritourism—and by implication, agri-heritage tourism—has been associated with opportunities to diversify income streams for farms that may be reliant upon falling and globally determined prices for their produce. Thus, agri-heritage tourism can help create employment and income security. While the growth of these benefits has been documented in the industrialised West, and indeed in many developing world destinations (e.g., Santeramo & Barbieri, 2017; Tew & Barbieri, 2012) the situation in Africa is less certain. And just as the capacity for agritourism to fulfil such promises (e.g., Adisa, 2013; Baykal, Yildiz, & Ataberk, 2020; Koutsouris et al., 2014; Rogerson, 2012) has been questioned, we need to also bear such warnings in mind for what benefits agri-heritage tourism may be able to bring to communities. This chapter considers the current position of agri-heritage tourism in Africa. In this chapter, we consider the current status of agri-heritage tourism with a focus on Sub-Saharan Africa (SSA) and outline some key challenges and opportunities for further development of this form of tourism.

Agriculture in Africa: The Context for Agritourism and Agri-heritage Tourism

To discuss the status of, and opportunities for, agri-heritage tourism in Africa, we first need to provide a broad context of agriculture in Africa, acknowledging the challenges of doing so for such a substantial and diverse continent. Africa is often perceived as a continent of drought and widespread poverty. However, the reality of Africa is as a home of diverse peoples that exhibit a remarkably long history of civilisation. This history has seen the development of a substantial and diverse range of agro-ecosystems that have evolved over many centuries. The transition from hunting and gathering to food-producing communities has been quite heterogeneous (Breunig & Neumann, 2002) and, consequently, Africa has developed diverse and locally adapted agricultural systems that conform to the landscape and environmental conditions.

African farming and agricultural systems are shaped by underlying indigenous farming knowledge, agro-ecological conditions, and socioeconomical dynamics. Our simple analysis of agri-heritage resources and tourism potential in selected African countries is designed largely to provide material for providing a broader understanding of the topic. Africa's agriculture is complex, so too is presenting a complete picture of Africa's agri-heritage resources and tourism potential because of variations between countries, as well as variations within countries due to differences in weather, history, culture, population dynamics, and geographical location.

African nations have diverse and rich agricultural landscapes and systems that provide ecological services and, of course, contribute to food production. Agriculture is the backbone economic activity of most African countries, and the agriculture sector is an integral part of life. Over 75% of the African population is rural dwelling, with agriculture being largely defined by small-scale holdings and family farming, and with a diversity of agro-ecosystems that form the livelihood basis of millions of farmers (FAO, 2008). The major farming systems include shifting cultivation, the bush fallow system or land rotation, the planted fallow system, compound or homestead farming, terrace farming, flood land cultivation, and transhumance pastoralism (Altieri & Koohafkan, 2008. Seemingly this diversity of farming itself suggests a potential for a range of agritourism, and by extension, agri-heritage tourism offerings.

However, the agricultural sector is not without its challenges, some of which may simultaneously provide incentives and disincentives for diversification into agritourism or agri-heritage tourism. As noted above, Africa is dominated by family farming and is reliant on family labour; Africa has 33 million farms of less than 2 hectares, accounting for 80% of all farms (NEPAD, 2013). Many farming systems are struggling to replenish soil fertility due to the lack of investment capacity and secure land tenure (NEPAD, 2013). Notwithstanding this small-holding pattern of land tenure, there has also been a growing issue of large-scale industrialisation of farming, leading to land insecurity in many countries (NEPAD, 2013).

One facilitating factor for the growth of agritourism in the West is the benign food security situation there. However, in Africa, one in four people are undernourished and food insecurity remains a substantial problem, especially in rural areas because the people producing food often do not make enough to feed their families due to the lack of adequate access to means of production (land, manure, and tools) (NEPAD, 2013). Africa faces the challenge of feeding a population that has doubled overall and tripled in urban areas over the last 30 years, yet support from critical stakeholders such as government and farmer organisations is limited, as is farmers' access to financial support, which 'hampers farmers' ability to take more risks and to increase investment' (NEPAD, 2013, p. 8).

In light of these challenges, it is not surprising then that the NEPAD (2013) Agency for the African Union's report, *African Agriculture, Transformation and Outlook*, fails to mention agritourism as an alternative means of rural livelihood for farmers. While emphasising the need for trans-sectoral dialogue and partnerships for agricultural development, the tourism sector is not identified as a potential partner. Likewise, despite NEPAD's acknowledgement of the diversity of farming systems found across Africa, with its unique regional and local production and consumption patterns, there is no mention of how such regional specialisations may be taken advantage of, for agri-heritage tourism or otherwise. This highlights the disconnect between the agriculture and tourism sectors, identified, for example, in relation to the development of agritourism in Rwanda (Anbalagan & Lovelock, 2014) and in Ethiopia (Degarege & Lovelock, 2021).

This lack of institutional recognition for the potential of agritourism (and by association agri-heritage tourism) remains, despite evidence from other developing world settings that such challenges may be overcome, and where a range of food crops have become key to the development of agritourism (Catalino & Lizardo, 2004; Eshun & Tettey, 2014; Hamilpurkar, 2012; Hamzah et al., 2012; Kunasekaran et al., 2011; Thomas-Francois & Francois, 2014). Such agritourism activities may include, for example, visits to farms to sample tropical fruits, tours to tea, sugar, coffee, rubber or cocoa plantations, silk producing facilities, as well as rice farms and spice-producing facilities (Berno, 2017; Rogerson & Rogerson, 2014).

Indigenous and Traditional Food Crops

Of special interest to this chapter is the potential role of indigenous and traditional food crops (ITFC) as specialty agri-heritage tourism products. Such crops may also contribute in multiple ways, having potential for on-farm agri-heritage tourism which involves the visitor in their production, harvesting, and processing, but ITFCs can also be a source of novelty food for restaurants catering to the tourist trade (Van Wyk, 2011). Investment in local cuisine utilising ITFCs and their culture of production may also enhance the local community's 'brand identity' (Mnguni & Giampiccoli, 2015). These food crops are fundamental to the cultural identities of diverse ethnic groups (Akinola et al., 2020), but, despite recognition in the West of their nutritional, sociocultural, environmental, and economic benefits, there has been an erosion of indigenous and traditional foods crops in Africa over the last 50 years (Akinola et al., 2020). In South Africa, for example, colonial and Apartheid rules have repressed traditional African agricultural practices as 'primitive' and 'inferior' (Cocks, Vetter, & Wiersum, 2018). Over time, this 'post-colonial displacement of these foods and their food ways' (Akinola et al., 2020, p. 2) has led to a decline in the knowledge and use of these food resources, thus also posing challenges for their adoption as agri-heritage products.

However, some conventional indigenous foods have survived and include crops such as sorghum, cowpeas, and sweet potatoes, while less conventional indigenous foods include those such as amaranth, pumpkin leaves, and calabash (van der Merwe, Cloete, & Van der Hoeven, 2016). In South Africa, the ITFCs that are actively cultivated for markets in South Africa include rooibos, buchu, pelargonium, and herbs (Blignaut et al., 2014). According to Ojiewo et al. (2013), more than 90% of all ITFCs in Africa originated in, or were ancient introductions to, the continent, and there are 400 well-defined plant species that encompass 53 botanical families, primarily used as vegetables in Africa. This suggests some potential to draw upon some of these species for agri-heritage tourism, particularly for international visitors who are not likely to be familiar with these plants and the foods created from them.

Ukpokolo, Okoye, and Lawuyi (2018) in their study of the social value of ITFCs found that in many cultures of Africa, ITFCs are linked with rituals, ceremonies, and traditional events. They cite the example of the 'new yam' festival in East Nigeria, originating from the ancient agrarian Igbo culture, and held to signal the harvest season. Thus, IFTCs provide opportunities not only for farm-based tourism activities but for event-based tourism involving the wider community. However, the extent to which IFTCs currently contribute to agri-heritage tourism is unclear. Indeed, there are multiple challenges for ITFC cultivation, including the lack of available land. In Zanzibar, Tanzania, for example, spice growing and associated spice tours have contributed to the development of the tourism sector as the leading economic sector. However, population growth and rapid expansion of settlements in various areas of Zanzibar are reducing the amount of available land for spice farming (Chami, 2019). Further to this, land use planning challenge, the growth of modern industrialised farming, and the influence of external market forces have led to the marginalisation of local agrobiodiversity more broadly (FAO et al., 2015; Shackleton, Pasquini, & Drescher, 2009). These factors provide a further challenge for the growth of agri-heritage tourism, as agricultural biodiversity has been identified as the core attraction of agritourism and the 'toured object' (Flanigan, Blackstock, & Hunter, 2014).

Agricultural Heritage Systems as Agri-heritage Tourism Attractions

Agrobiodiversity, along with local and traditional knowledge systems, are key selection criteria for GIAHS. GIAHS are a living, evolving system of human communities in an intricate relationship with their territory, cultural or agricultural landscape, or biophysical and wider social environment (FAO, 2021). Currently, around 60 GIAHS have been recognised by the FAO worldwide. Importantly, the FAO recognises the potential for promotion of agritourism as a benefit of establishing GIAHS, and an emergent body of work has identified the potential for agri-heritage tourism to arise from the creation of GIAHS and other associated agricultural heritage landscapes (Mitchell & Barrett, 2017; Sun et al., 2011; Sun et al., 2019; Vafadari, 2013).

The designation of agricultural heritage sites and landscapes as tourist destination sites is, however, underdeveloped in Africa. Despite the huge agricultural heritage resource potential, so far only three agricultural heritage systems from two countries are GIAHS-designated in SSA. These are the Oldonyonokie/Olkeri Maasai Pastoralist Heritage system in Kenya, the Engaresero Maasai Pastoralist Heritage Area, and the Shimbwe Juu Kihamba Agro-forestry Heritage Site in Tanzania. In North Africa, representation is more advanced, with seven agricultural heritage sites designated in four countries: one in Algeria, two in Morocco, three in Tunisia, and one in Egypt. Such agricultural heritage systems reflect long traditions of agricultural civilisation and environmental management and conservation. The Oldonyonokie/Olkeri Maasai Pastoralist Heritage that covers an area of 160,000 km^2 is located in southern Kenya. Maasai have developed and maintained agro-pastoral systems sustainably there for centuries. Such synergetic agro-pastoral systems, adapted to the arid lands of Kenya, are respectful to the wildlife and combine animals such as buffalos, goats, and sheep with endemic species and food plants such as maize and beans. Such agricultural landscapes have been managed, sustained, and conserved by local communities; it is unclear the extent to which these have been capitalised on for agri-heritage tourism in Africa. For example, in and around the Ngorongoro Conservation Area in Tanzania, the Maasai are involved in tourism development, and while this takes place within a pastoralist setting, tourism appears to focus on other cultural aspects such as song, dance, archaeological sites, and traditional housing. Indeed, tourism has actually been counterproductive to maintaining pastoralist traditions there (Melubo & Carr, 2019; Melubo & Lovelock, 2019).

Lack of Recognition of Agri-heritage Products

Notwithstanding such lost opportunities, there is a need to further inventory and document agricultural heritage resources and work with communities to convert them into tourism products where appropriate. In many parts of Africa, agritourism and agri-heritage tourism are unfamiliar concepts. Viljoen and Tlabela (2007) observe that agritourism in South Africa is a relatively new subsector and one that began with wine tourism and ostrich farming and has become more diverse over time. However, this diversification is yet to substantively include heritage components, particularly in terms of specialty agri-heritage products. In fact, Van Zyl's (2019) survey of South African farmers and the agritourism products they offered did not highlight any agri-heritage products. It appears that this form of tourism is developmental in many parts of Africa, with studies still largely focusing on identifying the potential for developing agri-heritage tourism products, with little evidence of products already in place. Lenao's (2016) Botswana study is a good example, where community members expressed optimism that some of their traditional foods, including seswaa (pounded meat) and phane worm (a caterpillar), would appeal to tourists. The following case from Ethiopia illustrates this focus on potential and some of the challenges associated with bringing this potential to agri-heritage tourism fruition.

Ethiopian Agri-heritage Tourism

Agricultural heritage landscapes are not a single location only but a system that includes both the landscape and human society which lives and utilises the resources therein. Agri-heritage tourism activities in agricultural landscapes include experiencing the history and heritage of agriculture. This emphasises the cultural and human side of agri-heritage resources worthy of recognition for tourism development. Equally important is the traditional knowledge and lifestyle of agri-heritage communities as a vital part of the system that can be developed into tourism products and experiences. Considering Ethiopia as a case, despite its long agricultural history (Butzer, 1981), there is limited effort and recognition in harnessing these cultural significances for tourism development purposes there. Ethiopia, like most destinations across Africa, has its regionally unique farming systems. These include seed farming, enset planting, shifting cultivation, and pastoral complexes (Westphal, 1975), all of which could be developed into different tourism products. The seed farming system includes the production of grain, particularly cereals, but also pulses and oilseeds. Seed-farming production systems are commonly observed throughout the central, northern, and eastern highlands. The enset-planting agricultural system provides the principal staple, a co-staple with cereals and/or tubers, or a minor component of cereal- or tuber-based systems. Enset serves as a staple food for a significant part of the Ethiopian population. The most common food items prepared from enset are kocho and bulla. Kocho is a starch-rich product obtained by fermenting the resulting pulp wrapped in enset leaves in an underground specially prepared pit within the enset home garden, and bulla is a by-product of the production of kocho and is prepared from the liquid extracted when the scrapings and pulp are squeezed (Borrell et al., 2020). As well as producing unique landscapes and food products, a range of heritage artefacts are associated with these agri-heritage systems; for example, the traditional plough culture uses a pointed, steel-tipped tine attached to a draught pole at an adjustable low angle (locally called maresha) that was introduced to Ethiopia between 1000 and 400 BC by Semitic tribes (Solomon, Snyman, & Smit, 2006). However, despite their unique nature, there has been little effort to develop enset-planting and seed farming systems as tourism products or experiences.

However, the Gedeo multi-storey agroforestry system, and the Konso Agroforestry and cultural landscapes have, to various extents, attracted some attention from tourism industry partners, and there are ongoing efforts to develop agri-heritage tourism. The Konso cultural landscape is an arid property of stone-walled terraces and fortified settlements in the Konso highlands of Ethiopia, recognised as a UNESCO World Heritage Site in 2011. Stone terracing is a very old kind of land management system in Ethiopia beginning probably earlier than 2,500 years ago (Michels, 2005). The Konso cultural landscape demonstrates a complex system of livelihood and landscape management, the dry-stone agricultural terraces that were created to protect the land from erosion and to retain soil for agriculture (Ferro-Vázquez et al., 2017). The maintenance of the terrace agricultural system reflects detailed knowledge of engineering, dynamics of the landscape, and hundreds of years of persistent human struggle to harness the hard, dry, and rocky environment. Due to its strategic location under the southern tourism routes of Ethiopia, the Konso cultural landscape has attracted the attention of tourists and tourism operators. Despite this huge potential, agri-heritage tourism is not specifically promoted to attract visitors. Generic cultural tourism that involves visiting terraces, landscapes and community is apparent.

The Gedeo multi-storey agroforestry system is located in the southern part of Ethiopia and is well known as an indigenous knowledge-based self-sustaining land use system (UNESCO, 2020). This 5,000-year-old sustainable land use system provides a unique opportunity for understanding the human–environment relationship maintained through indigenous institutions, values, and practices (Kanshie, 2002). A total of 195 plant species distributed across 155 genera and 66 families represent the plant biodiversity in Gedeo agriculture (UNESCO, 2020). The components of agroforestry are mainly coffee, enset, indigenous trees, root crops, and shrubs in which every plant occupies distinct layers of the vertical space of the plant community (Legese et al., 2014). The enset-coffee-associated woody species are the dominant groups in Gedeo agro-biodiversity and play a great role in the environmental and ecological sustainability of the region, but the tourism sector is yet to recognise the value of this unique agri-heritage system. Overall, tourism in the Gedeo area is less developed than in other parts of Ethiopia because of the lack of infrastructure, market-ready products, and promotional efforts. This could be one of the reasons for the underutilisation of the agri-heritage tourism potential of the Gedeo area and illustrates a problem likely to be found in many parts of Africa.

These cases demonstrate different agri-heritage resources and tourism development potential. However, they also illustrate the problem of assumptions that are based simply on the aggregation of agriculture, heritage, and tourism, and demonstrate the diverse and complex nature of agri-heritage. However, there are exceptions to this, where agri-heritage tourism has developed beyond the initial phases. The following two sections address two very important heritage crops in this regard: coffee and cocoa.

Coffee Tourism in Africa

The heritage of coffee grown around the world dates back to the 15th century in the forests of Ethiopia (National Coffee Association (NCA), 2021). Coffee is one of the most important agricultural crops in Africa (International Coffee Organization (ICO), 2012; Jolliffe, 2010). However, only destinations such as Ethiopia, Tanzania, and Kenya have been studied within the context of coffee. This is primarily due to their large share of coffee production in the continent and the cultural and historical significance of coffee associated with these countries (Anbalagan & Lovelock, 2014; Degarege & Lovelock, 2021). There are 22 coffee bean exporting countries in Africa. However, five of these countries (i.e., Ethiopia, Uganda, Ivory Coast (Cote d'Ivoire), Tanzania, and Kenya) produce 86% of the coffee in Africa (Mungai, 2015). Ethiopia at 39% has the highest share of coffee production in the continent (see Table 8.1).

While tourism is the main foreign exchange earner for many African countries, the coffee industry plays a vital role in their economic development, historically accounting for up to 60% of the total exports from countries like Ethiopia, Uganda, and Rwanda (Anbalagan & Lovelock, 2014; Degarege & Lovelock, 2021;

TABLE 8.1 Share of Africa's coffee production

Country	*Coffee production share (%)*
Ethiopia	39
Uganda	23
Côte d'Ivoire	13
Tanzania	6
Kenya	5
Madagascar	3
Cameroon	3
Other	8

(Adapted from Mungai, 2015).

Yun, 2014). Notably, tourism and coffee are two key development sectors, and coffee is a significant vehicle through which the development of the tourism sector, particularly rural tourism, may be realised. Although some countries (e.g., Ethiopia) have been able to produce synergistic results between the tourism and coffee sectors, there remains significant potential and opportunities to develop coffee tourism as a niche market within more African destinations.

Furthermore, it is assumed that coffee tourism as a component of rural tourism would be a source of income diversification, particularly for countries with a single dominant attraction, as in the case of Tanzania and wildlife tourism or Rwanda and gorilla tourism (Kumar & Woyesa, 2021). While coffee tourism is mainly discussed in the literature in relation to the role it plays in the development of the economy (Jolliffe, 2010; Yun, 2014), it is important to note that coffee is inextricably linked to the culture (customs and rituals) of many of the African coffee producing destinations. One such destination is Ethiopia, known as the birthplace of the Arabica coffee (Degarege & Lovelock, 2021; Jolliffe, 2010; Mitchell, 2019).

Coffee Tourism in Ethiopia

Firstly, it is important to mention that coffee tourism is not only limited to the visiting of coffee-growing landscapes, but also includes the cultural paraphernalia and practices relating to the consumption of coffee that has developed over the centuries in Ethiopia. Ethiopia as the birthplace of Arabica coffee has an abundant coffee culture. One of the key cultural practices in Ethiopia related to coffee consumption is the coffee ceremony—a ceremony also practised in Eritrea. The ceremony is a core cultural custom that involves a daily coffee serving routine, mainly for the purpose of getting together with family, neighbours, or other visitors (Degarege & Lovelock, 2021; Yun, 2014).

According to Mitchell (2019), the coffee ceremony is more than just drinking coffee, but also a ritual of daily life and a unique part of the Ethiopian coffee culture. In addition to the coffee ceremony, Yun (2014) states that Ethiopia's traditional cafes have an authenticity to them that piques the interest of many tourists. Aside from the traditional coffee culture relating to coffee consumption in Ethiopia, there are other coffee tourism attractions and activities, including visits to coffee farms and plantations (e.g., Kaffa), coffee museums, street coffee culture, and other tropical landscapes.

Challenges and Opportunities

While coffee has developed to become an important commodity in the Ethiopian economy and an integral part of the culture, it is faced with challenges. Interestingly, the challenges of coffee tourism are connected to the challenges of Ethiopia's tourism industry in general (Degarege & Lovelock, 2021). These challenges are not only limited to Ethiopia but are evident in other African coffee-producing destinations likewise (Anbalagan & Lovelock, 2014). One of the major challenges of coffee tourism is that of inadequate infrastructure (i.e., roads and irrigation systems) in the tourism industry. According to Yun (2014), the lack of expected amenities, such as paved roads, makes it problematic to encourage visits to certain coffee-growing areas and keep tourists for longer periods of time. This puts rural communities, particularly the underdeveloped coffee-growing communities, who have the potential to attract more tourists, at a disadvantage.

Another major challenge is that of a passive approach to tourism development, possibly a common issue in many African destinations, and prevalent both within government agencies and private tourism operators (Yun, 2014). This is expressed in various ways, including the lack of desire to create new products or programmes that could attract more tourists and little to no promotional activities. Despite the immense potential of Ethiopia as a home of Coffee Arabica, coffee tourism is currently underdeveloped due to the lack of proper tourism product development strategies (Degarege & Lovelock, 2021). Other challenges include low-quality souvenirs, lack of knowledge and standards, and illegal tourism businesses.

Despite the challenges, there are several opportunities for the development of coffee tourism in Ethiopia. Yun (2014) suggested the development of coffee routes/coffee trails. Ethiopia has quite divergent arid landscapes with not much significance placed on them both by the government and the local residents. Developing coffee trails could open up already existing but unnoticed landscapes. Coffee festivals with coffee resources could also be a viable coffee tourism attraction (Yun, 2014). To get the best out of these opportunities, it is important that strategic efforts are made to develop the infrastructure along with product development and promotional activities (Degarege & Lovelock, 2021; Kumar & Woyesa, 2021). One of the measures of success for coffee tourism would be the extent to which its benefits could be shared across the many thousands of small landowners who grow the crop, although admittedly this is a problem not only for agri-heritage tourism but an ongoing conundrum for many forms of contemporary tourism.

Cocoa Tourism in Africa

The cocoa bean is the seed of the cacao tree, popularly known for being the key ingredient in the production of chocolate. It is indigenous to the equatorial regions of the Americas but is now grown across many countries with a tropical climate including countries in SSA. Cocoa is one of the most important agricultural crops in the Sub-Saharan region of Africa and the main agricultural export for many West African countries. While there are 58 cocoa-producing countries globally (18 of which are in Africa), 70% of the world's cocoa beans come from four West African countries: Ivory Coast, Ghana, Nigeria, and Cameroon (see Figure 8.1) (Shahbandeh, 2021).

Unlike coffee production, the history of cocoa production in SSA is quite recent and only goes as far back as the late 19th century (Essegbey & Ofori-Gyamfi, 2012). However, its economic impact, particularly for those in the rural communities cannot be overemphasised. When one considers the fact that about 90% of

FIGURE 8.1 Map of Africa indicating the top four cocoa-producing countries.
Adapted from Crop Life, 2021.

cocoa in Africa is grown by small-scale farmers in rural communities, the economic significance placed on cocoa as the source of livelihood for millions becomes more evident (Wencong & Kwasi, 2019). The significance of cocoa in these communities and destinations is also reflected in their 'cocoa culture' which is expressed in songs and traditional rituals as in the case of the Ivory Coast (Cacao Barry, 2021).

Cocoa Tourism in Ivory Coast

The Ivory Coast is the largest supplier of cocoa in the world (Yock, 2021). Cocoa accounts for 40% of the destination's total exports (Yock, 2021). In an interview conducted by Adams (2019, n.p.) with the governor of a rural department of Adzopé, Abidjan, Ivory Coast, the governor, N'ZiKanga Rémi, was asked about the importance of cocoa in the country. Rémi responded by saying,

> It doesn't make sense to ask an Ivorian what cocoa means to him! It means everything! It's his first source of income! My education was funded by cocoa! Our houses are built with cocoa! The foundations of our roads, our schools, our hospitals is cocoa! Our government runs on cocoa! All our policy focuses on sustaining cocoa!

The significance of cocoa is also reflected in the culture of the Ivory Coast, particularly within her traditional religious context. For example, ceremonies and rituals are dedicated to mother nature, deities, or ancestors for all the elements allowing them to produce cocoa (Cacao Barry, 2021). This is also popular across other cocoa-producing destinations (Adams, 2019; Eshun & Tichaawa, 2020). In fact, in

Madagascar, there are traditional songs that actually speak of cocoa and the farmers who produce it (Cacao Barry, 2021). This is accompanied by the traditional dance called *Sacalava Antakarana* (Cacao Barry, 2021).

Challenges and Opportunities

Despite the economic significance of cocoa as an agricultural commodity, cocoa tourism remains underdeveloped and nearly non-existent in the Ivory Coast, as well as in other African countries. Unlike coffee tourism, which is quite established (although developing) in countries like Ethiopia, cocoa tourism is in its nascent state for African destinations. As such, the main challenge is for the governments of cocoa-producing destinations to consider developing cocoa tourism (e.g., agri-tours to cocoa farms) as an alternative business to diversify the tourism sector and support rural communities (Eshun & Tichaawa, 2020).

On a positive note, the government of the Ivory Coast has started working on improving market access for crops by upgrading transportation infrastructure and extending electricity provision to enhance value-added from cocoa (Yock, 2021). While these efforts target cocoa as an export cash crop and not cocoa tourism/agritourism, developed infrastructure would simultaneously aid the development of cocoa tourism or agritourism in general (Yock, 2021).

Furthermore, the existing cocoa culture and traditions (e.g., song and dance rituals as in the case of Madagascar), presents good opportunities for potential cocoa tourism products and activities such as cocoa festivals, exhibitions, museums, and street cafes. This would be beneficial for the economy of the Ivory Coast and help improve the standard of living for rural communities, particularly the small-scale cocoa farmers who work hard but get meagre pay (Adams, 2019).

Conclusion

The overall tone of the discussion in this chapter is one of cautious optimism about the future of agri-heritage tourism in Africa. To use an apt metaphor, only the tip of the plant has been harvested in terms of the potential of agri-heritage tourism in Africa. The background and cases discussed here suggest that there is substantial opportunity to grow agri-heritage tourism, with, optimistically, benefits arising for the communities involved, similar to those observed in Western settings. Undeniably there do appear to be barriers, and these are largely linked to generating inter-sectoral synergy between the agriculture and tourism sectors, such an outcome being more challenging in the African context compared to Western agritourism settings where the governance of both sectors is more advanced. Likewise, infrastructural, financing, product development, and marketing challenges exist, along with optimising livelihood benefits for the many small-holders who characterise the agriculture sector in Africa. There is also a need for destinations to undertake inventories of their agricultural heritage, accompanied by the crucial next steps of providing protection of these resources, and support for their integration into agri-heritage tourism products where desired. In the African context, such support will include product and service development, human resource development, marketing and promotion, and establishing market linkages (Spenceley, 2010).

Of course, it remains important to adopt a critical lens when considering the potential benefits of agri-heritage tourism, but it seems that it is likely to contribute to the concomitant benefits of providing a window into the diverse cultures found across Africa whilst also offering some livelihood diversification opportunities for farmers, their families, and communities where it is (sometimes desperately) needed. Further, there is potential for agri-heritage tourism, if economic sustainability can be demonstrated, to become a catalyst to rejuvenate interests among the youth towards agriculture, important for the sustainability of agriculture in Africa (Domfeh, 2013).

The current major tourism-generating regions in the West face a particular set of circumstances that ironically bode well in terms of producing demand for agri-heritage tourism in Africa. In particular, we believe that ITFC could feature strongly in agri-heritage tourism initiatives, providing strong attractions for visitors from parts of the world where modern food movements predicate a growing interest in moving away from industrialised farming and the industrial foods produced (Slocum & Curtis, 2017). Similarly, African heritage agricultural systems can teach us, by exposure through tourism, about traditional sustainable approaches to producing food, in an era where food production faces multiple challenges from climate change and worn-out agri-industrial systems.

References

Adams, T. (2019). From bean to bar in Ivory Coast, a country built on cocoa. *The Guardian*. Online at: https://www.theguardian.com/global-development/2019/feb/24/ivory-coast-cocoa-farmers-fairtrade-fortnight-women-farmers-trade-justice

Adisa, R.S. (2013). Agro-tourism in North-Central Nigeria: Perceptions, performance, problems, and prospects. *Journal of Travel and Tourism Research*, *13*(1/2), 54.

Akinola, R., Pereira, L.M., Mabhaudhi, T., de Bruin, F.M., & Rusch, L. (2020). A review of indigenous food crops in Africa and the implications for more sustainable and healthy food systems. *Sustainability*, *12*(8), 3493.

Altieri, M.A., & Koohafkan, P. (2008). *Enduring farms: Climate change, smallholders and traditional farming communities*. Penang, Malaysia: Third World Network.

Anbalagan, K., & Lovelock, B. (2014). The potential for coffee tourism development in Rwanda – neither black nor white. *Tourism and Hospitality Research*, *14*(1–2), 81–96.

Arroyo, C.G., Barbieri, C., & Rich, S.R. (2013). Defining agritourism: A comparative study of stakeholders' perceptions in Missouri and North Carolina. *Tourism Management*, *37*, 39–47.

Barbieri, C., & Mshenga, P. (2008). The role of firm and owner characteristics on the performance of agritourism farms. *Sociologia Ruralis*, *48*(2), 166–183.

Baykal, F., Yildiz, S., & Ataberk, E. (2020). The development potential of agritourism in the countryside of Izmir: The challenges and future prospects. *International Journal of Contemporary Tourism Research*, *4*(2), 143–162.

Berno, T. (2017). Bridging sustainable agriculture and sustainable tourism to enhance sustainability. In G.M. Mudacumura, M. Desta Mebratu, & S. Haque (Eds.), *Sustainable development policy and administration*, pp. 207–221. London: Routledge.

Blignaut, J., De Wit, M., Knot, J., Midgley, S., Crookes, D.J., Drimie, S., & Nkambule, N. (2014). *Sustainable agriculture: A viable option for enhanced food and nutritional security and a sustainable productive resource base in South Africa.* Pretoria: ASSET Research.

Borrell, J.S., Goodwin, M., Blomme, G., Jacobsen, K., Wendawek, A.M., Gashu, D., … & Wilkin, P. (2020). Enset-based agricultural systems in Ethiopia: A systematic review of production trends, agronomy, processing and the wider food security applications of a neglected banana relative. *Plants, People, Planet, 2*(3), 212–228.

Breunig, P., & Neumann, K. (2002). From hunters and gatherers to food producers: New archaeological and archaeobotanical evidence from the West African Sahel. In F.A. Hassan (Ed.), *Droughts, food and culture*, pp. 123–155. New York: Kluwer.

Butzer, K.W. (1981). Rise and fall of Axum, Ethiopia: A geo-archaeological interpretation. *American Antiquity, 46*(3), 471–495.

Cacao Barry (2021). *Cocoa culture.* Online at: https://www.cacao-barry.com/en-US/ep10#main-content

Catalino, A.H., & Lizardo, M. (2004). Agriculture, environmental services and agro-tourism in the Dominican Republic. *eJADE: Electronic Journal of Agricultural and Development Economics, 1*(1), 87–116.

Chami, A.A. (2019). Rapid spread of settlements endangers the future of historical touristic spice farming in Zanzibar. *Current Investigations in Agriculture and Current Research, 6*(3), 238.

Cocks, M., Vetter, S., & Wiersum, K.F. (2018). From universal to local: Perspectives on cultural landscape heritage in South Africa. *International Journal of Heritage Studies, 24*(1), 35–52.

Crop Life (2021). *Cocoa in West Africa.* Online at: https://croplife.org/trainingthroughlocalpartnerships/cocoa/img/cocoa-wa-info.pdf

Degarege, G.A., & Lovelock, B.A. (2021). Institutional barriers to coffee tourism development: Insights from Ethiopia - the birthplace of coffee. *International Journal of Culture, Tourism and Hospitality Research, 15*(3), 428–442.

Domfeh, K.A. (2013). *Farmer experiments: Ghana's agrotourism potential.* Online at: www.modernghana.com

Eshun, G., & Tettey, C. (2014). Agrotourism development in Ghana: A study of its prospects and challenges at Adjeikrom cocoa tour facility. In D. Szymańska, & S. Środa-Murawska (Eds.), *Bulletin of Geography, Socio-economic series*, No. 25, pp. 81–99. Toruń: Nicolaus Copernicus University Press.

Eshun, G., & Tichaawa, T.M. (2020). Developing agro-tourism tours on the cocoa sector in Africa: Emerging issues from Ghana. *EuroEconomica, 39*(1), 1–18.

Essegbey, G.O., & Ofori-Gyamfi, E. (2012). Ghana cocoa industry – an analysis from the innovation system perspective. *Technology and Investment, 3*(1), 276–286.

FAO (2008). *Supporting food security and reducing poverty in Kenya and the United Republic of Tanzania through dynamic conservation of globally important agricultural heritage systems (GIAHS).* Rome: Food and Agriculture Organization.

FAO (2021). Globally important agricultural heritage systems. Online at https://www.fao.org/giahs/giahsaroundtheworld/en/

FAO, IFAD & WFP (2015). *The state of food insecurity in the world 2015, meeting the 2015 international hunger targets: Taking stock of uneven progress.* Rome: Food and Agriculture Organization.

Ferro-Vázquez, C., Lang, C., Kaal, J., & Stump, D. (2017). When is a terrace not a terrace? The importance of understanding landscape evolution in studies of terraced agriculture. *Journal of Environmental Management, 202*, 500–513.

Flanigan, S., Blackstock, K., & Hunter, C. (2014). Agritourism from the perspective of providers and visitors: A typology-based study. *Tourism Management, 40*, 394–405.

Hamilpurkar, S. (2012). Agri tourism in Karnataka – Issues, constraints and possibilities. *International Journal of Research in Commerce, Economics & Management, 2*(7), 106–111.

Hamzah, A., Yassan, S., Samah, B., D'Silva, L., Tiraiyaei, N., Shaffril, H., & Uli, J. (2012). Socio-economic impact potential of agro tourism activities on *Desa Wawasan Nelayan* community in Peninsular Malaysia. *African Journal of Agricultural Research*, 7(32), 4581–4588.

International Coffee Organization (ICO) (2012). *About coffee*. Online at: http://www.ico.org/coffee_story.asp

Jolliffe, L. (ed.) (2010). *Coffee culture, destinations and tourism*. Bristol: Channel View Publications.

Kanshie, T.K. (2002). *Five thousand years of sustainability? A case study on Gedeo land use (Southern Ethiopia)*. Unpublished PhD thesis, Wageningen University, Netherlands.

Koutsouris, A., Gidarakou, I., Grava, F., & Michailidis, A. (2014). The phantom of (agri) tourism and agriculture symbiosis? A Greek case study. *Tourism Management Perspectives, 12*, 94–103.

Kumar, S., & Woyesa, T. (2021). Potential of coffee tourism for development in Ethiopia: A sustainable livelihood approach. *Environment, Development and Sustainability, 23*(1), 815–832.

Kunasekaran, P., Ramachandran, M., Yacob, M., & Shuib, A. (2011). Development of farmers' perception scale on agrotourism in Cameron Highlands, Malaysia. *World Applied Sciences Journal, 12*, 10–18.

Legese, G., Haile, A., Duncan, A.J., Dessie, T., Gizaw, S., & Rischkowsky, B. (2014). *Sheep and goat value chains in Ethiopia: A synthesis of opportunities and constraints*. Nairobi: ILRI.

Lenao, M. (2016). Rural cultural tourism development and agriculture: Evidence from residents of Mmatshumu Village in the Boteti region of Botswana. In H. Manwa, N. Moswete, & J. Saarinen (eds) *Cultural tourism in Southern Africa*, pp. 132–144. Bristol: Channel View Publications.

Melubo, K., & Carr, A. (2019). Developing indigenous tourism in the bomas: Critiquing issues from within the Maasai community in Tanzania. *Journal of Heritage Tourism, 14*(3), 219–232.

Melubo, K., & Lovelock, B. (2019). Living inside a UNESCO World Heritage Site: The perspective of the Maasai community in Tanzania. *Tourism Planning & Development, 16*(2), 197–216.

Michels, J.W. (2005). *Changing settlement patterns in the Aksum-Yeha region of Ethiopia: 700 BC-AD 850*. Cambridge: Cambridge Monographs in African Archaeology.

Mitchell, N.J., & Barrett, B. (2017). Exploring agricultural heritage landscapes: A journey across "terra incognita". *The George Wright Forum, 34*(2), 180–194.

Mitchell, S. (2019). Traditional Ethiopian coffee ceremony. *Culturally Ours*. Online at: https://culturallyours.com/2019/10/12/traditional-ethiopian-coffee-ceremony/

Mnguni, E.M., & Giampiccoli, A. (2015). Indigenous food and tourism for community well-being: A possible contributing way forward. *Mediterranean Journal of Social Sciences, 6*(3), 24–34.

Mungai, C. (2015). Which African countries produce the most coffee? *World Economic Forum*. Online at: https://www.weforum.org/agenda/2015/10/which-african-countries-produce-the-most-coffee/

National Coffee Association (NCA) (2021). *The history of coffee*. Online at: https://www.ncausa.org/About-Coffee/History-of-Coffee

NEPAD (2013). African agriculture, transformation and outlook. Online at https://www.nepad.org/caadp/publication/agriculture-africa-transformation-and-outlook

Ojiewo, C., Tenkouano, A., d'a Hughes, J., & Keatinge, J.D. (2013). Diversifying food and diets: Using indigenous vegetables to improve profitability, nutrition and health in Africa. In J. Fanzo, D. Hunter, T. Borelli, & F. Mattei (Eds.), *Diversifying food and diets: Using agricultural biodiversity to improve nutrition and health*, pp. 291–302. London: Routledge.

Rogerson, C.M. (2012). Strengthening agriculture-tourism linkages in the developing world: Opportunities, barriers and current initiatives. *African Journal of Agricultural Research*, 7(4), 616–623.

Rogerson, C.M., & Rogerson, J.M. (2014). Agritourism and local economic development in South Africa. In C.M. Rogerson & D. Szymańska (Eds.), *Bulletin of geography: Socio-economic series*, No. 26, pp. 93–106. Toruń: Nicolaus Copernicus University.

Santeramo, F.G., & Barbieri, C. (2017). On the demand for agritourism: A cursory review of methodologies and practice. *Tourism Planning & Development*, *14*(1), 139–148.

Shackleton, C., Pasquini, M., & Drescher, A. (2009). *African indigenous vegetables in urban agriculture*. London: Earthscan.

Shahbandeh, M. (2021). Cocoa bean production worldwide 2018/19 & 2020/21, by country. *Statista*. Online at: https://www.statista.com/statistics/263855/cocoa-bean-production-worldwide-by-region/

Slocum, S.L., & Curtis, K.R. (2017). *Food and agricultural tourism: Theory and best practice*. London: Routledge.

Solomon, T.B., Snyman, H.A., & Smit, G.N. (2006). Soil seed bank characteristics in relation to land use systems and distance from water in a semi-arid rangeland of southern Ethiopia. *South African Journal of Botany*, *72*(2), 263–271.

Spenceley, A. (2010). *Tourism product development interventions and best practices in sub-Saharan Africa: Part 1, Synthesis*. Washington, DC: World Bank.

Sun, Y., Jansen-Verbeke, M., Min, Q., & Cheng, S. (2011). Tourism potential of agricultural heritage systems. *Tourism Geographies*, *13*(1), 112–128.

Sun, Y., Timothy, D.J., Wang, Y., Min, Q., & Su, Y. (2019). Reflections on agricultural heritage systems and tourism in China. *Journal of China Tourism Research*, *15*(3), 359–378.

Tew, C., & Barbieri, C. (2012). The perceived benefits of agritourism: The provider's perspective. *Tourism Management*, *33*(1), 215–224.

Thomas-Francois, K., & Francois, A. (2014). Spices and agro-tourism on Grenada, the island of spice. In L. Jolliffe (Ed.), *Spices and tourism: Destinations, attractions and cuisines*, pp. 17–32. Bristol: Channel View Publications.

Tian, M., Min, Q.W., Jiao, W.J., Yuan, Z., Fuller, A.M., Yang, L., … & Cheng, B. (2016). Agricultural heritage systems tourism: Definition, characteristics and development framework. *Journal of Mountain Science*, *13*(3), 440–454.

Timothy, D.J. (2015). Cultural heritage, tourism and socio-economic development. In R. Sharpley, & D.J. Telfer (Eds.), *Tourism and development: Concepts and issues*, 2nd edn, pp. 237–249. Bristol: Channel View Publications.

Timothy, D.J. (2021). *Cultural heritage and tourism: An introduction*, 2nd edn. Bristol: Channel View Publications.

Torabi Farsani, N., Ghotbabadi, S.S., & Altafi, M. (2019). Agricultural heritage as a creative tourism attraction. *Asia Pacific Journal of Tourism Research*, *24*(6), 541–549.

Ukpokolo, F.C., Okoye, O.C., & Lawuyi, O.B. (2018). Cultural reconstruction of Iwa Ji Festival in Igbo-Ukwu, and fractured Igbo identity. *Synthesis Philosophica*, *65*, 149–164.

UNESCO (2020). Gedeo Cultural Landscape. Online at: https://whc.unesco.org/en/tentativelists/6448/

Vafadari, K. (2013). Tameike reservoirs as agricultural heritage: From the case study of Kunisaki Peninsula in Oita, Japan. *Journal of Resources and Ecology*, *4*(3), 220–230.

van der Merwe, J.D., Cloete, P.C., & Van der Hoeven, M. (2016). Promoting food security through indigenous and traditional food crops. *Agroecology and Sustainable Food Systems*, *40*(8), 830–847.

Van Wyk, B.E. (2011). The potential of South African plants in the development of new food and beverage products. *South African Journal of Botany*, 77(4), 857–868.

Van Zyl, C.C. (2019). *The size and scope of agri-tourism in South Africa*. PhD thesis, North-West University, South Africa.

Viljoen, J., & Tlabela, K. (2007). *Rural tourism in South Africa: Trends and challenges*. Pretoria: Human Sciences Research Council.

Wencong, L., & Kwasi, H.G.S.A. (2019). Transition of small farms in Ghana: Perspectives of farm heritage, employment and networks. *Land Use Policy*, *81*(1), 434 – 452.

Westphal, E. (1975). *Agricultural systems in Ethiopia*. Wageningen: Agricultural University.

Yock, E. (2021). Cocoa industry in the Côte d'Ivoire: The farming and production of cocoa beans in the West African country of Côte d'Ivoire. *ESRI*. Online at: https://www.arcgis.com/apps/Cascade/index.html?appid=dd19d4b6444e492da5b069d62d2c3ecb

Yun, O. (2014). *Coffee tourism in Ethiopia: Opportunities, challenges, and initiatives*. Unpublished doctoral thesis, University of Exeter, UK.

9

CULINARY TRADITIONS AND HERITAGE FOODS IN AFRICA

Roselyne N. Okech and Dallen J. Timothy

Introduction

The culinary tourism industry is now a significant force in the global economy (Long, 2018; World Tourism Organization, 2019). According to a 2016 summary, 88.2% of travellers consider food a significant component in their selection of a vacation destination, as well as in their vacation activities. A recent report by an industry organization claims that tourism contributes US$2.3 trillion to the global economy (World Travel and Tourism Council, 2017) which, given the importance of food as part of the travel experience, illustrates how culinary tourists can have a major monetary impact throughout the world (World Tourism Organization, 2019).

Food is an important aspect of cultural identity and an intricate crucible of human heritage. Culinary traditions have been passed on from generation to generation, making cuisines unique and encouraging the preservation of history, lifestyles, and beliefs through the growing, preparation, eating, and preservation of food. Culture, race, and ethnicity are intricately connected and important social identities, and the interplay of food and culture has had a huge global impact.

Food, perhaps more specifically cuisine, plays a prominent and integral role in the creation, maintenance, and promotion of regional and national identities (Bessiére, 2013; Bessière, 1998; Cusack, 2000; Highfield, 2017; Ramshaw, 2016; Smith, 1995). Palmer (1998) points out that food, along with landscape and the body, is an aspect of the modern material world that is important to both individual and collective identities. There is no doubt that food and the way various ingredients are combined and cooked, is an important element of a national cultural identity. This suggests a number of important dimensions of food experiences in tourism. For example, food experiences are often novel and contrasted with everyday eating. They often contrast between globalized and localized fare (Mak, Lumbers, & Eves, 2011), and there is an important role for 'authenticity' in distinguishing foods that

DOI: 10.4324/9781003153955-9

are presented to and consumed by tourists. There is a relationship between these dimensions in tourism since the 'local' is often equated with 'authentic'. In theory, therefore, what many tourists seek are local, authentic, and novel food experiences. Given the importance that cultural heritage has in society's economic development, some work has centred on its interpretation, planning, and management as a tourism resource in the context of food.

The purpose of this chapter is to highlight the importance of food heritage traditions in Africa. Examples of African cuisines are discussed, including how they are prepared and preserved. The economic, sociocultural, and environmental benefits and challenges of Africa's heritage foods are described, and the significance of traditional foods and health is also briefly discussed. Finally, reference is made to the influence of African food in diasporic communities, particularly in American culture. The chapter recognizes that more exposure and historical documentation of African food heritage needs to be explored.

Heritage and Food

Currently, food heritage constitutes one of the main attractions of tourist destinations, given that food is a marker of regional identity and a focal point of the tourist experience. As a result of this, a central theme in gastronomic tourism studies is the analysis of the impact that tourism practices have on the patterns of continuity and transformation of local food (Ron & Timothy, 2013; Thomé-Ortiz, 2018; Timothy, 2016b). Food can therefore provide the basis for the development of tourism experiences in a number of ways. First, it links culture and tourism. Food is often a gateway to local culture and can bring tourists and locals together in a shared cultural experience. Food can be linked to local customs, traditions, landscapes and systems of food production to provide engaging authentic experiences for tourists (Che, 2016; Sun et al., 2019). The second is developing the meal experience. Meals are a central part of the tourist experience and they provide an ideal setting for the creation and staging of memorable and meaningful experiences (Everett, 2016a; Ron & Timothy, 2013). The third way is producing distinctive foods, which can become distinguishing elements of the brand image of places and help to create uniqueness in a crowded marketplace (Che, 2006; Everett, 2016b). Fourth, tourism helps develop a critical infrastructure for food production and consumption. There is a growing critical infrastructure around the production, preparation, and consumption of food. This includes food producers, chefs, critics, other culinary trendsetters, journalists, bloggers, and information providers (Cappelen & Pedersen, 2020). Finally, food experiences can directly support local cultural development by providing the cultural capital necessary to create and sustain cultural production and consumption (Richards, 2012).

There is an innate human desire to preserve culture through food practices, especially when a culture is not the dominant one in a given society. However, what is perhaps less studied is the maintenance of cultural food practices and the sharing of cross-cultural gustatory practices by different cultural, racial, and ethnic groups within multicultural societies (Reddy & van Dam, 2020).

Currently, climate change is a grave challenge to the tourism sector since it increasingly affects the industry's core attributes, such as the length and quality of tourist seasons and environmental attributes of places. Natural and cultural sites coupled with authentic gastronomic products form a competitive advantage for a region. Since the length of visitor stays in most regions is quite short, culinary tourism may be properly blended as an integral part of a destination's tourism product to slow tourists down and help them appreciate the destination in greater depth. Destinations can take advantage of the unique features of culinary tourism, especially those elements that connect local gastronomy to a strong heritage narrative. This often involves cooking classes, farmland visits, and local food festivals, all of which can extend visitors' stay and enhance their experience (Donovan & De Bres, 2006; Everett, 2016a). Even though some catering establishments emphasize local culinary products, tour operators usually give inadequate attention to traditional food consumption, gastronomic events, and food-related festivals. This inattention potentially contributes to the underemphasis of heritage foods in the tourism product mix and the underperformance of food tourism which, in turn, creates a less than holistic heritage (or even general culinary) tourist experience and reduced visitor length of stay. This is a significant lost opportunity for enhancing a local sense of place, providing a stronger appreciation for local heritage, and lengthening tourists' stays in the region.

In recent years, the literature on the cultural heritage of food has grown in breadth and depth, but African food heritage continues to be underrepresented in popular and academic writing. Several cookery authors and historians have placed more focus on African cuisines (e.g., Duru, 2020; Hafner, 2002; Mallos, 2006; McCann, 2009), but the region's gustatory traditions continue to be unstudied in the academic arena, particularly from a heritage and tourism perspective.

African Cuisines: Ingredients and Cooking Methods

There are many food cultures in the world, some more distinctive than others but still just as significant to the communities they influence and the cultures they help to conserve. Food heritages are influenced by many factors, including specific ingredients and cooking techniques and methods. Each country in Africa has its own special and unique blend of cuisines. Most countries have national cuisines, and some have regional cuisines that are differentiated based on flavours, ingredients, preparation methods, and even eating cultures. The diverse nature of African cooking reflects local culinary traditions, together with different geographical and environmental conditions (e.g., rainforests, savannahs, and deserts) that typically determine the availability of vegetables, grains, and meat. There are, however, stark similarities in African cooking methods and ingredients. Each region of the continent has its own distinctive dishes, preparation techniques, and consumption procedures. Certain foods, however, are common to many regions (Table 9.1). African meals often contain one or more of these key ingredients.

TABLE 9.1 Common African culinary staples

Northern Africa	*Southern Africa*	*Eastern Africa*	*Central Africa*	*Western Africa*
Couscous	Braai	Ugali	Yam	Jollof rice
B'stilla	Boerewors	Tilapia	Palm Butter Soup	Yassa
Harira	Nsima	Irio	Foutou	Ndole
Matbucha	Game meat	Matoke	Bushmeat	Cachupa
Tajine	Fish	Chapati	Fish	Waakye

Source: authors' compilation.

These ingredients make African dishes unique and special. The central characteristic and focus of African cuisine is the starchy component (e.g., ugali, yams, rice, and couscous) (Chrones, n.d.), accompanied by stewed meats or vegetables, or a combination of both. These primary aliments, as in other parts of the world, reflect not only cultural traditions but also the environmental realities of what can be produced and consumed in Africa's deserts, savannahs, and rainforests. Starch filler foods, similar to the rice-based cuisines of Asia, are a hallmark. Cassava and yams are the main root vegetables. Steamed greens, mixtures of hot spices with root vegetables, and stew with or without meat, all are African-inspired. Peanuts, known as groundnuts in Africa, feature heavily in main dishes from garnish to peanut soups.

As previously noted, each region of the continent has its own specialty food items that reflect environmental conditions, endemic or indigenous plant and animal species, faith and religious practices, celebrations and family traditions, tastes, and foreign influences, among many other variables. The following sections highlight some of the specific culinary traditions in the main regions of Africa (see also Gheorghe & Bulin, 2014).

North Africa

The foods of Egypt and the Maghreb region, which includes the countries of Algeria, Morocco, Libya, and Tunisia, are some of the most recognizable throughout the world, with couscous being one of the most notable examples. Moroccan and Egyptian fares have spread throughout the world as popular ethnic food in Western countries, and the local Arab–Mediterranean cuisine is a significant part of the experience for tourists visiting the countries of North Africa (Belaidi & el Bouzidi, 2019; Mohamed et al., 2020; Qaddahat, Attaalla, & Hussein, 2016).

Common ingredients in this region include seafood, goat, beef, certain vegetables and fruits, legumes, and spices, including saffron and cumin. Couscous is the best-known and most fundamental food item in the Maghreb. In fact, it is such an important part of the region's culinary heritage and plays such a vital role in everyday life that it was inscribed by UNESCO on the Representative List of the Intangible Cultural Heritage of Humanity in 2020 as 'knowledge, know-how and the practices pertaining to the production and consumption of couscous' for Algeria, Mauritania, Morocco, and Tunisia.

There are also local names for certain kinds and styles of couscous, such as *burzqān* in Béja, Tunisia, where a fine-grain couscous is mixed with fresh butter, mutton, saffron, and chickpeas, sprinkled with hot milk, and garnished with raisins, almonds, pistachios, hazelnuts, and walnuts. *Malthūth* is a barley couscous prepared primarily by the poor. It is sorted, cleaned, and grilled on a platter known as a *ghanāʾā*. It is then pounded, sieved, and dried in the open air. A second sifting collects the barley. A further sieving through a finer sieve allows the larger grains to be retained. *Dashīsha*, as it is called in southern Algeria, is also the name of a porridge made of pounded wheat and butter. Smaller grains can be used for barley couscous or *ʿaṣīda* or *bazīn*, a kind of polenta with a sauce of chillies, bell peppers, onions, tomatoes, *harīsa*, and meat (Wright, 2021). In southern Tunisia, ground fenugreek is sprinkled on the couscous. Chilli pastes, such as the spicy Tunisian harissa, are also popular in the region. Tajine, is often served as a stew (as in Algeria) or a casserole (as in Tunisia). The Moroccan pastilla, or bastilla, is another regional favourite. Because the Maghreb is predominantly Muslim, halal meats are usually eaten (Wright, 2021), and much of the region's culinary landscape reflects its Mediterranean location and its long history of trans-Mediterranean trade with Europe, successive empires that controlled the area, and modern diasporic connections with former metropolitan colonizers.

West Africa

Nigeria and the coastal parts of West Africa are fond of chillies and hot spices. Coastal recipes include fish marinated in ginger, tomatoes, and cayenne, cooked in peanut oil. Peanuts (groundnuts) are a prominent ingredient in many soups and other dishes in West African fare, which is not surprising given that the region is one of the world's largest groundnut-producing areas, especially Nigeria, Cameroon, Chad, and Senegal.

The French colonial history in many countries of West Africa has influenced food traditions and cooking methods. Lime juice, chopped vegetables, peanuts, palm, and often coconut oils are also common in the region. The blackeyed pea is a staple of West Africa, okra is used in many dishes to thicken soups and stews, and plantains are a common food item. All of these reflect a tropical or subtropical climate that enables the growth of certain heat and humidity-loving plants. The cooking techniques of West Africa often combine fish and meat. Flaked and dried fish popularly known as 'crayfish' is combined with chicken, yams, chilli, onions, and water to make a flavourful stew.

Certain ingredients are more common in certain countries. In Ghana, for example, the most commonly used ingredients are hot peppers, ginger, and maize. Ghanaians' use of hot peppers reflects their belief that hot chillies will cool the body and cleanse or purify it. Fufu is usually made from cassava, and yams, and sometimes combined with cocoyam, plantains, or cornmeal. Today, there are many variations of fufu including cabbage, eggplant, almond, tigernut, and coconut, which are considered adequate substitutes for low-carbohydrate diets. Fufu can also be made from semolina, rice, or even instant potato flakes. Despite modern variations, fufu is still normally made by traditional methods: pounding the base substance in a mortar with a wooden spoon.

East Africa

East African foods are heavily inspired by Asian and Middle Eastern influences (Okech, 2014) owing to the region's long history of trade with East Asia, South Asia, and the Arabian Peninsula, as well as its widespread adoption of Islam (McCann, 2009). The diet is usually starch-based, with maize corn (*ugali*). This corresponds to the version of fufu in West and Central Africa. Barbecued meat (*nyama choma*) is popular but generally not eaten daily. Instead, dishes of beans and corn or *sukuma wiki* (sautéed leafy greens, often kale) are a large part of the daily aliment. Common spices include cumin, coconut, turmeric, tamarind, and coriander, especially in Swahili cooking. *Mutura* is a popular street food in Kenya, which is made of boiled goat sausage and goat blood, which is then barbecued (Great British Chefs, 2021). In Uganda, steamed bananas (*matoke*) provide starch fillers for many meals. Tsebhi and injera (stew and bread) are the commonest foods in Ethiopia and Eritrea. Further south, Tanzania has a completely different cuisine from those of Ethiopia and Eritrea. It is heavily influenced by Indian cooking, thanks to the nineteenth- and early twentieth-century migration of Indians into Tanzania. Likewise, Tanzanian cuisine includes abundant fish and seafood from its eastern coast and coconut milk in sauces and curries, which are heavily spiced.

The island of Zanzibar (part of Tanzania) is particularly well-known for *mchuzi wa samaki*, a fish and coconut curry spiced with cumin, garam masala, and coriander. Rice is more popular than beans and grains along the Tanzanian coast, including Zanzibar, but there is a distinct southern Indian influence throughout the country, with chapattis or pilau often prepared for celebrations. Djibouti is a small country on the Horn of Africa surrounded by Eritrea, Ethiopia, and Somalia, but it has its own food culture, in part due to its status as a former French colony and its relative nearness to the Arabian Peninsula across the Red Sea, sharing many characteristics with Yemen. French baguettes are still commonly eaten, and French mustard is a popular condiment, but probably the most well-liked dish is *fah-fah*, a stew made from goat, camel, or mutton and flavoured with an abundance of chillies (Great British Chefs, 2021). In addition to baguettes, flatbreads are common, especially with a biryani-like combination of rice and lamb.

Central Africa

Central Africans tend to stick to their food roots. Because they have been so centralized, many food trends have been unable to penetrate their cultures and cooking habits. People here use primarily local ingredients, except for cassava, peanuts, and chilli peppers. These ingredients have managed to permeate the otherwise rigid food culture of Central Africa. Food staples in this region include cassava (tapioca) and plantains, which are a multidimensional food. Plantains are eaten fried, baked, mashed, and used as an ingredient in many other dishes or simply eaten alone. Starchy foods are also made from fermented cassava roots and are served alongside meats. Stews are popular and include spinach stew and peanut stew, as well as African vegetable stew. Chicken and beef are the most common meats, but crocodile, warthog, monkey, and antelope are also prepared at times (SpicesInc, 2021).

Southern Africa

The food cultures of southern Africa are wide-reaching and play a very important part in tourism and local daily life. South African cuisine in particular is a unique fusion of many different external and internal cultural influences (SA-Venues, 2021). These include Indigenous, Dutch, French, German, Indian, and Malaysian flavours and techniques that continue to influence not only restaurant menus but also people's home cookery as well (Oktay & Sadıkoğlu, 2018). For example, the German-origin *Boerewors*, is a traditional sausage that is commonly consumed across the country.

The combination of these culinary influences has led to what is today known as the Afrikaans style of cooking. Because Afrikaaners traversed the country in search of places to settle without refrigeration, they became known in South Africa for their dried meats (e.g., biltong, which is similar to jerky, and dröewors, a dried sausage) and liberal use of spices and salts for the preservation of foods. The braai (barbecue) is one of the best-known 'cuisine' types in South Africa. Red meat, such as spiced sausages, chops, kebabs, steaks, and pork rashers, are put over hot coals, infusing them with smoky flavours (SA-Venues, 2021).

The most traditional meal of Zimbabwe is cassava which is milled and used as flour. Sadza is a staple corn meal similar to ugali in Kenya and among the Shona in Botswana. The general preference is for white maize rather than yellow maize, which is sometimes eaten during famine or hardships. Sadza can also be made from finger millet flour (Oktay & Sadıkoğlu, 2018).

African Culinary Influences

Culinary historian Diane Spivey (2018) describes Africa as a major contributor to the cuisine of North and South America, although this contribution has long been overlooked, trivialized, or denied. However, the discourse contained in volumes on American cooking is usually consistent in celebrating the European influences. When African American cuisine *is* recognized, as has increasingly been the case in the past few decades, credit for African food traditions and culinary legacies is usually relegated to the culinary and cultural shackle known as 'soul food', which specifically includes chitlins, corn bread, fatback, greens, and fried chicken. Despite the dearth of credit given to Africa's influence on gustatory traditions in the Americas, the story of the African culinary past actually began with 'Lucy' (the female hominin *Australopithecus afarensis* skeleton discovered in 1974) and East

Africa, the cradle of humankind and civilization. Over thousands of years, East African cuisine and culture slowly migrated to every other part of the continent, diversifying and establishing new concepts, while retaining basic aspects and characteristics of the old (Spivey, 2018).

Mitchell (1992) describes the slave trade that brought approximately half a million Africans to the United States. After the African slaves arrived, they were stripped of their material possessions but retained much of their distinctive cultural heritage, including the ability to prepare African-origin food well. Although the slave traders could remove material objects, they could not take away the intangible heritage of the slaves and the knowledge they carried with them. As a result of slavery and the slaves' resilience, the southern states of the United States have been heavily influenced by several African culinary traditions. For example, the inclusion of greens and vegetables in the daily diet, which some historians maintain added much-needed nutrients that 'saved white slaveholders from nutritional deficiencies' (Mitchell, 1992, n.p). The African diet mainly consists of starch (millet, yams, cassava) and meaty products. The effect of this food habit is today especially evident in Louisiana-style cookery in which chicken or seafood is served with a sauce over a bed of rice. Cajun dishes, such as gumbo and jambalaya, also demonstrate strong African influences (Mitchell, 1992).

The preservation and reinvention of culinary traditions and social patterns based on African heritage demonstrated strong cultural persistence and have left enduring legacies as is evident throughout the southern United States, Latin America, and the Caribbean (Donovan & De Bres, 2006; Goucher, 2014; Harris, 2011; Highfield, 2017; Houston, 2005; Niehoff, 2017; Twitty, 2018). Even though some of these nations could have adopted African culinary traditions and heritage as their own, in most cases, there is little recognition of the African roots of much of their modern food traditions. In general, especially in the United States, the African alimentary contribution is regularly ignored or undervalued and it is being subjected to racism and societal repression. For Africans and their descendants in the Americas, food and how it is prepared are deeply rooted and will always hold a fundamental place in the preservation and an eventual appreciation of African culinary heritage throughout the world (Highfield, 2017; Spivey, 2018). Despite its downtrodden history, in recent years, there has been a resurgence of interest among African Americans (and other Americans) in the culinary traditions of the South and the influence of African food heritage on the current foodscapes of the United States. Many cookbooks have been published in recent years to help preserve the recipes and histories of the slave South and its culinary heritage (e.g., Bailey, 2021; Lukas, Harris, & Grant, 2018; Tipton-Martin, 2019) and are indicative of a growing awareness and appreciation.

The Impact of Heritage Foods Outside of Africa

Most African foods have not been globally exposed and are not as widespread as an ethnic food choice, compared to other cuisines such as Indian, Chinese, and Italian.

Nevertheless, they still have a very significant impact and role to play in the culinary industry. However, it is not unusual that, as many people emigrate from Africa to the West, they carry with them their culinary traditions, which can help explain the many uniquely named African restaurants and eateries springing up in many Western countries. Some culinary fares, such as those of Ethiopia, Egypt, and Morocco, have seen a growing popularity in Europe and North America over the past few decades. Strong colonial linkages between Africa and some of its metropolitan colonizers have resulted in high levels of migration and ethnic eateries. For instance, in France, one can find many Algerian and Tunisian restaurants, especially in Mediterranean cities such as Marseille. In regions with a heavy African diasporic presence, many restaurants have been opened by immigrants to serve the general public, as well as to serve others of the same diaspora. Washington, DC, for example, has the largest Ethiopian and Eritrean immigrant population in the United States and one of the largest outside of Ethiopia and Eritrea. Food is one of the main unifiers of the Ethiopian and Eritrean communities and the city and its suburbs are home to dozens of Ethiopian restaurants (Chacko, 2003; Idris, 2020), which are one of the US capital's favourite ethnic foods. In London, restaurants featuring Ugandan, Ghanaian, and Kenyan cuisines are popular.

Besides restaurants, street markets and grocery stores outside of Africa often sell African food ingredients and spices. These help many immigrants carry on their traditions and food cultures wherever they live. For instance, there is a strong presence of West African shops in many countries in Europe and North America, which helps diasporic communities maintain their authentic or traditional gastronomic heritage. Preserving the traditional culinary heritage of diasporic people is not only about eating food. It is also about representing homeland and ethnic identity and creating an image, and it often has specific meanings for religious and cultural traditions (Frost & Laing, 2016; Lin, 2019; Timothy & Ron, 2016). Cuisines convey different meanings among food consumers and reveal the closeness of the relationships between individuals, their communities, their religions, and their ethnic identities. Immigrant cuisines are also valued by destinations because they improve the image of place and become part of the attraction in a destination's tourism system and contribute to local prosperity (Lin, 2019).

Economic Benefits and Challenges

There are many economic benefits of food tourism and, as alluded to earlier, many destinations in Africa have much to offer but they have not been adequately showcased. Many tourists seek authentic experiences during their travels. The quest for knowledge and education has spurred many destinations to include cooking shows, cookery classes, and food tours where visitors can practically engage with the locals in the preparation process from picking the ingredients to preparing and consuming the food. Since eating local foods can have significant impacts on the local economy, and especially with the recent COVID-19 challenges, it is vital for destinations to be creative with their culinary experiences. Few countries in Africa have dedicated food tourism as part of their overall tourism products. South Africa would be the exception, but elsewhere few visitors go into local homes to learn about the history of agricultural products, cooking ingredients, or food preparation. This part of Africa's heritage is clearly underdeveloped compared to other regions of the world. Hall and Sharples (2003) argue that local gastronomy brings value to the tourism industry in two ways: by supporting the sales of artisan food products to tourists and helping to preserve and cultivate landscapes (cited by Bonow & Rytkönen, 2012).

Food tourism also benefits farmers and rural communities (Telfer & Wall, 1996). Bonow and Rytkönen (2012) argue that gastronomy has become a vital part of preventing economic deprivation in late-modern societies. As a result, the advancement of gastronomy related to tourism may also act as a badge of quality for food-producing regions and localities (Bonow & Rytkönen, 2012). Local food producers are an important resource who need to be supported to strengthen their position in the local economy, keep tourist spending from leaking out (Bonow & Rytkönen, 2012; Hjalager, 2010), utilize local products more sustainably, and help protect the agricultural heritage of the region.

Sociocultural Benefits and Challenges

Food is a very integral part of culture and society in general. Therefore, the preservation of traditional foods can be a strong argument for preserving culture on the African continent. One significant challenge perhaps is the lack of documentation of some of the rich histories and heritage foods because of a fear of certain foods being accepted as 'authentic' recipes. Timothy (2016a) argues that in today's world, truly 'authentic' foods are nearly impossible to evaluate and appreciate, because every culture has now been influenced by others. He suggests that 'traditional' might be a better term to use rather than authentic, as traditional considers more broadly the influences of outside forces. For example, what would Italian food be today without tomatoes from the Americas?

Smith (2015) acknowledges an interesting balance between staunch traditionalists (purists) and people who are committed to preserving pure culinary heritage, and people who prefer fusion foods and argue that food is always in a constant state of evolution anyway. Smith (2015) argues that preserving traditional foods and ingredients is important for a variety of reasons—and that it can be gracefully integrated with responsible, interesting, and delicious fusion cuisine. The argument for preserving traditional cuisines, ingredients, methods of preparation, and recipes is not only about 'authenticity', which many people boil it down to. Authenticity, however, is not the issue. The issue is that food is an integral part of culture; it is necessary to sustain life. Food is life, but it is also a way humans express themselves, communicate, and relate to their own history. Food is woven into so many aspects of life, with the very act of cooking and food preparation marking a huge shift in our evolutionary history (Smith, 2015). Local cuisine represents the identity of a community or region. Sharing meals at a table or from the same plate and different customs surrounding how and on what occasions a certain meal is eaten may also lead to the integration of tourists to embrace local food cultures (Timothy & Ron, 2013). Consuming the food of other people and borrowing from their culinary practices may provide a rooted sense of the place visited and be representative of the symbolic consumption of a region or a place and facets of its climate, history, customs, and scenery (Bessiére, 2001).

Another significant benefit is that tourists can purchase food and beverage souvenirs such as wines (Algeria, Ethiopia, Morocco, South Africa), spices (Zanzibar), coffee (Kenya, Ethiopia, Ghana), and tea (Kenya, Tanzania, Zimbabwe). This may further reinforce the bond between the destination and the memories of those who visit. Another benefit is the passing of familial heritage in the form of recipes from one generation to another (Timothy, 2016c). Most traditional food recipes have remained in families for generations and are still being used on various occasions such as weddings, funerals, bride price negotiations, circumcision rites, and other celebrations. However, as technology evolves and the outmigration of younger generations continues, the interest in keeping cultural food heritage and traditions is increasingly difficult, as many young people are losing interest in their culture and embracing new traditions at their fingertips through globalization processes, social media, and other manifestations of modernity.

Environmental and Health Benefits and Challenges

Inevitably, food choices and preferences are directly linked to current and future environmental problems (Hall, 2020). For example, climate change is increasingly recognized as an issue of global concern, while food production and consumption along with dietary choices have a major impact on its dynamics. Around 10–12% of the global annual emissions of greenhouse gases and 75% of global deforestation come from agriculture. Therefore, climate change is an issue that the food system must play a part in mitigating, particularly with regard to the massive impacts of industrial-level food production. Climate change is believed to affect food security across a range of areas such as access to food, utilization, and price stability (Gomez-Zavaglia, Mejuto, & Simal-Gandara, 2020). It is increasingly important for global society to understand the concept of food security and to adopt behaviours that improve food security locally, regionally, and nationally by adopting sustainable ways of producing and consuming food. In this regard, much can be learned from Indigenous people and other traditional societies, such as many in Africa, whose time-honoured food-related practices were far more sustainable than today's industrial food systems (Dirar, 1993). People from different communities have different relationships with food, and their food choices are impacted by their socioeconomic and cultural backgrounds.

Both formal and informal education play a major role in supporting people to make informed decisions when it comes to choosing food. Indeed, studies by psychologists have recently underlined the importance of investigating the motives for, and functions of, eating in everyday life in order to understand what constitutes 'normal' eating behaviour. Human eating behaviour is regulated by multiple motives and is a complex function of biological, learned, sociocultural, and economic factors (Kapelari et al., 2020). Globally, food choices have a direct impact on the environment and overall health implications. Some staunch environmentalists advocate against the consumption of meat products from how the animals are captured, slaughtered, processed, packaged, and even consumed. Clark et al. (2019) argue that foods with lower impacts on one metric of environmental harm tend to have lower impacts on others. Additionally, of the foods associated with improved health (whole grain cereals, fruits, vegetables, legumes, nuts, olive oil, and fish), all except fish have among the lowest environmental impacts, and fish has markedly lower impacts than red meats and processed meats. Foods associated with the largest negative environmental impacts—unprocessed and processed red meat—are consistently associated with the largest increases in disease risk. Thus, dietary transitions towards greater consumption of healthier foods would generally improve environmental sustainability, although processed foods high in sugars harm health but can have relatively low environmental impacts (Clark et al., 2019).

A study conducted over a 20-year period in 187 countries concluded that diets from Sub-Saharan Africa are healthier compared to diets from other parts of the world. The main reason for this conclusion was that local diets in that region are rich in dietary fibre (fruits, vegetables, maize, fonio, millet, cocoyam, and beans) and low in added sugar, fat, and sodium (Adainoo, 2018). Based on these results, one can argue whether African foods are healthy or if they can improve the overall health of the world's population. Just as the 'Mediterranean Diet' is currently regarded as one of the healthiest in the world, an African diet could also lead to improvements in global health. Most nutritionists recommend certain dietary restrictions and sometimes the elimination of particular food categories that may have positive and negative benefits on a person's health and longevity. In the old days, many people lived very long and healthy lives because of physical fitness and rigorous activities such as tilling the ground, fetching water (sometimes from long distances), and walking to schools and markets. Contemporary food-related conveniences, such as fast food, ordering meals and groceries online, curbside pick-ups, home delivery, living in cities where there is little green space for planting vegetables, mobile phone apps, and other items have all contributed to poorer diets and sedentary living. Most diet-based diseases plaguing much of the world were far less common in earlier times when people ate traditional diets. Scientific studies show that conditions such as heart disease and diabetes skyrocketed when traditional dishes are abandoned in favour of convenient foods. The African Heritage Diet is being encouraged by many health specialists because it tends to have the following health benefits:

- Lower the risk of heart disease, high blood pressure, and stroke
- Avoid or help treat diabetes
- Fight certain cancers and many chronic diseases
- Reduce asthma, glaucoma, and kidney disease

- Nurture healthy babies
- Achieve a healthy weight and avoid obesity
- Reduce depression

(Oldways Cultural Food Traditions, 2021)

Based on traditional knowledge and food practices, there is significant potential for an African heritage diet to provide alternatives to unhealthy eating habits that have pervaded much of the industrialized world. In this way, the continent's culinary heritage could expand far beyond Africa, much the way other ethnic cuisines have done but for different reasons and through different means.

Conclusion

In the contemporary global tourism and hospitality sectors, food will continue to play an integral role in determining travel destinations and experiences. This chapter highlights the importance of food in cultures across Africa, provided examples from all areas of the region, and explained the economic, sociocultural, and environmental benefits of preserving traditional food heritages. Farming and hunting traditions, ingredients, preparation methods, and the ceremonies and festivities associated with gustatory heritage in Africa are vast and complex. They reflect a deep mix of indigeneity, colonialism, faith, globalization, and modernity, as well as Africans' struggles with nature, environmental determinism, poverty and wealth, climate change, patterns of land tenure, and other sociocultural, political, and economic forces.

Although African cuisines generally have not received widespread acclaim abroad, as many other ethnic foods have, there are certainly many unique opportunities that African countries can explore and utilize in the role of promoting food tourism globally, particularly with regard to the African diaspora(s) and its influences in Europe and the Americas. Indeed, culinary tourism can contribute to local and national economies, as well as stimulate a reimaging of the region's cuisines.

The unique characteristic of food is that people eat it every day even though some foods may be seasonal or scarce; yet, they are almost always available. Restaurant menus have evolved with descriptions of ingredients in the foods and the origins of the food. The African culinary landscape can capitalize on the opportunity to preserve the culinary heritage of the continent through restaurant offerings, culinary tourism, and a greater emphasis on the region's heritage cuisines as a tourism product, especially in light of the growing foodie and slow tourism markets, and an increased awareness of the significance of Indigenous knowledge and the value of protecting cultural heritage. It will be important to continue documenting the heritage cuisines of Africa to retain their authenticity and to protect both their tangible and intangible cultural attributes. Food is a central vestibule into the heritage traditions of Africa and a crucible of knowledge that has the potential to protect, promote, and provide a greater understanding of African traditions and identities.

References

Adainoo, B. (2018, April 11). Are African foods healthy? Online www.staywellnow.com

Bailey, P. (2021). *An African American cookbook: Exploring Black history and culture through traditional foods*. New York: Good Books.

Belaidi, S., & el Bouzidi, A. (2019). Food heritage as a mechanism to achieve economic development in the case of the High Atlas Mountains. (in Arabic) *Political Orbits, 3*(1), 46–58.

Bessière, J. (1998). Local development and heritage: traditional food and cuisine as tourist attractions in rural areas. *Sociologia Ruralis, 38*(1), 21–34.

Bessiére, J. (2001). The role of rural gastronomy in tourism. In L. Roberts, & D. Hall (eds) *Rural tourism and recreation: Principles to practice*, pp. 115–118. Wallingford: CABI.

Bessiére, J. (2013). Heritagisation, a challenge for tourism promotion and regional development: An example of food heritage. *Journal of Heritage Tourism, 84*(4), 275–291.

Bonow, M., & Rytkönen, P. (2012). Gastronomy and tourism as a regional development tool—the case of Jämtland. *Advances in Food, Hospitality and Tourism, 2*(1), 2–10.

Cappelen, S.M., & Pedersen, J.S. (2020). Inventing culinary heritage through strategic historical ambiguity. *Organization Studies, 42*(2), 223–243.

Chacko, E. (2003). Ethiopian ethos and the making of ethnic places in the Washington metropolitan area. *Journal of Cultural Geography, 20*(2), 21–42.

Che, D. (2006). Select Michigan: Local food production, food safety, culinary heritage, and branding in Michigan agritourism. *Tourism Review International, 9*(4), 349–363.

Che, D. (2016). Agricultural heritage, agritourism and rural livelihoods. In D.J. Timothy (Ed.), *Heritage cuisines: Traditions, identities and tourism*, pp. 77–87. London: Routledge.

Chrones, T.W. (n.d.). African food ingredients. Online: www.sallybernstein.com

Clark, M.A., Springmann, M., Hill, J., & Tilman, D. (2019). Multiple health and environmental impacts of foods. *Proceedings of the National Academy of Sciences of the United States of America, 116*(46), 23357–23362.

Cusack, I. (2000). African cuisines: Recipes for nationbuilding? *Journal of African Cultural Studies, 13*(2), 207–225.

Dirar, H. A. (1993). *The indigenous fermented foods of the Sudan: A study in African food and nutrition*. Wallingford: CABI.

Donovan, A., & De Bres, K. (2006). Foods of freedom: Juneteenth as a culinary tourist attraction. *Tourism Review International, 9*(4), 379–389.

Duru, M. (2020). An overview of African food historiography, 2013 to present. *Food and History, 18*(1–2), 211–221.

Everett, S. (2016a). *Food and drink tourism: Principles and practices*. London: Sage.

Everett, S. (2016b). Iconic cuisines, marketing and place promotion. In D.J. Timothy (Ed.), *Heritage cuisines: Traditions, identities and tourism*, pp. 119–131. London: Routledge.

Frost, W., & Laing, J. (2016). Cuisine, migration, colonialism and diasporic identities. In D.J. Timothy (Ed.), *Heritage cuisines: Traditions, identities and tourism*, pp. 37–52. London: Routledge.

Gheorghe, G., & Bulin, D. (2014). Cuisine: A regional tourism differentiation tool. *Knowledge Horizons—Economics, 6*(2), 194–199.

Gomez-Zavaglia, A., Mejuto, J. C., & Simal-Gandara, J. (2020). Mitigation of emerging implications of climate change on food production systems. *Food Research International, 134*, 109256.

Goucher, C. (2014). *Congotay! Congotay! A global history of Caribbean food*. London: Routledge.

Great British Chefs (2021). Everything you need to know about East African cuisine. Online: https://www.greatbritishchefs.com/features/east-african-food-guide

Hafner, D. (2002). *A taste of Africa: Traditional and modern African cooking*. Cambridge, MA: Wakefield Press.

Hall, C.M. (2020). Improving the recipe for culinary and food tourism? The need for a new menu. *Tourism Recreation Research*, *45*(2), 284–287.

Hall, C.M., & Sharples, L. (2003). The consumption of experiences or the experience of consumption? An introduction to the tourism of taste. In C.M. Hall, L. Sharples, R. Mitchell, N. Macionis, & B. Cambourne (Eds.), *Food tourism around the world: Development, management, and markets*, pp. 1–24. Oxford: Butterworth Heinemann.

Harris, J.B. (2011). *High on the hog: A culinary journey from Africa to America*. New York: Bloomsbury.

Highfield, J.B. (2017). *Food and foodways in African narratives: Community, culture, and heritage*. London: Routledge.

Hjalager, A-M. (2010). A review of innovation research in tourism. *Tourism Management*, *31*, 1–12.

Houston, L.M. (2005). *Food culture in the Caribbean*. Westport, CT: Greenwood Press.

Idris, M. (2020). Solidarity and social networks: The invisible backbone that Ethiopians and Eritreans in Washington DC used to transform Adams Morgan and U Street. *Journal of Immigrant & Refugee Studies*, *18*(4), 341–356.

Kapelari, S., Alexopoulos, G., Moussouri, T., Sagmeister, K.J., & Stampfer, F. (2020). Food heritage makes a difference: The importance of cultural knowledge for improving education for sustainable food choices. *Sustainability*, *12*(4), 1509.

Lin, Y.H. (2019). The transmission of traditional cuisines: A review. *Journal of Tourism and Hospitality*, *8*(2), 1–7.

Long, L.M. (2018). Cultural politics in culinary tourism with ethnic foods. *Journal of Business Management 58*(3), 316–324.

Lukas, A.G., Harris, J.B., & Grant, J. (2018). *Sweet Home Café cookbook: A celebration of African American cooking*. Washington, DC: Smithsonian Books.

Mak, A.H.N., Lumbers, M., & Eves, A. (2011). Globalisation and food consumption in tourism. *Annals of Tourism Research*, *39*(1), 171–196.

Mallos, T. (2006). *North African cooking*. Hong Kong: Periplus.

McCann, J.C. (2009). *Stirring the pot: A history of African cuisine*. Athens: Ohio University Press.

Mitchell, P.B. (1992). The African influence on Southern cuisine https://www.foodhistory.com/foodnotes/leftovers/african/infl/01/ accessed July 31, 2021

Mohamed, M.E., Hewedi, M., Lehto, X., & Maayouf, M. (2020). Marketing local food and cuisine culture online: A case study of DMO websites in Egypt. *International Journal of Tourism Cities*, *6*(4), 1045–1068.

Niehoff, A.H. (2017). *Caribbean transformations*. London: Routledge.

Okech, R.N. (2014). Developing culinary tourism: The role of food as a cultural heritage in Kenya. *Proceedings of the Second International Conference on Global Business, Economics, Finance and Social Science*, Chennai, India, 11–13 July, 2014.

Oktay, S. & Sadıkoğlu, S. (2018) The gastronomic cultures' impact on the African cuisine. *Journal of Ethnic Foods*, *5*, 140–146.

Oldways Cultural Food Traditions (2021). African heritage diet. Online: www.oldwayspt.org

Palmer, C. (1998). From theory to practice: Experiencing the nation in everyday life. *Journal of Material Culture*, *3*(2), 175–199.

Qaddahat, R., Attaalla, F., & Hussein, M. M. (2016). Halal tourism: Evaluating opportunities and challenges in the Middle East: Jordan and Egypt. *Journal of Faculty of Tourism and Hotels, Fayoum University*, *10*(2), 377–390.

Ramshaw, G. (2016). Food heritage and nationalism. In D.J. Timothy (Ed.), *Heritage cuisines: Traditions, identities and tourism*, pp. 53–64. London: Routledge.

Reddy, G., & van Dam, R.M. (2020). Food, culture and identity in multicultural societies: Insights from Singapore. *Appetite, 149*, 104633.

Richards, G. (2012). Food and the tourism experience: Major findings and policy orientations. In D. Dodd (Ed.), *Food and the tourism experience*, pp. 13–46. Paris: OECD.

Ron, A.S., & Timothy, D.J. (2013). The Land of Milk and Honey: Biblical foods, heritage and Holy Land tourism. *Journal of Heritage Tourism, 8*(2/3), 234–247.

SA-Venues (2021). South African food and cuisine. Online: https://www.sa-venues.com/cuisine.htm

Smith, I.F. (1995). *The case for indigenous West African food culture*. Dakar: UNESCO Regional Office.

Smith, S.E. (2015, August 28). Preserving food heritage, balancing new things. Online: https://meloukhia.net/

SpicesInc (2021). The culinary regions of African cuisine. Online: https://www.spicesinc.com/p-3808-the-culinary-regions-of-african-cuisine.aspx

Spivey, D. (2018). Transatlantic food migration: The African culinary influence on the cuisine of the Americas. Online: https://www.blackpast.org/global-african-history/perspectives

Sun, Y., Timothy, D.J., Wang, Y., Min, Q., & Su, Y. (2019). Reflections on agricultural heritage systems and tourism in China. *Journal of China Tourism Research, 15*(3), 359–378.

Telfer, D.J., & Wall, G. (1996). Linkages between tourism and food production. *Annals of Tourism Research, 23*, 635–653.

Thomé-Ortiz, H. (2018). Heritage cuisine and identity: Free time and its relation to the social reproduction of local food. *Journal of Heritage Tourism, 13*(2), 104–114.

Timothy, D.J. (2016a). Conclusion: Heritage cuisines and the patrimony of food. In D.J. Timothy (Ed.), *Heritage cuisines: Traditions, identities and tourism*, pp. 180–186. London: Routledge.

Timothy, D.J. (Ed.) (2016b). *Heritage cuisines: Traditions, identities and tourism*. London: Routledge.

Timothy, D.J. (2016c). Personal heritage, intergenerational food and nostalgia. In D.J. Timothy (Ed.), *Heritage cuisines: Traditions, identities and tourism*, pp. 65–76. London: Routledge.

Timothy, D.J., & Ron, A.S. (2013). Understanding heritage cuisines and tourism: Identity, image, authenticity and change. *Journal of Heritage Tourism, 8*(2/3), 99–104.

Timothy, D.J., & Ron, A.S. (2016). Religious heritage, spiritual aliment and food for the soul. In D.J. Timothy (Ed.), *Heritage cuisines: Traditions, identities and tourism*, pp. 104–118. London: Routledge.

Tipton-Martin, T. (2019). *Jubilee: Recipes from two centuries of African American cooking*. New York: Clarkson Potter.

Twitty, M.W. (2018). *The cooking gene: A journey through African American culinary history in the Old South*. New York: Harper Collins.

World Tourism Organization (2019). *Guidelines for the development of gastronomy tourism*. Madrid: UNWTO.

World Travel and Tourism Council (2017). Travel and tourism: Global economic impact. Online: https://www.wttc.org/-/media/files/reports/economic-impact-research/2017-documents/global-economic-impact-and-issues-2017.pdf

Wright, C. (2021). History of couscous. Online: www.cliffordwright.com

10

THE TRANSATLANTIC SLAVE TRADE

Africa's Epitomous Heritage Tourism

Aaron Yankholmes

Introduction

Drawing on Mazrui (1986), Teye (2009) argues that Africa's tripartite heritage (consisting of traditional or indigenous African heritage, Islamic heritage, and European colonization) could offer some economic and cultural benefits to host communities if properly developed, managed, presented, and interpreted to satisfy the personal, emotional, spiritual and educational needs of visitors. Consequently, slave heritage-based tourism initiatives helped put Ghana, Nigeria, Senegal, and The Gambia on the map of international tourism destinations, with Ghana in particular being an example of a successful slavery heritage destination (Engmann, 2021; Mensah, 2015).

The reasons for this success are easy to enumerate (Timothy & Teye, 2004). Ghana enjoys a considerable reputation as the country having the most documented slave trade landmarks. There are still numerous fortifications on Ghana's coast that illustrate the country's pivotal role in Europe's engagement with Africa. Moreover, a disproportionate number of African Americans trace their lineage to Ghana, and many of these people of the African diaspora desire to visit their 'ancestral homeland'.

This chapter examines the transatlantic slave trade between the sixteenth and mid-nineteenth centuries and the heritage it has left behind, particularly in West Africa but in other parts of the continent. It contextualizes slave heritage beyond Africa to highlight its manifestations in the Americas and Europe as an extension of the heritage of the transatlantic slave trade, which is commemorated in various ways in the United States, in many Caribbean islands, and in South America through tourism (Hernandez-Ramdwar, 2013). This global perspective on the slave trade is then utilized to set the tone for this broader manifestation of slave heritage-based tourism in Ghana, Senegal, The Gambia, and Nigeria.

DOI: 10.4324/9781003153955-10

The History and Legacies of Transatlantic Slavery

The Indian Ocean slave trade pre-dated the transatlantic slave trade substantially and continued much longer. Yet, until now, it has been less studied by scholars and less commercialized for tourism. In fact, the true scale of the transatlantic slave trade is far more known than its Indian Ocean counterpart (Hooper & Eltis, 2013). By using the transatlantic slave trade database as a model, scholars have uncovered areas that expand our knowledge of the connections between the two slave trades (Alpers, 2003; Campbell, 2003; Machado, 2004).

A large part of what has been written about the legacies of Africa's slavery has been portrayals of depopulation and its economic aspects: gains, losses, competition, volume, prices, supply, profits, and spatial distribution (Donkoh, 2007). Lovejoy (2012) points out that the major demographic and socioeconomic transformations that took place in Africa from the sixteenth to the nineteenth centuries were attributable to the scale of slave traffic to the New World, although the Muslim slave trade and internal developments also contributed to these changes. The sale of enslaved Africans became popular in the seventh century when Islam was gaining a foothold in North Africa. This happened seven centuries before Europeans explored the continent and ten centuries before enslaved West Africans were sold across the Atlantic to the Americas.

Curtin's (1969) pioneering census of the trade identified eight principal areas that were used by Europeans to buy and ship enslaved Africans to the New World and the western hemisphere. Although the distribution of slaves from each region varied, between 1650 and 1900, over ten million captives arrived in the Americas from several regions of Africa. Curtin's (1969) estimates are far less than those provided by the Slave Voyages (2021) of 12.5 million involuntary migrants from 1501 to 1867. More importantly, the Slave Voyages database suggests that the volume of slaves transported between the embarkation point and the transit areas after slavery's official abolition far exceeded those who were shipped across the Atlantic. The West African coast supplied over four million slaves in the seventeenth and eighteenth centuries in proportions that are shown in Table 10.1.

Table 10.1 suggests that the major sources of slaves from West Africa were the Bight of Benin (from the Rio Volta inclusive to just west of the Rio Nun), the Gold Coast (covering the area east of the Assini River up to and including the Volta River), and Bight of Biafra (extending from the Niger Delta in modern-day Nigeria to Cape Lopez in modern-day Gabon), in that order. The historiographical analysis by Slave Voyages has contributed to a considerable and sophisticated corpus of data for future research on slavery heritage, not only in West Africa, but also in Europe and the Americas. There is considerable work done in the English language literature on slavery heritage in Nigeria (e.g., Acey, 2019; Brown, 2003; Njoku, 2020), Ghana (e.g., Yankholmes et al., 2009; Yankholmes & McKercher, 2015a, 2015b), and Senegal and The Gambia (Bellagamba, 2009; Ebron, 2000; Gijanto, 2011; Parry, 2018).

TABLE 10.1 Slave exports from West Africa from 1650 to 1800

Region	*Total*	*Percent (%)*
Senegambia	445,000	9.5
Upper Guinea	207,000	4.4
Windward Coast	291,000	6.2
Gold Coast	1,121,000	24.0
Bight of Benin	1,545,000	33.1
Bight of Biafra	1,054,000	22.6
Total	4,663,000	100.0

Source: Lovejoy (2012).

As previously mentioned, slavery heritage-based tourism is heavily promoted in this handful of countries, although Ghana's efforts far exceed those of its neighbors (Timothy, 2021; Yankholmes & Akyeampong, 2010; Yankholmes & Timothy, 2017). A short description of each country's heritage assets is offered below. Despite the fact that the Bight of Benin (present-day western Nigeria) and the Bight of Biafra (present-day eastern Nigeria) were major slave trading emporiums (Lovejoy & Richardson, 1999), the main sites commemorating Nigeria's involvement in the trade are three museums in the coastal town of Badagry (Abaka, & Kumasenu, 2021). In contrast, even though Ghana (then Gold Coast) supplied 24% of captives exported in the eighteenth century, the country has turned its dark and contentious past into a major tourism asset. Many of the fortifications on Ghana's 500-km littoral (three castles, fifteen forts, four forts in partial ruins, four with visible structures, and two sites with evidence of former fortifications) are inscribed on UNESCO's World Heritage List (WHL) and have become flagship attractions for the tourism industry.

Senegal is the westernmost point of continental Africa. It became a battleground for several European metropoles including the Portuguese, British, Dutch, and French trying to control the Senegambia slave trade. France gained possession of the area in 1840 and Senegal became a French colony. Senegal has made significant efforts at harnessing the potential of its slavery past for tourism. The Island of Gorée, home to the eighteenth-century building said to have housed thousands of slaves was inscribed on the WHL in 1978, although some scholars dispute its role in the transatlantic slave trade (e.g., Austen, 2001; Eltis & Richardson, 2010; Thiaw & Wait, 2018). Yet, since the 1970s, Gorée has become a popular destination for African Americans seeking to retrace their roots (Tillet, 2009).

The Gambia is surrounded on the north, east, and south by Senegal and consists of strips of land stretching 350 km from west to east and 28 km wide along the banks of the Gambia River. Recognizing the importance of the Gambia River to the slave trade for its ability to penetrate deep into the interior of the continent, both Britain and France sought control of this area. Britain constructed a fort at James Island, while in 1681, the French established a slave station in Albreda two miles from the British quarters. Although slavery was legally abolished in 1894, the trade continued domestically until the 1930s because the British could not afford to antagonize local slaveholders, mostly chiefs. Although James Island and related sites on the Gambia River are visible remainders of the infamous slave trade, it was Alex Haley's famous book and docudrama series, *Roots*, that made The Gambia a major slave heritage destination in West Africa (Bellagamba, 2009; Gijanto, 2011).

Contemporary Expectations of the Transatlantic Slave Trade

The growing body of research on the transatlantic slave trade is not only related to the phenomenon's landmarks and legacies but also to expectations. Slavery heritage is both tangible and intangible (UNESCO, 2004). While gathering historical data and archaeological evidence is critical, what should be done to sites, buildings, and places linked to the 400 years of collective slavery memory is seen as being of utmost importance. The UNESCO Slave Route project was designed in 1994 with a particular emphasis on dealing with the repercussions of the transatlantic slave trade. Given the multiple interactions the international slave trade generated, UNESCO christened the project 'Slave Route' to address not only the universal phenomenon of slavery, but also, in an unequivocal way, how slavery shaped the continents and regions of Africa, the Arab World, Asia, the Indian Ocean, Caribbean, and the Americas and the dynamics of the movement of peoples, civilizations, and cultures involved (Teye, 2009; Timothy & Boyd, 2015; UNESCO, 2004). The objectives of the Slave Route project are twofold: to break the silence and make universally known the subject of transatlantic slavery, its causes, and modalities through scientific work, and to examine the consequences of the trade in a way that highlights the interactions between the peoples and civilizations concerned. There are three major areas of activity involving the Slave Route project, including multidisciplinary scientific research, education, and cultural tourism development. While these activities go hand in hand, the cultural tourism element is very visible and a major development focus of Ghana and some of its neighbors.

Despite its altruistic origins, the critical question that should be, but rarely is, asked in discussions of the Slave Route project is whether it truly captures the multitudes of contestations of slavery heritage, especially in West Africa. Either cultural tourism is the most viable option for optimizing the use and conservation of slavery heritage, in which case planning is necessary, or it is not, in which case the conceptualization and approach adopted by UNESCO must be interrogated further. This exercise then influences subsequent questions that are germane to government policy-making and the marketing of slavery heritage.

For obvious reasons, the promotion of slave heritage-based tourism in West Africa emphasizes the hope of socioeconomic development in former slave source communities. There is no gainsaying the fact that heritage tourism can be a source of revenue and is capable of creating jobs, providing new business opportunities, strengthening local economies, increasing property values, and growing opportunities for diversified local economies (McKercher & du Cros, 2002; Timothy, 2021). However, McKercher and du Cros (2002) caution against the arbitrary tourism use of cultural heritage assets with intrinsic cultural significance to local communities. If their caution is true, it would seem particularly sensitive in places associated with the slave trade.

Such places invoke many fundamental questions for local communities. Some of the most persistent questions are why indigenous and non-indigenous residents of former slave communities are interested in developing their slave assets for tourism, what makes a successful slave heritage-based tourism attraction, and what are the management, interpretation, and ethical issues involved. The rest of this section presents a brief review of research relevant to these questions, although it is not intended to be an exhaustive discussion of the extensive Anglophone literature on slave heritage-based tourism in Ghana. A more complete review of the literature may be found in Yankholmes and McKercher (2015a).

Though dated, Murphy's (1985) work lends credence to the claim that involving the local community in decision-making is crucial to sustainable development, which in turn justifies tourism as an economic development policy option. Interestingly, however, it does seem that when community residents are invited to participate in the tourism planning and decision-making process, they are also simultaneously being expected to be part of the product (Taylor, 1995). This is a contradiction to the traditional notion of community economic development where community-based planning efforts tend to be top-down and controlled by the government (Joppe, 1996). In the specific case of slavery heritage, empirical research on the expectations of former slave source communities in Ghana shows mixed results regarding slavery-oriented tourism development. Teye, Sönmez, and Sirakaya (2002) examine residents' perceived social, economic, and welfare impacts of tourism development in Cape Coast and Elmina, two coastal fort communities which are relatively well-established tourist destinations. Their study found a simmering discontent among residents against the backdrop of huge expectations from slave heritage-based tourism. Residents felt short-changed by the government regarding tourist revenue. Building on this study, Sirakaya, Teye, and Sönmez (2002) found that support for slavery-based heritage tourism was influenced by social-psychological factors, including perceptions of tourists, tourism's impacts, employment status, membership in community organizations, and awareness of tourism development projects in the community. In contrast, Boakye (2003) found that residents in Assin Manso, a major eighteenth-century slave market and famous for the Slave River, which is regarded as sacred, had more favorable attitudes about tourism. Their high level of support for tourism was based on perceived economic benefits, especially given the socioeconomic deprivation of the community. Consequently, Boakye (2003) suggested that any development of a community-based tourism project should aim not only to benefit the community but also to ensure the sacredness of the Slave River site. Yankholmes, Akyeampong, and Dei (2009) extended Boakye's (2003) study by analyzing slave heritage-based tourism in Danish-Osu, a former slave market in the seventeenth century controlled by the Danes for 200 years. The community hosts many international tourists but has other industries to support its economy. Results suggested residents were aware of the often-cited economic benefits of tourism and therefore were eager to see the promotion of tourism at slavery-related sites in the community.

Taken together, studies on community attitudes toward slave heritage-based tourism in Ghana suggest tourism is seen as a panacea for preserving the memory of the slave trade whose promotion and development should bring about concrete benefits for descendant slave communities. However, there are simmering contestations in the use of slave relics for tourism promotion both within and between communities and other geographical spaces where it occurs. Within former slave communities, apprehension still exists over the need for a tourism product based on their embarrassing past, especially given the presence of descendants of slaves, who never made it across the Atlantic, and who, because of a longstanding social stigma and discrimination, think the history of slavery should be forgotten (Yankholmes & McKercher, 2015b). Even more pronounced are the contestations between former slave communities over their historic role in slavery as either slave markets, slave camps, or European fort communities. While slave camps provided general welfare, such as food, shelter, and care for captives, purchases and raids took place at the inland and coastal markets (Perbi, 2004). As the name suggests, the fort communities were the fortified trading posts and residential enclaves of European settlers and trading firms (Hernæs, 1996). It should at this point be clearly stated that there were slave markets within the precincts or vicinity of the fort communities. Today, the distinction is quite useful because in the tourism marketplace it determines the type of slavery heritage offerings, the intricacies of collective slave memories and emotions attached to them, and ultimately how tourists react to the slavery attractions there. Schramm (2008) suggests that the perception of slavery is polarized between northern and southern Ghana and even among descendants of former slaves, descendants of enslavers, and descendants of preeminent slave profiteers. Even though the Ghanaian experience cannot be extrapolated to other former slave source communities in West Africa, the empirical research lends general support to the idea that at the community level, there is a deeper understanding of heritage that is not seen at the national or international level (Timothy, 1997). Likewise, it is not yet clear whether slavery-related heritage sites can be multi-sold and multi-consumed given the presence of different social groups and communities of interest.

The answer to the question of what makes a successful slave heritage attraction is complicated because many slave sites in West Africa and specifically in Ghana are derelict, remote, inaccessible, and, perhaps more importantly, poorly regarded by local and national authorities. There are many potentially viable sites associated with slavery, but they attract few people because of their isolation or lack of product development (Okoh, 2018). So far, it appears that the most successful slave heritage-based attractions are those designated by UNESCO to be of 'outstanding universal value'. As previously mentioned, the forts and related sites in Ghana, Senegal, and The Gambia have such recognition. However, while these places are deemed worthy of protection and preservation, and have brought in copious tourist dollars for their maintenance, they have also raised multiple, contested, and differentially embedded notions of slave heritage. For one, the World Heritage status of some slave sites suggests that some relics and places are imbued with special meaning and significance, while others are not. The UNESCO brand confers a certain level of legitimacy on aspects of history and the memory of slavery. Unfortunately, the UNESCO designation fails to recognize local and to some extent national heritages, as well as contestations in the use, interpretation, and commemoration of the slave trade.

Compared to many other cultural routes that have been developed into heritage products because of their contemporary intrinsic value (Timothy & Boyd, 2015), the Slave Route project as conceived by UNSECO is multinational in scope, has a larger geographical reach, and unquestionably exerts a degree of emotional influence on Africans and people of the African diaspora (Teye, 2009). Borrowing empirical historiography could not be the only explanation if one accepts that slave heritage sites are embodied symbols and icons of humankind's inhumanity. More questions about World Heritage designation are further complicated in the tourism realm, where residents of former slave source communities that are not UNESCO-inscribed think that the WHS brand automatically guarantees tourist arrivals and subsequent tourism development. Bruner (1996), Austin (2002), Kreamer (2004), and Teye and Timothy (2004) reached conclusions similar to those of Ashworth and van der Aa (2002, p. 447) that the 'idea of world heritage, and its manifestation in international tourism, may conflict with heritage used for local and national purposes'.

There, the matter would have to rest but for the fact that UNESCO's recognition of some slave attractions being worthy of World Heritage branding has exacerbated management problems in the realm of slave heritage-based tourism. A classic example relates to the haphazard management of the forts and castles in Ghana. While the Ghana Museums and Monuments Board is the statutory custodian of these assets, the Ghana Tourism Authority is responsible for marketing and ensuring quality control in heritage interpretation and representation of the slave trade theme. However, communication between these state agencies in finding ways to manage the sites in a truly sustainable manner has been elusive, triggering deep dissatisfaction among residents of both Cape Coast and Elmina regarding tourism's impacts (Teye, Sönmez, & Sirakaya 2002).

Indeed, much of Africa's cultural heritage assets are under increasing pressure and stress from varied natural, social, political, and environmental forces, namely erosion, vandalism, lax and inefficient legislative laws, insufficient trained heritage professionals, land management challenges, a lack of funds, and a laissez-faire attitude in host communities with regard to the value of their cultural heritage (Mabulla, 2000). It is not surprising, therefore, that there is no clearly defined cultural tourism market on the continent. This point is best appreciated in the larger context of tourism development in general, and the nature of cultural resources. From among many possible factors, Dieke (2003) argues that tourism development in Africa is influenced largely and directly by the nature of economic development. In this sense, the scale of development ranges from the dominant (i.e., most developed) to the least developed or late starters. Hence, Kenya, Mauritius, Seychelles, Morocco,

Tunisia, South Africa, Zimbabwe, Côte d'Ivoire, and Senegal are well-established 'successful' tourist destinations, whereas Nigeria, Cameroon, Eritrea, Sierra Leone, Angola, and Zambia, which, for a number of reasons, have limited tourism development but considerable potential. If Dieke's conclusions fail to convince skeptics, then reference can be made to Richards (2002a, p. 36), who suggests that even though much of Africa's cultural resources constitute 'a living heritage', many African countries 'are following the European model of cultural attraction development'. Beyond the differences in cultural resources offered in developed and developing countries in Africa, there are several reasons for doubting there is a clearly defined cultural tourism market, although such data are difficult to obtain.

Commodification of Slavery Heritage for Tourist Consumption

The extant literature establishes unambiguously that the commodification of culture for tourist consumption leads to a loss of authenticity; this is especially true in countries of the Global South. Cohen (1988, p. 380) defines commodification as

> a process by which things (and activities) come to be evaluated primarily in terms of their exchange value, in a context of trade, thereby becoming goods (and services); developed exchange systems in which the exchange value of things (and activities) is stated in terms of prices form a market.

This definition presupposes that commodifying cultural products is an optimal way of selling heritage places. McKercher and du Cros (2002) poignantly ask how much commodification can occur before an asset ceases to be authentic. They provide a distinction between cultural heritage assets and cultural tourism products:

> A cultural or heritage asset represents the uncommodified or raw asset that is identified for its intrinsic values. A cultural tourism product, on the other hand, represents an asset that has been transformed or commodified specifically for tourism consumption.
>
> *(McKercher & du Cros, 2002, p. 8)*

The robustness of the distinction suggests that slave heritage-based attractions are much less straightforward than is usually suggested in the contexts of authenticity and commodification, even though many of the relics of the slave trade lie in ruins and in a deplorable state of repair. A great deal of research has examined tourists' experiences at slave sites in Ghana. These studies tend to focus on the African diaspora, particularly African Americans. Austin (2002) alludes to the contention between African Americans and other groups of tourists over what constitutes an authentic representation and interpretation of slave history and what evokes slave memory at Cape Coast Castle. African American visitors raise concerns about the increasingly sanitized version of the slave trade as presented at the site. Their reaction is in part due to the desire to trace their ancestry and to experience a presumed 'homeland' even in the absence of any genealogical connection (Hosley, 2004), and in part because slave sites remind them of the trauma and humiliation their ancestors endured (Kemp, 2000; Mensah, 2015; Richards, 2002b; Timothy & Teye, 2004).

Not surprisingly, other types of tourists seek a qualitatively different experience and value features of slave heritage sites differently. For example, White American visitors to slave sites in Ghana tend to express feelings of shame, regret, sadness, and disgust for the evils perpetrated by their forebears (Amuquandoh & Brown, 2008; Teye & Timothy, 2004). The literature largely acknowledges that the slavery heritage market is heterogeneous. Based on visitors' connection to slavery and trip purpose, discrete market segments can be found to be racially, geographically, and experientially similar to one other and dissimilar from other segments (Yankholmes & McKercher, 2015b). Yet, there have been calls for the deliberate convincing of certain markets not to visit, visit selectively, or visit and 'consume' slave sites within a particular context in order to prevent conflict and ensure the respectfulness or sacredness of the site.

While demarketing has a place in upholding sustainability principles not only by protecting sites from overuse but also by discouraging distorted or otherwise inappropriate uses of the slave heritage attractions, it may not work for several reasons. In Ghana, such efforts may not be economically viable or sustainable in managing slave sites and former slave communities. Currently, most attractions, especially the forts and castles, have flexible pricing that permits discrimination by nationality and volume. Ghanaians pay a small fraction of the fees collected from foreign visitors, -whereas foreign visitors are made to pay per head in all cases, organized Ghanaians are entitled to a group discount. Additionally, because the majority of international visitors are Caucasians who come to see the slave forts and want highly commodified experiences, discouraging this market will have dire consequences. Even so, there appears to be stiff competition between Ghana, The Gambia, and Senegal to tap the diaspora heritage market (Araujo, 2011). Spurred by the recent killing of George Floyd by Minneapolis police, diaspora tourism to West Africa in the form of spiritual quests, roots discovery, personal pilgrimages, or education has taken on a new meaning (Dillette, 2021).

Demarketing slavery heritage tourism would be a tough balancing act because attracting the minority market segment will not only inevitably displace the majority elsewhere, but it will also affect the overall quality of the experience. Even though African Americans are emotionally and personally connected to slave sites, contrary to their Caucasian counterparts, the latter feel their experiences are more 'authentic' because of the presence of the former. There is, therefore, an increasing need for destination managers in the region to match products and services to customer needs to create a unique selling point.

Discussion and Conclusion

This chapter examined several issues concerning slave heritage-based tourism in West Africa. Along with the growth of tourism has been an awakening of interest in the region's heritage borne out of the long history of contact with Europeans through the transatlantic slave trade. The slave trade as a critical factor in the world economy between 1501 and 1867 saw millions of African captives transported to Europe and the Americas. Present-day Ghana, Nigeria, Senegal, and The Gambia were key sources and embarkation points for the slave-export trade. While societal blame continues to ignite international controversy as to who should ultimately bear the greater guilt and shame for perpetuating this heinous phenomenon, contemporary use, conservation, and interpretation of slave monuments and sites is also fraught with dissonance, given the varied interest groups and multiple heritages connected to the slave trade. Therefore, there is a need to unravel the subtleties both within and between former slave trading outlets in West Africa or geographical spaces where dissonance occurs. Regrettably, much of the Anglophone research to date on slave heritage-based tourism has concentrated on Ghana with few studies focused on other countries in the region.

Even in the context of Ghana, several lingering questions remain about how various interest groups in former slave source communities recollect and forget the history and memory of slavery. While the silence and shame that hitherto engulfed public discourse on slavery have fizzled away with the launch of the UNESCO Slave Route project, slavery still acts to legitimatize social hierarchies and political structures in Ghana given the unequal power relations between and among different stakeholder groups.

A similarly complex picture emerges from visitors' perspective. Although slavery heritage sites cater to the needs of different types of visitors, the symbolic meanings attached to the sites evoke qualitatively different experiences and engagement for different visitor groups. As a result, people who feel a personal connection to slavery and seek to trace their lineage envision their trips to be a personal pilgrimage, whereas those visiting purely for pleasure and sightseeing think they are engaged in the mundane act of touring. Each group reconciles its trip purpose and proximity to slavery in ways that mediate their experiences at slave commemorative locales.

Accompanied by the scenarios enumerated above are the growing spatial and political contestations posed by UNESCO's designation of some slave sites as World Heritage. The issues this designation raises are of obvious consequence for former slave sites, particularly those in West Africa. This branding can be of decisive importance whether or not what is selected and remembered at the global scale reflects the narratives and heritage interpretations at the local level. No doubt, dissonance emerges if the messages vary by heritage designation status. Equally important are the consequences of how individuals and different social groups project their history and memory of slavery against each other. The clamoring by other sites for WHL designation brings into sharp focus not only the way visitors and local communities remember and commemorate slavery but also the role of government in resolving the contested spaces and dissonant heritages at those places. There is, in fact, evidence to suggest that the governments of Ghana, Senegal, The Gambia, and Nigeria have not delivered on the promise of socioeconomic development through slave heritage-based tourism promotion. Thus, in none of these countries can the heritage of slavery guarantee continuing success without considering the different interest groups and collective interests that lay claim legitimately or not to the history and memory of the transatlantic slave trade and multiple heritages created and shared by them.

References

Abaka, E., & Kumasenu, G.X. (2021). Slavery, remembrance, and sites of historical memory: The case of Badagry. *African Economic History*, *49*(1), 104–126.

Acey, C. (2019). Heritage preservation as survival: Mediating social and ecological risk and resilience at the Slave Port of Badagry, Nigeria. In V. Bharne & T. Sandmeier (Eds.), *Routledge companion to global heritage conservation*, pp. 345–363. London: Routledge.

Alpers, E. (2003). The African diaspora in the Indian Ocean: A comparative perspective. In S. Jayasuriya, & R. Pankhurst (Eds.), *The African diaspora in the Indian Ocean*, pp. 20–26. Trenton, NJ: Africa World Press.

Amuquandoh, F.E., & Brown, D.O. (2008). A content analysis of sentiments expressed by visitors to cultural heritage sites at the Elmina and Cape Coast former slave castles in Ghana. *The Consortium Journal of Hospitality and Tourism*, *12*, 77–90.

Araujo, A.L. (2011). Welcome home the diaspora: Slave trade heritage tourism and public memory of slavery. *Ethnologies*, *32*(2), 145–178.

Ashworth, G.J., & van der Aa, B.J.M. (2002). Bamyan: Whose heritage was it and what should we do about it? *Current Issues in Tourism*, *5*(5), 447–457.

Austen, R.A. (2001). The slave trade as history and memory: Confrontations of slaving voyage documents and communal traditions. *The William and Mary Quarterly*, *58*(1), 229–244.

Austin, N.K. (2002). Managing heritage attractions: Marketing challenges at sensitive historical sites. *International Journal of Tourism Research*, *4*(6), 447–457.

Bellagamba, A. (2009). Back to the land of roots: African American tourism and the cultural heritage of the River Gambia. *Cahiers d'Etudes Africaines*, *49*(1–2), 453–476.

Boakye, K.A.A. (2003). Breaking the silence: Community perceptions and attitudes towards the Atlantic slave trade: A case study of Assin Manso. Unpublished master's thesis. University of Cape Coast, Cape Coast, Ghana.

Brown, C.A. (2003). Identity and the contested history of slavery in southeastern Nigeria: An oral history project. In S.A. Diouf (Ed.), *Fighting the slave trade: West African strategies*, pp. 219–226. Athens, OH: Ohio University Press.

Bruner, E.M. (1996). Tourism in Ghana: The representation of slavery and the return of the Black diaspora. *American Anthropologist*, *98*(2), 290–304.

Campbell, G. (2003). Introduction: Slavery and other forms of unfree labour in the Indian Ocean world. *Slavery & Abolition 24*(2), ix–xxxii.

Cohen, E. (1988). Authenticity and commoditization in tourism. *Annals of Tourism Research*, *15*, 371–386.

Curtin, P.D. (1969). *The Atlantic slave trade: A census*. Madison: University of Wisconsin Press.

Dieke, P.U.C. (2003). Tourism in Africa's economic development: Policy implications. *Management Decision*, *41*(3), 287–295.

Dillette, A. (2021). Roots tourism: A second wave of double consciousness for African Americans. *Journal of Sustainable Tourism*, *29*(2/3), 412–427.

Donkoh, W.J. (2007). Legacies of the Transatlantic Slave Trade in Ghana: Definitions, understandings and perceptions. In J.K. Anquandah, N.J. Opoku-Agyemang, & M.R. Doortmont (Eds.), *Transatlantic slave trade: Landmarks, legacies, expectations*, pp. 305–325. Accra: Sub-Saharan Publications, Proceedings of the International Conference on the Historic Slave Route.

Ebron, P. (2000). Tourists as pilgrims: Commercial fashioning of transatlantic politics. *American Ethnologist*, *26*(4), 910–932.

Eltis, D., & Richardson, D. (eds) (2010). *Atlas of the Transatlantic slave trade*. New Haven: Yale University Press.

Engmann, R.A.A. (2021). Coups, castles, and cultural heritage: Conversations with Flight Lieutenant Jerry John Rawlings, former president of Ghana. *Journal of Heritage Tourism*, *16*(6), 722–737.

Gijanto, L.A. (2011). Competing narratives: Tensions between diaspora tourism and the Atlantic past in The Gambia. *Journal of Heritage Tourism*, *6*(3), 227–243.

Hernæs, P. (1996). European fort community on the coast in the era of the slave trade. In J. Everaet & J. Parmentier (Eds.), *Shipping, factories and colonialism*, pp. 167–180. Brussels: Koninklijke Academic Voor Overzeese Wetenschappen.

Hernandez-Ramdwar, C. (2013). African traditional religions in the Caribbean and Brazil: Models of religious tourism and impacts of commodification. *Journal of Heritage Tourism*, *8*(1), 81–88.

Hooper, J., & Eltis, D. (2013). The Indian Ocean in transatlantic slavery. *Slavery & Abolition*, *34*(3), 353–375.

Hosley, B. (2004). Transatlantic dreaming: Slavery, tourism and diasporic encounters. In F. Markowitz & A. Stefansson (Eds.), *Homecomings: Unsettling paths of return*, pp. 166–182. Lanham, MD: Lexington Books.

Joppe, M. (1996). Sustainable community tourism development revisited. *Tourism Management*, *17*(7), 475–479.

Kemp, R. (2000). Appointment in Ghana: An African American woman unravels the mystery of her ancestors. *Modern Maturity*, July–August, 1–17.

Kreamer, C.M. (2004). The politics of memory: Ghana's Cape Coast Castle museum exhibition "Crossroads of People, Crossroads of Trade". *Ghana Studies*, 7, 79–91.

Lovejoy, P.E. (2012). *Transformations in slavery* (3rd ed.). Cambridge: Cambridge University Press.

Lovejoy, P.E., & Richardson, D. (1999). Trust, pawnship, and Atlantic history: The institutional foundations of the old Calabar slave trade. *The American Historical Review*, *104*(2), 333–355.

Mabulla, A.Z. (2000). Strategy for cultural heritage management (CHM) in Africa: A case study. *African Archaeological Review*, *17*(4), 211–233.

Machado, P. (2004). A forgotten corner of the Indian Ocean: Gujarati merchants, Portuguese India and the Mozambique slave-trade, c. 1730–1830. *Slavery & Abolition*, *24*(2), 17–32.

Mazrui, A. (1986). *The Africans: A triple heritage*. London: BBC Publications.

McKercher, B., & du Cros, H. (2002). *Cultural tourism: Partnership between tourism and cultural heritage management*. New York: Haworth Press.

Mensah, I. (2015). The roots tourism experience of diaspora Africans: A focus on the Cape Coast and Elmina Castles. *Journal of Heritage Tourism*, *10*(3), 213–232.

Murphy, P (1985). *Tourism: A community approach*. London: Routledge.

Njoku, R. (2020). *West African masking traditions and diaspora masquerade carnivals: History, memory, and transnationalism*. Rochester, NY: University of Rochester Press.

Okoh, L. (2018). How Nigeria is preserving the legacy of its slave ports. Accessed from https://theculturetrip.com/africa/nigeria/articles/how-nigeria-is-preseving-the-legacy-of-its-slave-ports/ (Retrieved May 27, 2021).

Parry, T.D. (2018). What is Africa to me now? African-American heritage tourism in Senegambia. *Journal of Contemporary African Studies, 36*(2), 245–263.

Perbi, A.A. (2004). *A history of indigenous slavery in Ghana from the 15th to the 19th century*. Accra: Sub-Saharan Publications.

Richards, G. (2002a). Satisfying the cultural tourist: Challenges for the new millennium. In J. Akama, & P. Sterry (Eds.), *Cultural tourism in Africa: Strategies for the new millennium*, pp. 36–41. Arnhem, Netherlands: Association for Tourism and Leisure Education.

Richards, S.L. (2002b). Cultural travel to Ghana's slave castles: A commentary. *International Research in Geographical and Environmental Education, 11*, 372–375.

Schramm, K. (2008). Slave Route projects: Tracing the heritage of slavery in Ghana. In F. de Jong & M. Rowlands (Eds.), *Reclaiming heritage: Alternative imaginations in West Africa*, pp. 71–98. Walnut Creek, CA: Left Coast Press.

Sirakaya, E., Teye, V.B., & Sönmez, S. (2002). Understanding residents' support for tourism development in the Central Region of Ghana. *Journal of Travel Research, 41*, 57–67.

Slave Voyages (2021). Explore the origins and forced relocations of enslaved Africans across the Atlantic world. Accessed August 2, 2021. Online: https://www.slavevoyages.org/

Taylor, G. (1995). The community approach: Does it really work? *Tourism Management, 16*(7), 487–489.

Teye, V.B., Sönmez, S., & Sirakaya, E. (2002). Residents' attitudes toward tourism development. *Annals of Tourism Research, 29*(3), 668–688.

Teye, V.B. (2009). Tourism and Africa's tripartite cultural past. In D.J. Timothy & G.P. Nyaupane (Eds.), *Cultural heritage and tourism in the developing world: A regional perspective*, pp. 165–185. London: Routledge.

Teye, V.B., & Timothy, D.J. (2004). The varied colours of slave heritage in West Africa: White American stakeholders. *Space and Culture*, 7(2), 145–155.

Thiaw, I., & Wait, G. (2018). Presenting archaeology and heritage at a UNESCO World Heritage site: Goree Island, Senegal. *Advances in Archaeological Practice, 6*(3), 238–247.

Tillet, S. (2009). In the shadow of the castle: (trans)nationalism, African American tourism, and Gorée Island. *Research in African Literatures, 40*(4), 122–141.

Timothy, D.J. (1997). Tourism and the personal heritage experience. *Annals of Tourism Research, 34*(3), 751–754.

Timothy, D.J. (2021). *Cultural heritage and tourism: An introduction*. 2nd ed. Bristol: Channel View Publications.

Timothy, D.J., & Boyd, S.W. (2015). *Tourism and trails: Cultural, ecological and management issues*. Bristol: Channel View Publications.

Timothy, D.J., & Teye, V.B. (2004). American children of the African diaspora: Journeys to the motherland. In T. Coles & D.J. Timothy (Eds.), *Tourism, diasporas and space*, pp. 111–123. London: Routledge.

UNESCO (2004). *From slavery to freedom. Special edition on 2004 International Year to commemorate the struggle against slavery and its abolition*. Paris: UNESCO.

Yankholmes, A., & Akyeampong, O.A. (2010). Tourists' perceptions of heritage tourism development in Danish-Osu, Ghana. *International Journal of Tourism Research, 12*(5), 603–616.

Yankholmes, A., Akyeampong, O.A., & Dei, L.A. (2009). Residents' perceptions of Transatlantic Slave Trade attractions for tourism development in Danish-Osu, Ghana. *Journal of Heritage Tourism, 4*(4), 315–329.

Yankholmes, A., & McKercher, B. (2015a). Rethinking slavery heritage tourism. *Journal of Heritage Tourism, 10*(3), 233–247.

Yankholmes, A., & McKercher, B. (2015b). Understanding visitors to slavery heritage in Ghana. *Tourism Management, 5*, 22–32.

Yankholmes, A., & Timothy, D.J. (2017). Social distance between local residents and African–American expatriates in the context of Ghana's slavery-based heritage tourism. *International Journal of Tourism Research, 19*(5), 486–495.

11

AFRICAN DIASPORA TOURISM

Concepts, Issues, and Prospects beyond Slavery-Oriented Heritage

Emmanuel Akwasi Adu-Ampong and Ishmael Mensah

Introduction

In recent years, the concept of diaspora has become invested with new dreams and possibilities while remaining a contested academic concept. Traditionally, diaspora was the outcome of forced dispersals. However, ongoing forces of globalisation have resulted in voluntary dispersals of communities who have become constituted diasporas. Diaspora commonly refers to groups of people living outside their place of origin and who define themselves in relation to this distant place of origin. With a long history of migration and contemporary dispersals, African diasporas have emerged in many parts of the world, defining themselves by and through a strong link to the continent. Identifying with a distant place of origin, diasporic dreams and possibilities of return travel, resettlement, and reengagement are nurtured and realised. In this context, African diasporic people often undertake trips between their current place of residence and their defined place of origin. These homecomings of the African diaspora to their places of origin have increasingly been fused with slavery-related roots tourism—the return travel of the descendants of enslaved Africans in search of their roots.

In this chapter, we examine African diaspora tourism as a sociocultural phenomenon involving travel from one's place of residence to a place of origin with the objective of searching for one's roots, maintaining heritage links, visiting family, friends, and relatives, undertaking religious pilgrimages, exploring business opportunities and for leisure or other purposes. All of these constitute various perspectives on heritage, as they involve diasporic people traveling back to their homeland, which in itself reflects types of heritage consumption regardless of their motives. This chapter offers a broad discussion and conceptualisation of diasporas beyond slavery heritage-driven roots tourism to include diaspora tourism emanating from economic migration and colonial connections between

DOI: 10.4324/9781003153955-11

Africa and Europe. We start by conceptualising the emergence of the African diaspora before linking this to tourism. We then hone in specifically on the characteristics and features of African diasporic tourism. In particular, we examine diaspora tourism promotion in Africa and the role of diaspora tourism in the socioeconomic development of the region. We then focus on the relationship between hosts (Africans) and guests (diaspora Africans) before drawing out key conclusions.

African Diasporas: Emergence and Conceptualisation

Diaspora, as a word and as an academic concept remains fluid and contested notwithstanding its long and rich historical and etymological lineage. Choi (2003) argues that historically, the concept has three meanings: original, classical and contemporary. In its original meaning, diaspora can be traced back to the Greek words *speiro dia*. These words were used by the ancient Greeks with positive connotations to refer to migration and colonialisation through military conquest and expansion. The Greek words were then used in translating the Old Testament from the original Hebrew. Thus, the *speiro dia* became translated into English as to 'sow or scatter over' (Shuval, 2000, p. 42) or 'dispersal' (Helmreich, 1992, p. 245). It is through this biblical evolution of the word and close alignment with Jewish history that diaspora came to take on the negative connotations of forced dispersals. This represents the classical meaning of diaspora with a very specific meaning and reference to 'the exile of the Jews from their historic homeland and their dispersion throughout many lands, signifying as well the oppression and moral degradation resulting from that dispersion' (Saffran, 1991, p. 89). Diaspora in its contemporary meaning is seen as a social construct with a more general meaning relating to a group of people who have settled away from their ancestral homelands for a variety of push and pull reasons. The reasons for dispersals can include seeking refuge as victims, colonial settlers from imperial expansion, indentured labour migrants, and voluntary business migrants among others (Choi, 2003; Cohen, 1997; Williams, 2019).

The discourse of diaspora has travelled and hybridised into new forms within critical and popular discourse, facilitated by processes of globalisation and transnationalism (Clifford, 1994). Thus, in addition to the notion of forced dispersal through persecution, war, and slavery with its negative connotation of collective trauma, contemporary meanings of diaspora referring to groups whose migratory patterns are the result of (economic) globalisation forces and/or domestic economic stresses have a more positive connotation. A common working definition of diaspora is consequently used to refer to groups of people living outside their place of origin and who define themselves in relation to this distant place of origin through the collective consciousness of their original homeland and sharing common daily experiences in the new country (Tettey & Puplampu, 2005). This includes expatriates; migrant workers; immigrants; political, economic, and climate-change refugees; ethnic and racial minority groups, as well as overseas communities. Metaphorically, diaspora is seen as displaced groups who feel a connection with a prior home, even if they have never visited their ethnic homeland, and take steps to maintain such links (Saffran, 1991; Shuval, 2000). It is in this context that the notion of the Black and/or African diaspora has emerged.

'African diaspora' as a term only emerged in the 1950s and 1960s, but African diasporas have long existed in different parts of the world and mobilised under different terms such as Pan-Africanism, Garveyism, and the Negritude movement (Shepperson, 1993; Zeleza, 2005). The notion of an African diaspora is however doubly contested because of the common conflation of 'African' with 'Black' and the fact that not all Africans are Black and not all Black people consider themselves Africans (Essien, 2010; Zeleza, 2005, 2008). Thus, while the concept of diaspora developed from the central idea of dispersal, it is important to note that not all dispersals or migrations result in diasporas. Diasporas are formed through a process of collective consciousness of origin/original homeland (there) and shared lived socio-cultural experience in a new place (here). Diaspora is an unending process of being and becoming, and what constitutes the African diaspora, therefore, is an unending journey of movement and of being and becoming both spatially and discursively (Essien, 2010; Zeleza, 2008).

Studies on the African diaspora tend to be centred on, and biased towards, the dispersals in the 'Black Atlantic' and the African American experience originating from the transatlantic slave trade (TAST). Indeed, while Brazil for instance has the largest African diaspora in the world, research on the topic disproportionately places heavy importance on African Americans (diaspora in the United States) (Zeleza, 2005), to the neglect of other African diaspora formations. This has overshadowed other forms of African diaspora movements and counter-movements such as the exodus of former enslaved Africans from Brazil who (re)settled in Ghana, Nigeria, Benin, and other parts of West Africa from the early 1800s through voluntary return after manumission, through deportation after slave rebellions, or after the abolishing of slavery. This reverse African diaspora has unique longings and experiences as outlined by Essien's (2010) study of those who settled in Ghana, who become the *Tabom* people—Ghanaian–Brazilians. He shows the diversities of identity among this group and how some occupy the liminal space of being in their Ghanaian homeland while longing to visit their Brazilian homeland.

Historically, there is a rich diversity of African diaspora (counter) movements involving Africans of Asian, European, and Arabic descent on the continent (Koser, 2003). To reflect this rich diversity, there is a need to de-centre the spatial and racial preoccupations with the 'Black Atlantic' world and 'blackness', respectively. In doing this, Zeleza (2005) argues that there are four dominant geographical dimensions of what might be termed a global African diaspora: Intra-Africa; Indian ocean; Mediterranean; and Atlantic. The lack of research has been a key challenge to a full appreciation of these dimensions with the bulk of existing research focused on the Atlantic dimension. Indeed, Palmer (2000) identifies at least six key great African dispersals and movements within and outside the continent:

- The great exodus of about 100,000 years ago;
- The movement of Bantu-speaking people from contemporary Cameroon and Nigeria to other parts of the African continent and to the Indian Ocean around 3,000 BCE;
- The movement of traders, soldiers, merchants, and slavers to parts of Asia, Europe, and the Middle East from around the 5th century BCE;
- The transatlantic slave trade beginning in the mid-15th century;
- The East African slave trade to Asia; and
- The contemporary movement of Africa and peoples of African descent to various societies throughout the globe.

While these various dispersals form the basis for the development of the African diaspora in a wider sense, there is a tendency in research to privilege modern dispersals and diasporic formations. We need to move beyond the presentist orientation of contemporary scholarship in order to historicise and pluralise the multiple dispersals and formations of African diasporas. This chapter proceeds from this ambition of a wide conceptualisation of African diaspora, which (Zeleza, 2005, p. 8) describes as 'those peoples dispersed from the continent in historic and contemporary times, who have constituted themselves or been constituted into diasporas' and who maintain their connection to the continent through the material forms of economic remittances, business development, and tourism. These African diasporas are found in various parts of the world and have multiple identities and differing levels of identification with Africa (Zeleza, 2005).

Fluid Formations and Motives of African Diaspora Tourism

If diasporas are an unending fluid state of being and becoming based on a process of collective consciousness of origin, then this fluid state is anchored in regular (and stories of) travel back to the homeland. The 'mobilities turn' in tourism studies specifically and the social sciences, in general, have provided a framework in which to consider the various ways transnational dispersals and migration intersect with tourism (Coles & Timothy, 2004). Travel and tourism help diaspora communities to engage reflexively in the processes of learning and self-discovery of their diasporic identities both in their adopted countries and in their original homeland (Hollinshead, 1998; Iorio & Corsale, 2013; Marschall, 2017). Diaspora tourism then is simply the sociocultural phenomenon of people visiting a country of origin (real or imagined) from where they or their ancestors had to leave (voluntarily or involuntarily) due to certain social, economic, political, and environmental factors, in order to maintain and/or re-establish connections (Corsale & Vuytsyk, 2016; Marschall, 2015; Tichaawa, 2017).

The intersection of diasporas with travel and tourism takes several forms and has spurred a set of related typologies such as 'genealogical tourism' (Higginbotham, 2012; Santos & Yan, 2010; Timothy, 2008), 'ancestral tourism' (Alexander, Bryce, & Murdy, 2017), 'ethnic tourism' (King, 1994), 'legacy tourism' (McCain & Ray, 2003), 'visiting friends and family tourism' (Asiedu, 2005), 'homesick tourism' (Marschall, 2015), 'personal heritage tourism' (Timothy, 1997), homeland trips (Kibria, 2002), and 'roots tourism' (Mensah, 2015; Tichaawa, 2017; Timothy, 2008). While subtle differences exist among these typologies of diasporic tourism in terms of motivations, there is a shared underlining motivation stemming from the emotional attachment to, and nostalgic search for, identity and belonging in the place of origin (Coles & Timothy, 2004). Thus, African diaspora tourism can be seen in its varied forms stemming from the six great historical and contemporary African dispersals and movements identified by Palmer (2000) and outlined in the previous section.

Consequently, there is a need to pluralise the flows of African diaspora tourism from around the world to the continent. In this way, the over-emphasis on the African American and Black Atlantic experience can be decentred to make visible other forms of African diaspora tourism flows from other dispersals. For instance, we can begin to map out the features of African diaspora tourism that follows from the immigration of Africans to Europe and the formations of diaspora based on the historical and linguistic trails of colonialism (Zeleza, 2008), asylum-seeking (Koser, 2003), and economic migration (Mangala, 2017). In addition to more recent economic and conflict-induced refugee migration patterns out of Africa, there are still strong colonial connections with European countries. Thus, we can see immigrants from British colonies going to Britain and those from French colonies going to France in the post-independence era, as well as a sizeable diaspora of Eritreans in Italy (Arnone, 2011). There are also regional patterns, with a number of other Africans based in South Africa, such as the Nigeria diaspora (Marschall, 2017) and the Zimbabwean diaspora (Serumaga-Zake & Unisa, 2017). These patterns of migration and diaspora formation result in peculiarities in how return travel and tourism to their 'homelands' are organised and performed.

A key distinguishing feature of diaspora tourists compared to other forms of tourists is their possession of a shared cultural and ancestral identity with the place of visit. Thus, from a demand perspective, the main motivation of diaspora tourists is the push factor of a shared social identity with a diasporic community and a pull factor of personal and/or social identity with the homeland. The desire to trace one's roots and/or to rediscover one's heritage act as both a push and pull factor in diaspora tourism. In a recent study, Otoo, Kim, and Choi (2021) identify six main motivational pull factors for diaspora tourism: nostalgia-seeking desire to relive past experiences (Huang, Ramshaw, & Norman, 2016), (re)connecting to ancestral roots (Mensah, 2015), personal emotional connections to a homeland (Timothy, 1997; Yankholmes & McKercher, 2015), a quest for identity discovery (Arnone, 2011; Huang, Haller, & Ramshaw, 2013), a sense of pride in and desire for material contributions to the homeland (Afrifah, 2014; Asiedu, 2005; Yankholmes, 2016), and the quest for family reunions (Huang, Ramshaw, & Norman, 2016).

In pluralising African diaspora tourism and motivations beyond slavery heritage-related roots tourism, a number of key differences and effects of this form of tourism become evident. For instance, the diaspora tourism of Zimbabwean or Nigerian residents in South Africa and those travelling to the continent from outside involve regular repeat visitations to the homeland. Such travels in addition to being motivated by emotional entanglements of personal heritage are also for practical reasons including exploring business opportunities, maintaining property, and fulfilling family obligations such as attending weddings, funerals, and other community social events (Marschall, 2017; Masiiwa & Doroh, 2011; Rogerson, 2015). Thus, there is a wider socioeconomic importance attached to Africa's diaspora tourism because it is seen as yielding longer stays, repeat visits, spreading of benefits away from tourism hotspots, forging deeper links between tourism and local economic development, and increasing financial transfers and remittances (Tichaawa, 2017). The result is that a supply-side response from African governments has emerged through the establishment of diaspora liaison institutions and increased promotion of African diaspora tourism in its various forms.

Diaspora Tourism Promotion in Africa

Many African countries have recognised the importance of the diaspora travel and tourism market and have instituted a number of promotional strategies geared towards attracting their diaspora. Among the strategies employed by African governments are the establishment of institutions to promote diaspora tourism, diaspora business development, staging of special events and programmes, easing entry restrictions, and preserving sites and monuments such as those connected to the TAST. First, a number of African countries have established diaspora institutions to help streamline the engagement with their diaspora and to help facilitate travel by diaspora Africans to their ancestral homeland. These are usually state or quasi-state institutions that have been tasked with strengthening the relationships between a state and its dispersed population by addressing the specific needs of the diaspora, as well as coordinating the initiatives of government targeted at the diaspora.

The establishment of diaspora promotion institutions has been described by Agunias and Newland (2012) as formal state offices in executive or legislative branches of government that are dedicated to the affairs of emigrants and their descendants. For instance, there is Ethiopia's Diaspora Coordinating Office, Ghana's Diaspora Affairs Office, and Zambia's Diaspora Liaison Office. In some instances, a high-level institution such as a ministry is established purposely for diaspora affairs. Ghana's Ministry of Tourism, in 2003, launched a Strategic Action Plan focused on diaspora tourism with a central aim being to establish Ghana as the 'homeland' for Africans in the diaspora (Kleist, 2013). Subsequently, in 2006, the ministry's portfolio was enlarged and renamed the Ministry of Tourism and Diaspora Relations (MOTDR). This was in anticipation of the 2007 golden jubilee celebration of Ghana's independence and the launch of the Joseph Project, all geared towards attracting diaspora Africans, especially African Americans and Afro-Caribbeans to

Ghana (Adu-Ampong, 2019). Similarly, the Senegalese government has established a Ministry of Foreign Affairs and Senegalese Diaspora (MAESE), which has the primary responsibility of managing the rights of the Senegalese diaspora and promoting its interests (Mouthaan, 2019).

The diaspora affairs office in most African countries tends to be a department under a ministry or even under the presidency as is currently the case in Ghana. In Somalia, the Office for Diaspora Affairs is a department within the Ministry of Foreign Affairs and Investment Promotion (Gamlen et al., 2013). In Tunisia, the Office of Tunisians Abroad (Office des Tunisiens à l'Etranger or OTE) was established by law under the Ministry of Social Affairs and Solidarity in 1988. In some other countries, different institutions deal with diaspora-related issues, such as in Ethiopia where there are three different institutions in charge of Ethiopian diaspora affairs, namely the General Directorate for Ethiopian Expatriates (Ministry of Foreign Affairs), the Ethiopian Diasporas Coordinating Office (Ministry of Capacity Building), and the Diasporas Desk in the Addis Ababa City Government Investment Authority (MIDA, 2004). The promotion of diaspora tourism forms a key component of the activities of these diaspora institutions.

Moreover, many African countries have instituted a number of measures that seek to ease entry restrictions for members of their diasporas, ensuring an easy entry and facilitating long-term stays. These measures are mainly composed of legal instruments that seek to address specific challenges or remove obstacles to the integration of the diaspora into the societies of their ancestral homeland. For instance, the Ethiopian government enacted a law in 2002 to permit Ethiopian migrants with foreign citizenship to be treated just like Ethiopian nationals by issuing them a Person of Ethiopian Origin card, popularly known as the 'yellow card'. Among the privileges enjoyed by bearers of the card are entry into Ethiopia without a visa and the right to live and work in the country (Plaza & Rotha, 2011). Easing visa restrictions undoubtedly helps boost arrivals in the country from its diasporas abroad. Since Ghana identified the African diaspora as a niche market, a number of legislations have been passed to facilitate their movement to the country. These include the Right of Abode (2000) law, which gives people of African descent the right to apply and be granted the right to stay in Ghana indefinitely (Yankholmes & Timothy, 2017). Subsequently, the government in public ceremonies granted Ghanaian citizenship to 34 diaspora Africans in 2016 and another 126 in 2019. The Dual Citizenship Act of 2002 was passed to extend dual citizenship to people who have been naturalised as Ghanaians but live in other countries. Members of the Ghanaian diaspora also have the right to vote during general elections through the Representation of People's Act (ROPAB) (Act 699), although the country's Electoral Commission has not yet fully implemented this. All these legal instruments are meant to facilitate a deeper attachment and easy access to the homeland for African diasporas.

Some African governments make use of events to acknowledge and welcome their diasporas back home (Sim & Leith, 2013). Government initiatives have focused on the promotion of events based on the tangible and intangible cultural heritage connected with the TAST which are found in coastal towns of countries such as Senegal, Ghana, and Benin (Araujo, 2010; Pierre, 2009). In West African countries in particular, Black heritage festivals have been staged to attract the African diaspora that was separated from their kith and kin as a result of the TAST. For instance, as far back as 1966, Senegal, under the leadership of Léopold Sedar Senghor, held the first World Festival of Black Arts to promote Gorée Island as a site of memory of the TAST and as a tourist destination for the African diaspora. This attracted a number of prominent African Americans to Senegal. Also, the historical Island of Saint-Louis in Senegal has been hosting its famous Jazz Festival, which attracts hundreds of diasporic Africans every year since 1993. In Benin, efforts to develop tourism by attracting African American tourists to the country led to the creation of a landscape dedicated to Ouidah's role in the slave trade (Landry, 2011). This landscape comprises the Ouida Museum of History, Fort of São João Batista da Ajuda, and the 'Door of No Return' monument. In addition, the government embarked on a project of restoration and conservation of the royal palaces as sites of visitation.

Ghana is another country that has employed events to boost diaspora tourism, most prominently through the yearly Emancipation Day and Pan African Historical Theatre Festival (PANAFEST). These events have been used by the government to encourage returns to Ghana from the diaspora, to promote tourism, and to deepen the government's engagement with both the general African and Ghanaian diaspora (Hasty, 2002; Mensah, 2015; Yankholmes & McKercher, 2015; Yankholmes & Timothy, 2017). Travel by the African diaspora to Ghana reached a crescendo in 2019 when the government launched a year-long programme of events under the banner of 'the year of return 2019' to commemorate the 400 years since the first enslaved Africans arrived in Jamestown, Virginia, in the United States. Activities included musical concerts, carnivals, food and trade bazaars, fashion shows, and other reverential events fronted by a number of prominent African Americans and celebrities visiting Ghana. Other notable events that have been created to attract the African diaspora are the Black Heritage Festival in Badagry, Nigeria, and the biannual International Roots Festival in Banjul, The Gambia, which attracts thousands of African Americans seeking to reconnect with their African roots and to experience The Gambia's culture and heritage.

African diaspora tourism promotion has also been facilitated through educational and birthright tour packages developed by governments, usually in collaboration with international and local tour operators. According to Newland and Taylor (2010), a number of tour operators, regional associations, intergovernmental organisations, and individual countries actively develop and promote cultural, historic, and natural heritage sites under the umbrella of African diaspora tourism. Due to the huge business potential of diaspora tourism, a number of specialised travel agents and tour operators offer dedicated products to diaspora tourists from different backgrounds (Corsale & Vuytsyk, 2016). As a result, diaspora tour packages have been tailored to the needs of people of African descent. Tour packages have included activities like homecoming ceremonies, naming ceremonies, and other rites of passage. For example, the African American travel agent Meredith

McCleary has been instrumental in promoting diaspora tourism in southern African destinations as a specialist in travel to Botswana and southern African spa destinations (Newland & Taylor, 2010). In Ghana, Land Tours, a business owned by Mona Boyd, who is herself a returnee married to a Ghanaian, has played an instrumental role in bringing many African Americans back to Africa through their innovative tour packages. The Eritrean government has been very active in its diasporic people to travel back to Eritrea through a programme dubbed 'Know-Your-Country-Tour' (Zuran Hagerka). According to Conrad (2010), second-generation diasporic Eritreans who embark on the tour learn about their country of origin and its history. These tours are organised by the NGO, the National Union of Eritrean Youth and Students, with support from the government. In many cases in Africa, roots-seeking travel packages are arranged by both local and international tour agencies with assistance from the immigration associations and other private business owners (Li, McKercher, & Chan, 2020). Morocco initiated a project known as 'Operation Marhaba' to welcome and provide assistance to members of the Moroccan diaspora returning home from Europe for the summer holidays. Marhaba, which means 'welcome and attraction' in Arabic, has resulted in large numbers of the Moroccan diaspora visiting their ancestral homeland for a short period every year. Furthermore, it has been estimated that the Algerian diaspora accounts for about 70% of that country's tourist arrivals (International Centre for Migration Policy Development, 2011).

Diaspora Tourism and the Socioeconomic Development of Africa

The African diaspora has been very instrumental in the socioeconomic development of the continent through remittances, investments, knowledge transfer, as well as technology and skills transfer. As a result, many African countries and the African Union (AU) have begun to recognise the important role the diaspora plays in the economic development of the region (Chand, 2016). In this light, the AU, in 2001, recognised the African diaspora as the sixth region of Africa. The AU further endorsed and supported the establishment of the Diaspora African Forum (DAF) in 2007 during the AU Summit in Accra, as a diplomatic mission for the African diaspora which seeks to promote and encourage the involvement of Africans abroad in the development of the continent. In spite of this, there is a recognition that the historical linkages between the African diaspora, which comprises more than 140 million people, and the continent have not yet been fully explored (African Development Bank, 2011).

Tourism provides a good medium for exploring the historical linkages and harnessing the socioeconomic potential of the African diaspora. This is mainly due to the fact that members of the diaspora, unlike conventional international tourists, can be actively involved in the tourism industry of the destination in different ways, such as serving as investors, remitters, philanthropists, volunteers, promoters, and informal ambassadors (Seraphin, Korstanje, & Gowreesunkar, 2020). Teye, Alhassan, and Setrana (2017) referred to Ghana's longstanding history of diaspora engagement through the medium of tourism. They argue that one of the best ways to initiate engagement with people in the diaspora is to develop tourism programmes to attract them to their natal states. Diaspora tourism goes beyond a mere pleasure vacation. Due to its fluidity, the African diaspora in particular tends to have different motivations for travel including visiting friends and relatives, searching for roots, and fulfilment of other sociocultural obligations. The diaspora market, therefore, tends to have greater economic potential and has been projected to serve as the engine of growth of the tourism industry in the future (Huang, Haller, & Ramshaw, 2013).

There are some peculiar features of diaspora tourism that confer greater economic gains on a destination. Firstly, diaspora tourists tend to stay longer in the homeland than other tourists (Coles & Timothy, 2004; Corsale & Vuytsyk, 2016; Huang, Haller, & Ramshaw, 2013). Also, diaspora tourism is less characterised by seasonality than other types of tourism, and diaspora tourists are more likely to repeat their visits due to the fact that they have personal connections with the homeland and their families and may be obliged to attend festivals, family reunions, and other social gatherings (Iarmolenko & Kerstetter, 2015; Pérez-López, 2007). Diaspora tourism tends to have a greater economic impact on African countries because diaspora tourists usually stay in smaller locally owned accommodation facilities, visit rural areas and consume local goods and services (Corsale & Vuytsyk, 2016; Newland & Taylor, 2010; Roberts, 2012; Seraphin, Korstanje, & Gowreesunkar, 2020). These help in enhancing the economic fortunes of the destination by growing local businesses, providing off-season employment opportunities, and encouraging local entrepreneurial activities. By patronising local goods and services diaspora tourists also help to minimise economic leakages (Asiedu, 2005).

The socioeconomic benefits of diaspora tourism are often far-reaching and transcend the tourism industry. Diaspora tourism has enormous economic potential in Africa owing to the availability of rich traditions in music, dance, art, and architecture, which are of interest to the diaspora and has contributed to the development of growing tourism markets in countries like Ethiopia, Cape Verde, and Mali (World Bank, 2011). Diaspora tourism can be used to generate foreign exchange and foreign direct investment, as well as technology and knowledge transfer. This is exemplified by Ghana's 'Beyond the Return' project under the theme 'A Decade of Renaissance: 2020–2030', which seeks to strategically engage the African diaspora. This ten-year project is anchored on seven pillars and will galvanise the resources and potentials of the diaspora towards tourism development, investment, and socioeconomic growth. It seeks to build on the success of the 'year-of-return 2019' campaign, which reportedly generated some $1.9 billion through economic activities and resulted in an increase in tourist arrivals of over 200,000 (Daily Graphic, 2019).

Many African countries, including Ghana, Eritrea, Cape Verde, and Ethiopia, continue to profit from the promotion of diaspora tourism. The nostalgic trade has been identified as one of the areas where Cape Verde has the greatest potential (Resende-Santos, 2016). Also, the quest for Eritrean identity among the Eritrean diaspora has ignited a wave of travel to Eritrea from Europe. People of the Eritrean diaspora who travel back to their homeland, show off their wealth by acquiring properties and injecting money into the local economy and by so doing, contribute to the economic development of the country (Arnone, 2011). Similarly, apart from their personal travels to Ethiopia, emigrant Ethiopians and their progeny are also among the main investors in larger hotels, some of which are managed under international brands such as Sheraton, Hilton, Radisson, Marriott, Golden Tulip, Best Western, and Ramada (Altes, 2018). In Ghana, a study by Afrifah (2014) found that African Americans who visit the country contribute significantly to the incomes of local craftsmen, which enables them to support their families and sustain their businesses. Asiedu (2005) also found that Ghanaians abroad returning home for visits spent an average of £2,769 on international travel, of which a significant amount (£585) was spent on incidentals such as contributions to community development funds and expenses for funerals; another £433 was used to purchase food and entertainment for friends and relatives.

Relations between Hosts (Africans) and Guests (Diaspora Africans)

When people of the African diasporas travel to their ancestral homelands, they expect to reconnect with their kith and kin and to be accepted and treated as locals. However, in reality, this is not always the case. The relationship between Africans on the continent and those in the diaspora has not always been cordial (Mensah, 2015; Mwakikagile, 2005; Zachary, 2001). This has been attributed to the fact that Africans and diasporic Africans have had different historical and cultural experiences (Appiah, 1992). Though Africans on the continent and those in the diaspora have some common identity features, they grew up in different cultural contexts. Their expectations of diaspora tourism are therefore shaped by their varied experiences, and as such, their expectations do not always coincide (Yankholmes & Timothy, 2017).

Diasporic tourists seek to cultivate social bonds with their hosts based on their shared African identity, Africans on the other hand are more interested in the potential economic benefits they can derive from the host–guest relationships (Reed, 2013). This has often resulted in misunderstandings and contention in the framing of the events targeted at the African diaspora. Pierre (2013) reports of the disappointment of African Americans regarding the lack of Ghanaian engagement in the emotional and spiritual aspects of events like the Emancipation Day celebrations (Ebron, 2002; Pierre, 2013).

Though most diasporic Africans want to be treated as brothers and sisters on the continent, they are often regarded as foreigners and treated differently, and in some cases, they suffer racial abuse. In Eritrea, for instance, there is an intra-generational conflict between second-generation Eritreans abroad and their peers who live in Eritrea. This stems from Eritrea's policy of graduated citizenship, which exempts diasporic Eritreans from many of the citizenship duties that their peers at home have to fulfil. The experiences of being disapproved by Eritreans sometimes force visiting Eritreans to re-negotiate their sense of belonging (Tecle, 2012). In Ghana, people of other nationalities, including the African diaspora, are required to pay higher entrance fees to attractions and sites than their local counterparts. This has often been met with protests by the diasporic tourists (Bruner, 1996; Yankholmes & Timothy, 2017).

Diaspora Africans sometimes find the very racial prejudices that confront them in their adopted homelands waiting for them when they arrive in the motherland (Mensah, 2015). Diaspora Africans are subjected to subtle forms of racism by their kinsmen when they are labelled foreigners or tourists. In Ghana, an African American is referred to as *obroni*, an Akan Twi word for a white person (Reed, 2013). According to Bruner (1996), the irony of the term 'obroni' is that it labels African Americans as both White and foreign, whereas they see themselves as Black and at home. African Americans have been confronted with similar situations in East Africa where they are referred to as *muzungu* a Kiswahili word for Whites or foreigners. For many diaspora Africans, the journey to their ancestral land is sacred (Timothy, 2008; Timothy & Teye, 2004); therefore, referring to them as tourists only trivialises the experience, adds insult to injury, and invokes the profane rather than the sacred (Reed, 2005). They would rather prefer to be called returnees or pilgrims. As a result of these conflicts and misunderstandings, tourists from the African diaspora usually leave the ancestral homeland having established a sense of connection to their imagined ancestors but without having established a sense of connection to their contemporary African brothers and sisters (Holsey, 2004; Yankholmes & Timothy, 2017).

Conclusion

This chapter has sought to decentre the overemphasis on Black Atlantic slavery heritage-induced diaspora tourism experiences to look at other heritage-related travel patterns in various African diasporas. We have outlined the travel motivations and experiences of intra-African diaspora tourism flows vis-à-vis Nigerian and Zimbabwean economic diasporas in South Africa who undertake regular travel back to their home countries. In addition to being motivated by emotional entanglements of personal heritage, such travels also involve exploring business investment opportunities and the need to fulfil family obligations. The chapter also addressed the dynamics of diaspora tourism of first- and second-generation Eritreans, Ethiopians, and Moroccans living in Europe who embark on regular travel to their homelands. In highlighting these cases, this chapter has sought to historicise and pluralise the varied flows and patterns of African diaspora tourism. On the supply side, this chapter examines governmental strategies and programmes, as well as private sector enterprises, in promoting African diaspora tourism.

African diaspora tourism is a sociocultural phenomenon for which this chapter has sought to offer a broad conceptualisation and discussion. Beyond slavery heritage-driven roots tourism, this chapter has traced flows emanating from economic migration and colonial connections between Africa and Europe. In so doing a number of issues and questions have been raised, which can serve as standing points for future research. In recognition of the socioeconomic contributions of their diasporic communities, many African countries have established agencies and ministries to promote and facilitate diaspora tourism and related investments in the homeland. The mechanisms involved in these and often tense relationships between diaspora returnees and residents all require further research. We hope this chapter will serve as an inspiration for wider engagement with African diaspora tourism in all its varied dynamic flows.

References

Adu-Ampong, E.A. (2019). Historical trajectories of tourism development policies and planning in Ghana, 1957–2017. *Tourism Planning & Development, 16*(2), 124–141.

Africa Development Bank (2011). *The role of the diaspora in nation building: Lessons for fragile and post-conflict countries in Africa*. Addis Ababa: Africa Development Bank.

Afrifah, M. (2014). *Diaspora tourism and homeland development: exploring the impacts of African American tourists on the livelihoods of local traders in southern Ghana*. Unpublished doctoral dissertation, King's College London, University of London.

Agunias, D., & Newland, K. (2012). *Developing a road map for engaging diasporas in development: A handbook for policymakers and practitioners in home and host countries*. Washington, DC: International Organization for Migration.

Alexander, M., Bryce, D., & Murdy, S. (2017). Delivering the past: Providing personalized ancestral tourism experiences. *Journal of Travel Research, 56*(4), 543–555.

Altes, C. (2018). *Analysis of tourism value chain in Ethiopia: Final report*. The Hague: Ministry of Foreign Affairs, CBI.

Appiah, K.A. (1992). *In my father's house: Africa in the philosophy of culture*. Oxford: Oxford University Press.

Araujo, A. (2010). Welcome the diaspora: Slave trade heritage tourism and the public memory of slavery. *Ethnologies, 32*(2), 145–178.

Arnone, A. (2011). Tourism and the Eritrean diaspora. *Journal of Contemporary African Studies, 29*(4), 441–454.

Asiedu, A. (2005). Some benefits of migrants' return visits to Ghana. *Population, Space and Place, 11*, 1–11.

Bruner, E.M. (1996). Tourism in Ghana: The representation of slavery and the return of the black diaspora. *American Anthropologist, 98*(2), 290–304.

Chand, M. (2016). Leveraging the diaspora for Africa's economic development. *Journal of African Business, 17*(3), 273–290.

Choi, I. (2003). Korean diaspora in the making: Its current status and impact on the Korean economy. In C.F. Bergsten & I. Choi (Eds.), *The Korean diaspora in the world economy*, pp. 9–17. Washington, DC: Institute for International Economics.

Clifford, J. (1994). Diasporas. *Cultural Anthropology, 9*(3), 302–338.

Cohen, R. (1997). *Global diasporas*. London: Routledge.

Coles, T., & Timothy, D.J. (2004). 'My field is the world': Conceptualizing diasporas, travel and tourism. In T. Coles & D.J. Timothy (Eds.), *Tourism, diasporas and* space, pp. 1–29. London: Routledge.

Conrad, B. (2010). *"We are the prisoners of our dreams": Long-distance nationalism and the Eritrean diaspora in Germany*. Unpublished doctoral dissertation, Staats-und Universitätsbibliothek Hamburg Carl von Ossietzky.

Corsale, A., & Vuytsyk, O. (2016). Long distance attachments and implications for tourism development: The case of western Ukrainian diaspora. *Tourism Planning and Development, 13*(1), 88–110.

Daily Graphic (December, 16, 2019). Year of Return generates $1.9 billion. Online at: https://www.graphic.com.gh/news/general-news/year-of-return-generates-1-9-billion.html

Ebron, P. (2002). *Performing Africa*. Princeton: Princeton University Press.

Essien, K. (2010). *African diaspora in reverse: The Tabom people in Ghana, 1820s-2009*. Unpublished doctoral dissertation, University of Texas at Austin.

Gamlen, A., Cummings, M., Vaaler, P.M., & Rossouw, L. (2013). *Explaining the rise of diaspora institutions*. Oxford: Oxford University, International Migration Institute.

Hasty, J. (2002). Rites of passage, routes of redemption: Emancipation tourism and the wealth of culture. *Africa Today, 49*(3), 47–76.

Helmreich, S. (1992). Kinship, nation, and Paul Gilroy's concept of diaspora. *Diaspora: A Journal of Transnational Studies, 2*(2), 243–249.

Higginbotham, G. (2012). Seeking roots and tracing lineages: Constructing a framework of reference for roots and genealogical tourism. *Journal of Heritage Tourism*, 7(3), 189–203.

Hollinshead, K. (1998). Tourism and the restless peoples: A dialectical inspection of Bhabha's halfway populations. *Tourism Culture & Communication, 1*(1), 49–77.

Holsey, B. (2004). Transatlantic dreaming: Slavery, tourism, and diasporic Encounters. In F. Markowitz & A.H. Stefansson (Eds.), *Homecomings: Unsettling paths of return*, pp. 166–182. Lanham, MD: Lexington Books.

Huang, W., Haller, W.J., & Ramshaw, G. (2013). Diaspora tourism and homeland attachment: An exploratory analysis. *Tourism Analysis, 18*, 285–296.

Huang, W., Ramshaw, G., & Norman, W.C. (2016). Homecoming or tourism? Diaspora tourism experience of second generation immigrants. *Tourism Geographies, 18*(1), 59–79.

International Centre for Migration Policy Development (2011). *Enhancing diaspora engagement*. Vienna: International Centre for Migration Policy Development.

Iarmolenko, S., & Kerstetter, D. (2015). Potential predictors of diaspora tourism for Ukrainian immigrants in the USA. *World Leisure Journal, 57*(3), 221–234.

Iorio, M., & Corsale, A. (2013). Diaspora and tourism: Transylvanian Saxons visiting the homeland. *Tourism Geographies, 15*(2), 198–232.

Kibria, N. (2002). Of blood, belonging, and homeland trips: Transnationalism and identity among second-generation Chinese and Korean Americans. In P. Levitt & M.C. Waters (Eds.), *The changing face of home*, pp. 295–311. New York: Russell Sage Foundation.

King, B. (1994). What is ethnic tourism? An Australian perspective. *Tourism Management 15*(3), 173–176.

Kleist, N. (2013). Flexible politics of belonging: Diaspora mobilisation in Ghana. *African Studies, 72*(2), 285–306.

Koser, K. (Ed.) (2003). *New African diasporas*. London: Routledge.

Landry, T.R. (2011). Touring the slave route: Inaccurate authenticities in Bénin, West Africa. In H. Silverman (Ed.), *Contested cultural heritage: Religion, nationalism, erasure, and exclusion in a global world*, pp. 205–231. New York: Springer.

Li, T.E., McKercher, B., & Chan, E.T.H. (2020). Towards a conceptual framework for diaspora tourism. *Current Issues in Tourism, 23*(17), 2109–2126.

Mangala, J. (Ed.) (2017). *Africa and its global diaspora: The policy and politics of emigration*. Cham, Switzerland: Springer.

Marschall, S. (2015). 'Homesick tourism': Memory, identity and (be)longing. *Current Issues in Tourism, 18*(9), 876–892.

Marschall, S. (2017). Transnational migrant home visits as identity practice: The case of African migrants in South Africa. *Annals of Tourism Research, 63*, 140–150.

Masiiwa, M., & Doroh, B. (2011). *Harnessing the diaspora potential for socio-economic development in Zimbabwe: Investment, trade and participation in political rocesses*. Harare: Institute of Development Studies, National Association of NGOs.

McCain, G., & Ray, N.M. (2003). Legacy tourism: The search for personal meaning in heritage travel. *Tourism Management, 24*(6), 713–717.

Mensah, I. (2015). The roots tourism experience of diaspora Africans: A focus on the Cape Coast and Elmina Castles. *Journal of Heritage Tourism, 10*(3), 213–232.

MIDA (2004). *Return of the qualified Ethiopian health care professionals from Sweden to Ethiopia*. Addis Ababa: IOM MIDA Pilot Project.

Mouthaan, M. (2019). Unpacking domestic preferences in the policy-'receiving' state: The EU's migration cooperation with Senegal and Ghana. *Comparative Migration Studies*, 7(1), 1–20.

Mwakikagile, G. (2005). *Relations between Africans and African Americans: Misconceptions, myths and realities*. Washington, DC: National Academic Press.

Newland, K., & Taylor, C. (2010). *Heritage tourism and nostalgia trade: A diaspora niche in the development landscape*. Washington, DC: Migration Policy Institute, USAID.

Otoo, F.E., Kim, S., & Choi, Y. (2021). Developing a multidimensional measurement scale for diaspora tourists' motivation. *Journal of Travel Research, 60*(2), 417–433.

Palmer, C. (2000). The African diaspora. *The Black Scholar, 30*(3–4), 56–59.

Pérez-López, J.F. (2007). The diaspora as a commercial network for Cuban reconstruction. *Annual Proceedings of the Association for the Study of the Cuban Economy, 17*, 399–409.

Pierre, J. (2009). Beyond heritage tourism, race and the politics of African-diasporic interactions. *Social Text, 27*(1), 59–81.

Pierre, J. (2013). *The predicament of Blackness: Postcolonial Ghana and the politics of race*. Chicago: University of Chicago Press.

Plaza, S., & Rotha, D. (2011). Harnessing diaspora resources for Africa. In S. Plaza & D. Rotha (Eds.), *Diaspora for development in Africa*, pp. 1–54. Washington, DC: World Bank.

Reed, A. (2005). *Slave trade memory: Ghana's heritage tourism and the promotion of African identity*. Paper presented to the *Mid-America Alliance for African Studies Conference*, Springfield, MO.

Reed, A. (2013). Diaspora tourism. In A. Quayson & G. Daswani (Eds.), *A companion to diaspora and transnationalism*, pp. 524–538. Chichester: Wiley Blackwell.

Resende-Santos, J. (2016). Cape Verde: Rethinking diaspora in development policy. *International Migration, 54*(2), 82–97.

Roberts, S. (2012). Assessing the potential of diaspora tourism. *Journal of Eastern Caribbean Studies, 37*(3/4), 115–131.

Rogerson, C.M. (2015). Unpacking business tourism mobilities in sub-Saharan Africa. *Current Issues in Tourism, 18*(1), 44–56.

Saffran, W. (1991). Diasporas in modern societies: Myths of homeland and return. *Diaspora 1*(1), 83–99.

Santos, C.A., & Yan, G. (2010). Genealogical tourism: A phenomenological examination. *Journal of Travel Research, 49*(1), 56–67.

Seraphin, H., Korstanje, M., & Gowreesunkar, V. (2020). Diaspora and ambidextrous management of tourism in post-colonial, post-conflict and post-disaster destinations. *Journal of Tourism and Cultural Change, 18*(2), 113–132.

Serumaga-Zake, P.A., & Unisa, S.B.L. (2017). Migration and tourism: The challenges of Zimbabwean diaspora in South Africa. *African Journal of Hospitality, Tourism and Leisure, 6*(4), 1–20.

Shepperson, G. (1993). African diaspora: Concept and context. In E. Joseph & E. Harris (Eds.), *Global dimensions of the African diaspora*, pp. 41–50. Washington, DC: Howard University Press.

Shuval, J.T. (2000). Diaspora migration: Definitional ambiguities and a theoretical paradigm. *International Migration, 38*(5), 41–56.

Sim, D., & Leith, M. (2013). Diaspora tourists and the Scottish Homecoming. *Journal of Heritage Tourism, 8*(4), 259–274.

Tecle, S. (2012). The paradoxes of state-led transnationalism: Capturing continuity, change and rupture in the Eritrean transnational social field. Unpublished master's thesis, York University, Toronto, Canada.

Tettey, W., & Puplampu, K.P. (2005). Continental Africans in Canada: Exploring a neglected dimension of the African-Canadian experience. In W. Tettey & K.P. Puplampu (Eds.), *The African diaspora in Canada: Negotiating identity and belonging*, pp. 3–24. Calgary, AB: University of Calgary Press.

Teye, J.K., Alhassan, O., & Setrana, M.B. (2017). Evolution and nature of diaspora engagement policies in Ghana. In J. Mangala (Ed.), *Africa and its global diaspora: The policy and politics of emigration*, pp. 143–173. Cham, Switzerland: Palgrave Macmillan/Springer.

Tichaawa, T.M. (2017). The nature of diasporic tourism in Cameroon: An opportunity for tourism development. *African Journal for Hospitality, Tourism and Leisure, 6*(4), 1–13.

Timothy, D.J. (1997). Tourism and the personal heritage experience. *Annals of Tourism Research, 24*(3), 751–754.

Timothy, D.J. (2008). Genealogical mobility: Tourism and the search for a personal past. In D.J. Timothy & J. Kay Guelke (Eds.), *Geography and genealogy: Locating personal pasts*, pp. 115–135. Aldershot: Ashgate.

Timothy, D.J., & Teye, V.B. (2004). American children of the African diaspora: Journeys to the motherland. In T. Coles & D.J. Timothy (Eds.), *Tourism, diasporas and space*, pp. 111–123. London: Routledge.

Williams, A.M. (2019). Migration, tourism, and globalisation. In D.J. Timothy (Ed.), *Handbook of globalisation and tourism*, pp. 66–88. Cheltenham: Edward Elgar.

World Bank (2011). *Africa region tourism strategy: Transformation through tourism— harnessing tourism for growth and improved livelihoods*. Washington, DC: World Bank.

Yankholmes, A. (2016). A look at the consumption behaviours along Ghana's Slave Routes. *Tourism Recreation Research, 41*(3), 272–282.

Yankholmes, A., & McKercher, B. (2015). Understanding visitors to slavery heritage sites in Ghana. *Tourism Management, 51*, 22–32.

Yankholmes, A., & Timothy, D.J. (2017). Social distance between local residents and African–American expatriates in the context of Ghana's slavery-based heritage tourism. *International Journal of Tourism Research, 19*(5), 486–495.

Zachary, P. (2001). Tangled roots: For African-Americans in Ghana, the grass isn't always greener. *Wall Street Journal*, March 14.

Zeleza, P.T. (2005). Rewriting the African diaspora: Beyond the Black Atlantic. *African Affairs, 104*(414), 35–68.

Zeleza, P.T. (2008). The challenges of studying the African diasporas. *African Sociological Review/Revue Africaine de Sociologie, 12*(2), 4–21.

12

INDUSTRIAL HERITAGE TOURISM IN AFRICA

Prospects and Challenges

Clinton D. van der Merwe and Dallen J. Timothy

Introduction

Industrial tourism is an important niche of heritage tourism. The best-known and developed industrial heritage attractions are located in the Global North. Amidst the economic downturns in the last two decades, as well as the challenges of the COVID-19 pandemic, many countries are trying to diversify their tourism economies and offerings to encourage economic development (Gössling, Scott, & Hall, 2020), including developing their heritage assets for increased tourism growth. Africa has a rich history of industrialization associated with ancient sites and later colonialism and could therefore benefit enormously by developing and more effectively marketing its industrial heritage assets to attract tourists. Nevertheless, the region overall is far behind other parts of the world in this regard. This chapter outlines certain trends related to industrial heritage-based tourism in Africa and examines several ways in which the region can leverage its industrial past and present to help grow its tourism economy.

Industrial Heritage Tourism

For millennia, humankind has utilized various mechanisms and processes to sustain and simplify life. Some of the world's earliest industrial activities included the production of microliths (stone tools), metallurgy, and fish processing. Canals, roadworks, bridges, viaducts, dams, and other industrial complexes were built in ancient days. With the Industrial Revolution and the widespread construction of railway lines in the 19th century, places and economies industrialized rapidly, and people became dependent on the processes of primary (extractive) and secondary (manufacturing) economic activities, and the world, for the most part, became reliant upon industrial economies, where manufacturing and other heavy production processes

DOI: 10.4324/9781003153955-12

dominated growing employment opportunities and the products used in everyday life. In the mid-20th century, however, as a growing service and technology-based economy began to surpass that of the extractive and manufacturing sectors, and as many natural resources diminished, many heavy industries began a steep decline. This frequently resulted in a proliferation of derelict industrial landscapes and the abandonment of many resource-dependent communities.

This post-Fordist turn also included a greater environmental consciousness and a realization of the need to protect the industrial heritage and renew many of the brownfields and landscapes of dereliction that heavy industry and its decline had brought about. Efforts began to beautify industrial areas and to turn brownfields, derelict harbors, and industrial wastelands into usable recreation and tourism spaces for local communities and tourists. There are multitudes of examples of these changes, with some of the most prominent ones being the waterfronts in Sydney, Toronto, Baltimore, Singapore, and Cape Town (Babalis & Townshend, 2018; Chang & Huang, 2011; Ferreira & Visser, 2007). Common outcomes of these movements include parks and greenspaces located on rehabilitated industrial areas, the preservation of industrial sites and other means of production, urban industrial zone (especially waterfront) redevelopment, the repurposing of former industrial buildings and areas into modern uses (e.g., apartment blocks, office spaces, conference and events centers, shopping centers, and leisure complexes), and the establishment of industrial museums and interpretive centers.

In response to economic decline, outmigration of young people, and the threat of extinction, many small resource communities reinvented their economic sectors toward tourism. In many Western countries, this has manifested in what Getz (1993; see also Murphy et al., 2011) called 'tourist-shopping villages' where the industrial past (and often ethnic immigration related to industrial development) provides a unique heritage backdrop for a contemporary focus on shopping and rural amenities. Similarly, many abandoned resource settlements have become ghost towns, or nearly ghost towns, which often also create unique selling points as part of the heritage milieu of industrial regions that appeal to many tourists (Prideaux & Timothy, 2011).

Thus, industrial tourism, like other tourisms, provides salient socioeconomic benefits for industrial communities, not only for areas where heavy industry has ceased but also where it remains in operation (Castillo Canalejo, López-Guzmán, & Millán Vázquez de la Torre, 2010). Like farmers often adopt agritourism as a source of supplemental income, many mines and other industrial complexes now see tourism as an additional income source. However, mines and other heavy industry sectors have also begun using tourism as a goodwill gesture for outreach to the community, as well as a source of 'propaganda' in an effort to satiate public dissatisfaction with the environmental and social problems that heavy industry often causes. Tourism has been used by some industrial sectors, particularly mining, to show the benefits and benevolence of heavy industry for the surrounding community (e.g., jobs and social welfare) and the natural environment, downplaying its negative ecological effects and highlighting what the company has magnanimously done to mitigate ecological problems (Rudd & Davis, 1998). For example, local authorities in the Potosí mining area in Bolivia have decided to utilize mine-based tourism as a means of narrating Indigenous stories and constructing a colonial-indigenous discourse that aims to elevate the socioeconomic position of the area's native population (Pretes, 2002). In some cases, former miners (or other industrial sector workers) are re-employed as guides and interpreters at such sites to provide a more intimate look at industrial life and as part of community outreach efforts (Pretes, 2002).

The importance of protecting and promoting industrial heritage as a tool for understanding the human experience and humankind's socioeconomic development is widely recognized (Alfrey & Putnam, 1992; Wicke, Berger, & Golombek, 2018). Industrial heritage is now a prominent sector of the heritage and cultural industries and is seen as being both an economic driver and a crucible of knowledge about humankind's struggle to survive in unusually harsh environments and the development of rural identities. It is, in fact, a strong manifestation of 'ordinary heritage', or the past of the everyday life of ordinary people—something that has traditionally been ignored in mass tourism narratives and itineraries but which is gaining currency in special-interest and alternative forms of heritage tourism (Timothy, 2014, 2018). Since its inception in 1972, the UNESCO World Heritage List (WHL) has notoriously focused on the world's grandest and most opulent heritage places because these are the most iconic localities that demonstrate 'outstanding universal value'. This has meant that simpler and ordinary forms of heritage, including heavy industry, have not been well valued or commemorated by such global inventories (Cleere, 2001; Falser, 2001; Timothy, 2014). However, UNESCO's recent efforts to emphasize intangible heritage, holistic cultural landscapes, non-ancient heritage places, and the heritage of regular working-class populations recognize the importance of ordinary heritage and have seen an increase in inscriptions of industrial sites on the WHL or the Tentative WHL.

In addition to UNESCO, other international organizations have begun to recognize and appreciate industrial heritage on a global and regional scale. For example, in 1987, the Council of Europe established its Cultural Routes program as a means of highlighting and connecting Europe's cultural heritage through long-distance and interconnected networks of trails, locations, communities, and heritage sites along common European themes. The program's aim is to stimulate tourism and to emphasize a common European past. Industrial heritage is the primary or secondary focus of at least seven of the Council's 45 established routes, including the Routes of the Olive Tree, the European Route of Ceramics, and the European Route of Industrial Heritage (Council of Europe, 2022).

There are generally three temporal contexts of industrial heritage resources: extinct or disbanded industrial complexes that are valued as archaeological sites; recently closed sectors that are either derelict, torn down, or renovated and repurposed for other uses; and functioning industrial projects that continue to produce their product while simultaneously acting as a tourist attraction. Castillo Canalejo and her colleagues (2010) distinguish industrial heritage assets in three areas: industrial relics that can be converted into museums; transportation elements (e.g., roads, bridges, and railways); and sociocultural features that link social organizations and communities to their industrial past, including workers' housing compounds. There are several types of industrial heritage assets, including but not limited to, all elements of places, people, and events associated with the primary and secondary industrial sectors. Table 12.1 illustrates many of the common themes and industrial heritage types that are counted as important heritage assets based upon a wide range of industrial activities.

Industrial heritage-based tourism entails the development of leisure activities and services around sites, buildings, and landscapes that originated with industrial resource processes (Edwards & Llurdés i Coit, 1996; ICOMOS, 2003; Xie, 2015), such as mining, manufacturing, and other extractive or production sectors (Timothy, 2021a; van der Merwe & Rogerson, 2013, 2018). Industrial archaeology, or the 'evidence, material and immaterial, of documents, artefacts, stratigraphy and structures, human settlements and natural and urban landscapes, created for or by industrial processes' (ICOMOS, 2003, cited in Clarke & Fisher, 2018, p. 16), includes derelict industrial complexes, as well as functioning heavy industry that is an important part of the extant heritage complex of a region (Timothy, 2021a; Timothy & Tahan, 2020). Thus, industrial heritage tourism includes visits to derelict industrial complexes, restored industrial landscapes, and currently operating industrial spaces as previously indicated.

Many mining areas throughout the world, especially in the Global North, have turned to tourism with their focus being on telling the story of mining, demonstrating the process, and leading guided tours of industrial relics (Conlin & Jolliffe, 2011; Perfetto & Vargas-Sánchez, 2018; Rudd & Davis, 1998; Timothy, 2021a; Vargas-Sánchez, 2015). Most European countries and parts of North America, where the Industrial Revolution first unfolded, have rich heritagescapes of industries related to railways, shipping, canals, mining, waterworks, forestry, and manufacturing. Touring mines, dams and factories has long been a strong focus of industrial tourism in the Global North (Beer, Rybár, & Kaľavský, 2018; Firth, 2011; Hasimoto & Telfer, 2017; Hollick, 2011; Perfetto & Vargas-Sánchez, 2018; Pretes, 2002; Różycki & Dryglas, 2017; Timothy, 2021a; Xie, 2015).

TABLE 12.1 Common themes of interest in industrial heritage

Agriculture & food	Fishing and fish processing	Metallurgy
Art & industry	Glass	Mining & collieries
Bridges	Hydroelectricity	Museums
Canals & waterways	Iron & steel	Oil & petrochemicals
Ceramics & clay	Industrial communities	Paper
Chemicals	Land transportation	Railways
Communications	Leatherworks	Textiles
Conservation	Marine transportation	Water
Energy & power	Mechanical engineering	Wood

Source: TICCIH (2022).

Industrial Heritage Tourism in Africa

Relatively few African countries have functioning industrial heritage sites that are utilized for tourism purposes, and only very little research has been undertaken on tourism in these contexts. Likewise, few sites have received significant global recognition as centers of industrial heritage-based tourism (Khalil & Elgohary, 2020; Nassar & Sharaf Eldin, 2013; van der Merwe & Rogerson, 2018). Nevertheless, Africa has a strong industrial past and present, which has the potential to grow this sector of heritage tourism. Table 12.2 illustrates some prominent African examples of industrial heritage that have either been developed for tourism or have the potential to become developed. Some of them are designated UNESCO World Heritage Sites, while others have yet to realize their full potential.

Although the number of industrial heritage sites is growing on the UNESCO WHL and in other inventories, the representation of Africa remains far below that of other regions (Breen, 2007; Ndoro, 2017; Yang, 2017). Industrial heritage is far less pervasive as a tourism resource in Africa compared to its role in North America, Europe, Australia, and parts of Asia, and its recognition as a heritage asset in the region is not widespread (Minchinton, 1983). Nonetheless, it is an important part of the continent's heritage that has considerable potential to be tapped for tourism. Industrial heritage plays a more significant tourism role in the developed countries of Africa than it does in the lesser developed states, and some sites are an iconic part of the tourism offerings of certain countries, including the Aswan Dam and the Port Said Lighthouse in Egypt (Komesaroff, 1980; Nasir & Ndoro, 2018), as well as the many functioning mines of South Africa and Botswana (Mahachi & Ketshabile, 2013; Najafi, Hamzeh, & Moqimi, 2014; van der Merwe & Rogerson, 2013, 2018). Except for a few notable exceptions (e.g., Buyeke & Njoroge, 2015; Demewoz & Abebe, 2018; Najafi, Hamzeh, & Moqimi, 2014; Tseane-Gumbi & Ani, 2018; van der Merwe & Rogerson, 2013, 2018), relatively few researchers have discussed the notion of industrial heritage tourism on the continent.

Africa is home to many different industries, from ancient quarrying and metallurgy to medieval shipping ports and modern-day diamond and gold mining. Most countries of Africa have some type of heavy industry, the following being among the most common in the region: mining, quarrying, manufacturing, milling, utilities, and power generation (e.g., dams, power stations, and alternative energy), transportation (e.g., railways, canals, and marine infrastructure), metallurgy and smelting, oil drilling, trade routes, and salt extraction (see Table 12.2). Although Africa produces many natural resources, perhaps the most famous of these are diamonds and gold. Diamonds are particularly plentiful in southern Africa (e.g., Botswana, DR Congo, South Africa, and Angola), while gold is found all over the continent, as exemplified by the rise and fall of many ancient African empires in connection with gold mining, processing, and trade. The top five gold-producing countries of Africa are listed in Table 12.3, with other relevant information.

TABLE 12.2 Selected industrial heritage sites in Africa

Country/ territory	*Site*	*Type of industrial heritage*	*Other remarks*
Algeria	Oued Meswar Dam	Waterworks, dam	Currently functioning for water supply and irrigation
Botswana	Jwaneng Diamond Mine and other mines	Mining	Functioning mines for diamonds, gold, copper, nickel, and coal
Burkina Faso	Ancient ferrous metallurgy sites	Metalworks, iron production, furnaces, and mines	Ancient period, UNESCO World Heritage Site (2019)
Burkina Faso	Ruins of Loropéni	Mining	Medieval gold mining site, UNESCO World Heritage Site (2009)
Cape Verde	Saltworks of Pedra de Lume	Salt processing	Active from 18th to 20th century, on UNESCO's Tentative List of World Heritage Sites
DR Congo	Grottes de Matupi	Microlithic industry	Ancient stone works, up to 40,000 years ago, on UNESCO's Tentative List of World Heritage Sites
DR Congo	Mining in Lubumbashi	Mining	Many active mining operations, especially copper and cobalt
Egypt	Aswan Dam	Waterworks, dam	Currently functioning for water supply, irrigation, and hydropower
Eswatini	Ngwenya Mine	Mining	Iron mining—site of world's earliest mining operations, on UNESCO's Tentative List of World Heritage Sites
Ethiopia	Gafat Industrial Village	Metalworks and cannon factory	Heyday in the 1800s
Kenya	Railway system and infrastructure	Railway and transportation	Currently functioning but also utilized as a heritage asset
Madagascar	Hydraulic canal system for rice production	Waterworks	Historic system, currently still in use and undergoing renovation
Mauritania	Tegdaoust archaeological site	Metallurgy site	Gold, iron, and copper processes site, on UNESCO's Tentative List of World Heritage Sites
Mayotte (France)	Sugar cane mills	Sugar production	No longer in operation but several mills in ruins

(*Continued*)

TABLE 12.2 (Continued)

Country/ territory	*Site*	*Type of industrial heritage*	*Other remarks*
Namibia	Benguela Current Marine Ecosystem	Guano/fertilizer production and fishing	On UNESCO's Tentative List of World Heritage Sites
Namibia	Kolmanskop	Ghost town and mining village	Former diamond mining area, abandoned since 1956 but now a tourist attraction
Niger	Cultural Routes of the Sahara Desert: Salt Route	Salt processing and salt trade	Ancient salt production and caravan trade, on UNESCO's Tentative List of World Heritage Sites
Réunion (France)	Sugar cane mills	Sugar production	Two mills continue to function, producing sugar and rum
Sierra Leone	National Railway Museum	Railways, industrial complex	Government railway operated between 1897 and 1974. Much of its material culture is on display
South Africa	Gold Reef City	Mining	Amusement park on remnants of a goldmine
South Africa	Cullinan diamond mines	Mining	Still functioning. Source of largest gem-quality diamond ever found
South Africa	Turbine Hall	Power station	Repurposed for conferences and events
South Africa	Rand Steam Laundries	Laundries associated with gold mining	Repurposed as a shopping center
South Africa	Soweto Cooling Towers	Power station	Repurposed venue for public art and bungee jumping
South Africa	NZASM railway lines	Railway	1887–1908, now defunct with various sites being conserved
Tanzania	Ruins of Kilwa Kisiwani and Ruins of Songo Mnara	Coastal shipping ports	Ports for shipping gold, silver, and pearls, 9th–19th centuries, UNESCO World Heritage Site (1981)
The Gambia	Wassu Stone Circles quarry	Stone quarry	3rd century BC to 16th century AD, source of stones for the Wassu Stone Circles, on UNESCO's Tentative List of World Heritage Sites
Togo	Granaries of the Caves of Nok and Mamproug	Food production and food storage	17th–19th centuries, on UNESCO's Tentative List of World Heritage Sites

(*Continued*)

TABLE 12.2 (Continued)

Country/ territory	*Site*	*Type of industrial heritage*	*Other remarks*
Tunisia	Ancient quarries of numidic marble, Chimtou	Marble quarries	Source of much of the Roman Empire's marble, on UNESCO's Tentative List of World Heritage Sites
Tunisia	The Roman hydraulic complex of Zaghouan-Carthage	Waterworks and storage tanks	Masterpiece of Roman engineering, on UNESCO's Tentative List of World Heritage Sites
Uganda	Kibiro	Salt production village	900 years of salt production, and a women-only industry. On UNESCO's Tentative List of World Heritage Sites
Zimbabwe	Naletale Cluster of Dzimbabwes	Mining	Gold, copper, and iron mining and metalwork, 16th–18th centuries, on UNESCO's Tentative List of World Heritage Sites
Zimbabwe	Ziwa National Monument	Iron smelting sites	Ancient smelting site with terraced agriculture, on UNESCO's Tentative List of World Heritage Sites

Source: Authors' compilation.

TABLE 12.3 Top gold-producing countries of Africa, 2019

Rank	*Country*	*Tonnage (metric tonnes)*	*Other observations*
1	Ghana	142.4	Large companies and 'artisanal miners' contribute to Ghana's mining success. Gold comprises 90% of Ghana's mineral exports, with estimates suggesting the country has a reserve of 1,000 metric tonnes
2	South Africa	118.2	Gold extraction has declined in SA in recent years. It is estimated that SA still has some 6,000 tonnes of gold reserves
3	Sudan	76.6	Most of the country's gold is located in the Eriab area of the Nuba Mountains, near the Nile River and the Blue Nile

(*Continued*)

TABLE 12.3 (Continued)

Rank	*Country*	*Tonnage (metric tonnes)*	*Other observations*
4	Mali	71.1	Large companies and 'artisanal miners' contribute to Mali's mining success. The country is estimated to have 800 tonnes of gold reserves. After cotton and livestock, gold is Mali's third largest export item
5	Burkina Faso	62.0	Gold is one of Burkina Faso's biggest export products, and it is estimated that the country has approximately 154 million tonnes of gold in reserve

Source: NS Energy (2020).

The continent is also a major producer of crude oil. In 2021, Libya produced 28.4 billion barrels, Nigeria 36.9 billion, Algeria 12.2 billion, Angola 7.8 billion, and Sudan 5 billion (Benson, 2022). Africa's oil industry has resulted in a massive growth of business tourism but has so far had little to do with industrial heritage-based tourism as other industries have done in the region. Although oil drilling is an important part of some countries' socioeconomic heritage, it is not often recognized as such, the way mining and other extractive sectors often are.

Railways also have a rich history and heritage in many regions of Africa, most notably in the northern and southern parts of the realm, having much to do with the mining industry, trade, and colonial expansionism. Africa's railway systems began in the mid-1800s in Alexandria, Egypt, and expanded from there to other areas until about the 1930s (BBC, 2022). Significant lines were built in all regions of Africa in the 1800s and early 1900s (see Table 12.4), and many countries have built railroads in recent years, but these have less to do with their industrial heritage and are meant to satisfy modern transportation needs. The biggest rail networks today are in South Africa, Morocco, Egypt, and Algeria but they are growing in other countries as well. The industrial archaeology of the Netherlands-South African Railway Company (NZASM) in South Africa and some of the historical lines in Kenya have a great deal of potential for the country's tourism sector if it is properly conserved, developed, and managed sustainably (Clarke & Fisher, 2018; Njunguna, Wahome, & Deisser, 2018), such as ensuring that these efforts empower communities and provide employment opportunities for those whose heritage is being consumed.

As noted earlier, the abandonment of industrial settlements has been a major problem for many industry-dependent communities, particularly mining, forestry, and plantation towns. This abandonment has resulted in the proliferation of ghost towns in many parts of the world, which are an important post-industrial landscape feature in places such as North America, Latin America, Australia, Europe, and Asia and have become salient tourist attractions in their respective regions (Ballantyne, 2001; Hasimoto & Telfer, 2017; Prideaux & Timothy, 2011; Timothy, 2021a). Far less has been written about deindustrialization-caused ghost towns

TABLE 12.4 Early (historical) railways in Africa

Region and country	*Starting point*	*Finishing point*	*Year completed*
North Africa			
Egypt	Alexandria	Cairo	1856
Sudan	Wadi Halfa	Khartoum	1898
Algeria	Algiers		1919
Tunisia	Tunis		1919
Morocco	Casablanca	Rabat	1923
West Africa			
Ghana	Sekondi	Kumasi	1903
Nigeria	Lagos	Kano	1912
Sierra Leone	Freetown		1909
Congo	Brazzaville	Pointe Noire	1932
East Africa			
Kenya–Uganda	Mombasa	Lake Victoria	1901
Tanganyika (Tanzania)	Tanga	Usambara Hills	1905
Djibouti–Ethiopia	Djibouti	Addis Ababa	1917
Southern Africa			
South Africa–Zimbabwe	Cape Town	Bulawayo	1897
Congo–Angola	Copper belt	Benguela	1931

Source: Based on information in BBC (2022).

in Africa, but there is potential for resurrecting local economies through tourism in such places. Prominent examples include Dallol, Ethiopia; La Güera, Western Sahara; Kolmanskop, Namibia; and many localities in South Africa and Zimbabwe (Alexander, 2010; Kamete, 2012; Lohnert, 2017; Mediero, 2017; Nel et al., 2003). In all of Africa, industrial heritage-based tourism is particularly prevalent in Egypt and in South Africa, with the country's oldest mine, Kromdraai, for example, being a massively popular tourist attraction, as are the other mines and industrial complexes in the country.

Like other parts of the world, many of the mines in southern Africa are closing as mineral resources become depleted. Scholars have argued that the industrial heritage of such places should not be ignored in favor of other heritages but should become the focus of attracting tourism to the affected regions to compensate for lost jobs and economic stagnation (Tseane-Gumbi & Ani, 2018). In addition to ghost towns and derelict industrial areas, diamond museums and similar attractions might be a useful focus of tourism development (Mahachi & Ketshabile, 2013), much the same way the Gold Museum in Bogota, Colombia, represents that country's gold mining heritage and has achieved considerable success as one of Bogota's main tourist attractions (Moreno, 2013).

Repurposing is another recent trend that is catching on in less-developed countries where, in most cases, old was seen as something to dismantle or neglect. In the more developed countries of Africa, we are seeing a wave of repurposing derelict industrial spaces and buildings (i.e., former factories, assembly plants, or mining-related structures) into tourism services, such as restaurants, lodging, and museums, through private and public–private enterprises. This is particularly prominent in South Africa (Läuferts & Mavunganidze, 2009). The Turbine Boutique Hotel and Spa in Knysna, Western Cape, was once the Thesen Power Station and is a good example of repurposing industrial archaeology into modern tourism uses. It is generally regarded as a success story and 'illustrates how adaptive reuse can be used as a strategy to preserve industrial architectural heritage' (Edwards, 2017, p. vi).

Although rehabilitating industrial areas, including harbors and waterfronts, as well as former industrial brownfields, is a common practice in the Global North, it is not yet a common practice in Africa. However, there are a few examples, such as the recent and current beautification of industrial brownfields associated with former mining landscapes in Kénadsa, Algeria, for industrial tourism and recreational purposes (Mostadi & Biara, 2019).

Like other heritages, industrial heritage may be an important component of a country's identity and international image. Some countries' global image is at least partially linked to their industrial past and present. For example, in Cuba, a country well known for its sugar production, '…the Hershey sugar mill represents much of Cuba's history, heritage, culture, and economy' (Salinas Chavez et al., 2018, p. 426). Many African countries have similar rich histories associated with mining and processing (e.g., sugar, peanuts, cacao, rubber, and coffee) that could be tapped for heritage tourism. Gold and diamond mines, oil wells, rubber plantations, historic railways, ghost towns, and repurposed historic buildings and sites are just a few of the many latent assets that might help the countries of Africa develop their tourism sectors.

Industrial Heritage Challenges

As the earlier discussion makes clear, there is much potential for industrial heritage-based tourism in many African countries (Buyeke & Njoroge, 2015; Demewoz & Abebe, 2018; Läuferts & Mavunganidze, 2009; Newman & Page, 2017; Tseane-Gumbi & Ani, 2018), but there are also major constraints, both in the past and at present. Many of these constraints are similar to those of other regions, such as a lack of funding and trained specialists, security concerns (image problems), and a lack of access to the rural areas where many sites can be found. However, there are also significant barriers that are unique to Africa's context. These include, among others, resentment over colonialism, a lack of definition or appreciation of industrial heritage, corruption and a lack of institutional support, and pervading legal and policy frameworks that target extraordinary heritage over the protection of industrial archaeology and ordinary heritage (Njunguna, Wahome, & Deisser, 2018).

In their zeal to exploit the resources of their colonial claims in Africa, European metropoles made mining and other industries a top priority (Seif-El-Islam, 2020). Although the Portuguese made some early efforts to establish gold mining in the interior of Africa in the 15th and 16th centuries, their efforts failed, and gold mining was truly established only in the 1880s, after copper (1854) and diamonds (1867) had already been harvested (Geography, 2022). From the mid-1800s until the early 20th century, European colonizers established mines throughout the continent, becoming wealthy on the backs of African laborers by extracting gold, diamonds, tin, coal, iron, copper, bauxite, and many other minerals that enriched the Europeans. Even after African territories became sovereign states, the mining sector remained largely under the control of foreign companies and expatriate managers. Although the revenue from mineral exports was shared between the foreign mining companies and African governments, the latter often lined their personal pockets at the expense of the citizenry (Geography, 2022).

The brutality with which some colonial powers treated their African subjects is bewildering (Hochschild, 2019). A prominent example is the case of Belgium's King Leopold II, who personally owned the Congo Free State (now DR Congo) (1885–1908) and all of its natural resources, including rubber. Under his rule, the Congolese people were subject to beatings, wanton limb amputations, false imprisonment, and executions for not filling rubber quotas and other prescribed infractions (Rutz, 2018). The colonial legacies of slavery and violence are not easily forgotten in Africa (Arowolo, 2010; Ayittey, 1992).

Because mining and other industries are so often associated with the colonial hold on Africa, its brutality and exploitative nature, and with greater demands to decolonize tourism on the continent, there have been calls to decolonize mining and other industrial activities. For many people, the use of colonial-era (including modern-day) industry for tourism purposes is a distasteful prospect (Pwiti & Ndoro, 1999), although there is a growing interest in using disused industrial heritage as a tool for educating the public about the negative effects of colonialism (Ray, 2018; Turunen, 2020; Ugwuanyi, 2021).

Another salient impediment is the lack of recognition of the importance and value of industrial patrimony because, as noted earlier, it is often associated with colonialism, but also because different cultures value heritage differently and some might not recognize heavy industry as true heritage if it is not closely related to traditional cultural heritage (Timothy, 2021b). The meaning of heritage is often ingrained in language and social practices and values. Likewise, it is often seen as something grandiose, extraordinary, and having ancient roots. This is especially the view in non-Western global contexts and keeps many societies from valuing their own 'insignificant' traits and characteristics as a heritage asset that outsiders might appreciate (Timothy, 2021b).

The International Committee for the Conservation of the Industrial Heritage (TICCIH) is a global organization that aims to promote international cooperation in protecting, researching, documenting, interpreting, and promoting education related to industrial heritage (TICCIH, 2022). The organization works closely with ICOMOS on matters related to industrial archaeology and has 46 member states. Of these members, only three are in Africa: Egypt, Morocco, and South Africa. This dearth of African representation reflects the lack of appreciation of the continent's industrial archaeology as a heritage asset (Seif-El-Islam, 2020).

In many countries, corruption and state capture are hugely problematic both in terms of heritage protection and heritage promotion, as much-needed revenue and funding are stolen by politicians instead of being invested into appropriate state and cultural heritage bodies (Breen, 2007; Nomishan, Tubi, & Gubam, in press; Yap & Saha, 2013). There is also corruption in the issuance of new mining licenses and infrastructure development that destroys areas of high heritage value (Leonard & Lebogang, 2018). Likewise, favoritism and nepotism are rampant, with certain communities, regions, and cultures receiving much greater financial and policy support than other areas where industrial heritage might prevail.

Conclusion

Although industrial heritage-based tourism is far less developed in Africa than it is in many other regions, especially in the Global North, the potential for increased industrial heritage tourism in Africa is immense. The lack of attention to this subsector of heritage is largely a result of poor governance, lack of funding, corruption, inappropriate legislation, insecurity, a lack of political will, and a general unawareness of what constitutes heritage—all conditions that mar the process of leveraging these resources and assets in the heritage tourism marketplace (Timothy, 2021b).

Like other forms of heritage tourism, industrial tourism must be linked to communities and tourism services. This is particularly true at lesser-known industrial heritage complexes. Considerations for destination planners should include the location of the site in relation to other attractions and services, accessibility, provide well-trained staff to deliver quality interpretation, ensure historical accuracy, and develop in relation to other services to provide for the needs of tourists (Hollick, 2011). In addition to these normative planning needs, industrial heritage protection and tourism require unique planning approaches and paradigms. First, it is vital to building a greater understanding of industrial heritage—that it is indeed an important part of a nation's patrimony—and the importance of protecting and promoting it. There needs to be a strong realization that, despite their colonial and exploitative roots, post-industrial landscapes are worth preserving because they provide knowledge about the past and the present and might help heal society and move it forward. Second, it is imperative to create solidarity among community members, the tourism industry, and government officials to stimulate stakeholder buy-in so that they will work together to avoid the destruction of this heritage even if they aim to decolonize it. An important part of this process may involve enacting and implementing legislation and policies that protect different sorts of heritage that reflect the complexity of Africa's checkered history. Third, to provide balanced perspectives and a truer understanding of the past, planners and site managers must provide a holistic interpretive program that emphasizes both African and European perspectives. Finally, Africa's tourism market needs to be re-segmented to include broader perspectives on heritage consumers and then provide for their needs by ensuring safer and more accessible environments. Tourism planners must also realize that 'ordinary heritage' is as important as extraordinary heritage, and there is a growing niche market that will pay to demonstrate this importance.

There is a notable research lacuna regarding the industrial heritage tourism sector in Africa. More work needs to be done to identify the extent and nature of industrial heritage and industrial archaeology assets that exist in this region and how these resources can be leveraged to promote a sustainable heritage tourism industry. Likewise, additional research is needed to understand the constraints to industrial heritage-based tourism, and other heritage tourisms, and how these might be overcome through appropriate planning, destination management, and policy-making.

References

Alexander, N. (2010). *Kolmanskop: An industrial heritage resource or only a tourist attraction?* Unpublished master's thesis, University of Cape Town.

Alfrey, J., & Putnam, T. (1992). *The industrial heritage: Managing resources and uses*. London: Routledge.

Arowolo, D. (2010). The effects of Western civilization and culture on Africa. *Afro Asian Journal of Social Sciences, 1*(1), 1–13.

Ayittey, G.B.N. (1992). *Africa betrayed*. New York: Palgrave Macmillan.

Babalis, D., & Townshend, T.G. (Eds.) (2018). *Urban waterfronts and cultural heritage: New perspectives and opportunities*. Florence: Altralinea Edizioni.

Ballantyne, P. (2001). Unsettled country: Reading the Australian ghost town. *Journal of Australian Studies, 25*(70), 33–38.

BBC (2022). The story of Africa: Railways. Online at: https://www.bbc.co.uk/worldservice/africa/features/storyofafrica/11chapter12.shtml#:~:text=Africa's%20network%20of%20railways%20was,were%20government%20owned%20and%20run

Beer, M., Rybár, R., & Kaľavský, M. (2018). Renewable energy sources as an attractive element of industrial tourism. *Current Issues in Tourism, 21*(18), 2139–2151.

Benson, E.A. (2022). Here are 10 African countries with the largest crude oil production output in 2021. *Business Insider Africa*, 14 January 2022. Online at: https://africa.businessinsider.com/local/markets/10-african-countries-with-the-largest-crude-oil-production-output-in-2021/qbj8bj3

Breen, C. (2007). Advocacy, international development and World Heritage Sites in sub-Saharan Africa. *World Archaeology, 39*(3), 355–370.

Buyeke, E., & Njoroge, J.M. (2015). Soap stone mining, challenges and opportunities for tourism development: Case study of soapstone in Tabaka, Kisii County, Kenya. *African Journal of Tourism, Hospitality and Leisure Studies, 1*(1), 1–10.

Castillo Canalejo, A.M., López-Guzmán, T.J., & Millán Vázquez de la Torre, G. (2010). Industrial mining tourism activities. *Anatolia, 21*(2), 1–9.

Chang, T.C., & Huang, S. (2011). Reclaiming the city: Waterfront development in Singapore. *Urban Studies, 48*(10), 2085–2100.

Clarke, N.J., & Fisher, R. (2018). The industrial archaeology of the NZASM in South Africa and some of its unique artefactual residue. *Architecture SA*, September/October, 16–23.

Cleere, H. (2001). The uneasy bedfellows: Universality and cultural heritage. In R. Layton, P.G. Stone, & J. Thomas (Eds.), *Destruction and conservation of cultural property*, pp. 38–45. London: Routledge.

Conlin, M., & Jolliffe, L. (Eds.) (2011). *Mining heritage and tourism: A global synthesis*. London: Routledge.

Council of Europe (2022). Cultural Routes of the Council of Europe programme. Online at: https://www.coe.int/en/web/cultural-routes

Demewoz, E., & Abebe, F. (2018). Investigating the tourism potential of Gafat industrial village: South Gondar Administrative Zone, North Western Ethiopia. *African Journal of Hospitality, Tourism and Leisure*, 7(4), 1–15.

Edwards, J.A., & Llurdés I Coit, J.C. (1996). Mines and quarries: Industrial heritage tourism. *Annals of Tourism Research, 23*(2), 341–363.
Edwards, R. (2017). *The adaptive reuse of the former Thesen Island Power Station: A case study.* Unpublished MTech Dissertation, Cape Peninsula University of Technology, South Africa.
Falser, M. (2001). *Industrial heritage analysis: World Heritage List and Tentative List.* Bangkok: UNESCO World Heritage Centre.
Ferreira, S., & Visser, G. (2007). Creating an African Riviera: Revisiting the impact of the Victoria and Alfred Waterfront development in Cape Town. *Urban Forum, 18*(3), 227–246.
Firth, T.M. (2011). Tourism as a means to industrial heritage conservation: Achilles heel or saving grace? *Journal of Heritage Tourism, 6*(1), 45–62.
Geography (2022). Africa: Minerals and mining. Online at: https://geography.name/minerals-and-mining/
Getz, D. (1993). Tourist shopping villages: Development and planning strategies. *Tourism Management, 14*(1), 15–26.
Gössling, S., Scott, D., & Hall, C.M. (2020). Pandemics, tourism and global change: A rapid assessment of COVID-19. *Journal of Sustainable Tourism, 29*(1), 1–20.
Hasimoto, A., & Telfer, D.J. (2017). Transformation of Gunkanjima (Battleship Island): From a coalmine island to a modern industrial heritage tourism site in Japan. *Journal of Heritage Tourism, 12*(2), 107–124.
Hochschild, A. (2019). *King Leopold's ghost: A story of greed, terror and heroism in colonial Africa.* London: Macmillan.
Hollick, M. (2011). Sustainaing the benefit of heritage mining for the site, city and region: Exploring the success of Sovereign Hill outdoor museum. In M. Conlin & L. Jolliffe (Eds.), *Mining heritage and tourism: A global synthesis*, pp. 108–128. London: Routledge.
ICOMOS (2003). The Nizhny Tagil Charter for the Industrial Heritage. Online at: https://www.icomos.org/18thapril/2006/nizhny-tagil-charter-e.pdf
Kamete, A.Y. (2012). Of prosperity, ghost towns and havens: Mining and urbanisation in Zimbabwe. *Journal of Contemporary African Studies, 30*(4), 589–609.
Khalil, N.I., & Elgohary, A.G. (2020). The role of industrial heritage in sustainable tourism development Case study of Mina EL-Basal district in Alexandria–Egypt. *International Journal of Heritage, Tourism and Hospitality, 14*(3), 217–250.
Komesaroff, M.B. (1980). The industrial archaeology of lighthouses. *The Journal of Navigation, 33*(1), 131–137.
Läuferts, M., & Mavunganidze, J. (2009). Ruins of the past: Industrial heritage in Johannesburg. *WIT Transactions on The Built Environment, 109*, 533–552.
Leonard, L., & Lebogang, T. (2018). Exploring the impacts of mining on tourism growth and local sustainability: The case of Mapungubwe Heritage Site, Limpopo, South Africa. *Sustainable Development, 26*, 206–216.
Lohnert, B. (2017). *Migration and the rural-urban transition in Sub-Saharan Africa.* Berlin: Centre for Rural Development, Humboldt Universität zu Berlin.
Mahachi, D., & Ketshabile, L. (2013). Diamond mining as a possible strategy for tourism product diversification in Botswana. *Journal for Development and Leadership, 2*(2), 35–50.
Mediero, B.A. (2017). La industria de las algas en el antiguo Sáhara Español. *Vegueta: Anuario de la Facultad de Geografía e Historia, 17*, 323–340.
Minchinton, W. (1983). World industrial archaeology: A survey. *World Archaeology, 15*(2), 125–136.
Moreno, D. (2013). Les objets d'art dans les nouvelles interfaces numériques. *Interfaces Numériques, 2*(2), 281–295.

Mostadi, A., & Biara, R.W. (2019). Sustainable development of brownfield site for a new landscape perception of an industrial heritage in the city of Kenadsa. *International Journal of Conservation Science, 10*(1), 165–176.

Murphy, L., Benckendorff, P., Moscardo, G., & Pearce, P.L. (2011). *Tourist shopping villages: Forms and functions*. London: Routledge.

Najafi, N., Hamzeh, F., & Moqimi, S. (2014). The contribution of 'Kimberley' as a rehabilitated mine to South Africa's tourism income. *African Journal of Hospitality, Tourism and Leisure, 3*(2), 1–10.

Nasir, A.A.M., & Ndoro, W. (2018). Heritage and energy development issues: A controversial complex relationship. In W. Ndoro, S. Chirikure, & J. Deacon (Eds.), *Managing heritage in Africa: Who cares?*, pp. 193–208. London: Routledge.

Nassar, D.M., & Sharaf Eldin, S. (2013). Towards adaptive reuse of the industrial heritage in Minet El-Bassal district at Alexandria. *Architecture and Planning Journal, 22*(1), 1–17.

Ndoro, W. (2017). World Heritage and development in Sub-Saharan Africa. *American Anthropologist, 119*(1), 129–131.

Nel, E.L., Hill, T.R., Aitchison, K.C., & Buthelezi, S. (2003). The closure of coal mines and local development responses in Coal-Rim Cluster, northern KwaZulu-Natal, South Africa. *Development Southern Africa, 20*(3), 369–385.

Newman, C., & Page, J. (2017). *Industrial clusters: The case for special economic zones in Africa*. Helsinki: United Nations University World Institute for Development Economics Research.

Njunguna, M.B., Wahome, E.W., & Deisser, A.M. (2018). Saving the Iindustry from itself: A case of the railway industrial Heritage in Kenya. *The Historic Environment: Policy & Practice, 9*(1), 21–38.

Nomishan, T.S., Tubi, P-K., & Gubam, D.S. (in press). Cultural heritage management and the effect of corruption in Nigeria: Hampering sustainable development via cultural heritage destruction. *Journal of Cultural Heritage Management and Sustainable Development*.

NS Energy (2020). Top five gold mining countries of Africa: From Ghana to Burkina Faso. August 28, 2020. Online at: https://www.nsenergybusiness.com/news/top-gold-mining-countries-africa/#

Perfetto, M.C., & Vargas-Sánchez, A. (2018). Towards a smart tourism business ecosystem based on industrial heritage: Research perspectives from the mining region of Rio Tinto, Spain. *Journal of Heritage Tourism, 13*(6), 528–549.

Pretes, M. (2002). Touring mines and mining tourists. *Annals of Tourism Research, 29*(2), 439–456.

Prideaux, B., & Timothy, D.J. (2011). From mining boom towns to tourist haunts: The ghost town life cycle. In M. Conlin & L. Jolliffe (Eds.), *Mining heritage and tourism: A global synthesis*, pp. 227–238. London: Routledge.

Pwiti, G., & Ndoro, W. (1999). The legacy of colonialism: Perceptions of the cultural heritage in Southern Africa, with special reference to Zimbabwe. *The African Archaeological Review, 16*(3), 143–153.

Ray, H.P. (Ed.) (2018). *Decolonizing heritage in South Asia: The global, the national and the transnational*. London: Taylor & Francis.

Różycki, P., & Dryglas, D. (2017). Mining tourism, sacral and other forms of tourism practices in antique mines—analysis of the results. *Acta Montanistica Slovaca, 22*(1), 58–66.

Rudd, M.A., & Davis, J.A. (1998). Industrial heritage tourism at the Bingham Canyon copper mine. *Journal of Travel Research, 36*(3), 85–89.

Rutz, M.A. (2018). *King Leopold's Congo and the "Scramble for Africa": A short history with documents*. Indianapolis: Hackett.

Salinas Chavez, E., Delgado Mesa, F.A., Henthorne, T.L., & Miller, M.M. (2018). The Hershey sugar mill in Cuba: From global industrial heritage to local sustainable tourism development. *Journal of Heritage Tourism, 13*(5), 426–439.

Seif-El-Islam, R. (2020). Héritage industriel en Algérie, entre reconnaissance et abandon: Le cas du patrimoine minier et agricole de la ville d'Annaba. *Ethnologies, 42*(1–2), 167–185.

TICCIH (2022). Guiding principles and agreements. Online at: https://ticcih.org/about/about-ticcih/

Timothy, D.J. (2014). Views of the vernacular: Tourism and heritage of the ordinary. In J. Kaminski, A. Benson, & D. Arnold (Eds.), *Contemporary issues in cultural heritage tourism*, pp. 32–44. London: Routledge.

Timothy, D.J. (2018). Making sense of heritage tourism: Research trends in a maturing field of study. *Tourism Management Perspectives, 25*, 177–180.

Timothy, D.J. (2021a). *Cultural heritage and tourism: An introduction*, 2nd Edn. Bristol: Channel View Publications.

Timothy, D.J. (2021b). Democratising the cultural past: Western values, the Global South and cross-cultural perspectives in heritage tourism. In J. Saarinen & J.M. Rogerson (Eds.), *Tourism, change and the global south*, pp. 33–47. London: Routledge.

Timothy, D.J., & Tahan, L.G. (2020). Archaeology and tourism: Consuming, managing and protecting the human past. In D.J. Timothy & L.G. Tahan (Eds.), *Archaeology and tourism: Touring the past*, pp. 1–25. Bristol: Channel View Publications.

Tseane-Gumbi, L.A., & Ani, K. (2018). The political economy of mining tourism: A strategic nation building opportunity for the North West Province, South Africa. *African Renaissance, 15*(3), 125–138.

Turunen, J. (2020). Decolonising European minds through heritage. *International Journal of Heritage Studies, 26*(10), 1013–1028.

Ugwuanyi, J.K. (2021). Time-space politics and heritagization in Africa: Understanding where to begin decolonization. *International Journal of Heritage Studies, 27*(4), 356–374.

van der Merwe, C.D., & Rogerson, C.M. (2013). Industrial heritage tourism at the 'Big Hole', Kimberley, South Africa. *African Journal for Physical, Health Education, Recreation and Dance, 19*(3), 155–171.

van der Merwe, C.D., & Rogerson, C.M. (2018). The local development challenges of industrial heritage in the developing world: Evidence from Cullinan, South Africa. *GeoJournal of Tourism and Geosites, 21*(1), 186–199.

Vargas-Sánchez, A. (2015). Industrial heritage and tourism: A review of the literature. In E. Waterton & S. Watson (Eds.), *The Palgrave handbook of contemporary heritage research*, pp. 219–233. London: Palgrave Macmillan.

Wicke, C., Berger, S., & Golombek, J. (Eds.) (2018). *Industrial heritage and regional identities.* London: Routledge.

Xie, P.F. (2015). *Industrial heritage tourism*. Bristol: Channel View Publications.

Yang, X. (2017). Industrial heritage tourism development and city image reconstruction in Chinese traditional industrial cities: A web content analysis. *Journal of Heritage Tourism, 12*(3), 267–280.

Yap, G., & Saha, S. (2013). Do political instability, terrorism, and corruption have deterring effects on tourism development even in the presence of UNESCO heritage? A cross-country panel estimate *Tourism Analysis, 18*(5), 587–599.

13

DARK TOURISM AND 'PAINFUL PASTS' IN AFRICA

Concepts, Contexts, and Challenges

Philip R. Stone

> *When death finds you, may it find you alive!*
> *African Proverb*

Introduction

The world is sadly strewed with traumascapes of death and disaster that disturb our collective consciousness. These sites of atrocities and adversity display our painful pasts and (re)frame catastrophes and calamities. In turn, places *of* or *associated with* fatality attract contemporary tourists, resulting in a mediation of 'mortality moments' as well as contemplation of our difficult heritage. Indeed, our significant dead wander amongst us through engineered remembrance. Souls are memorialised and memories are orchestrated by a visitor economy that trades on tragedy. However, with the dead now returning to the touristic feast, tourists encounter reimagined 'ghosts' that warn us of our fights, follies, and misfortunes. It is here that the term 'dark tourism' – visiting sites of death or the seemingly macabre – has entered academic parlance (Stone, 2006). In turn, dark tourism has shone a critical light on the thanatological condition of contemporary society. Moreover, dark tourism raises profound questions of how and why we selectively recall our dead and inherent tourist experiences thereof. However, while dark tourism is a global occurrence, much of the academic literature over the past 20 years or so has focussed on dark tourism within European, American, or Australasian contexts. Consequently, there has been limited scrutiny of dark tourism from an African perspective.

The purpose of this chapter, therefore, is to address this gap in the literature and to outline key themes, issues, and challenges of dark tourism in continental Africa. In so doing, I offer an exploratory post-disciplinary discussion of dark tourism (after

DOI: 10.4324/9781003153955-13

Stone, 2011) with pan-African perspectives. Of course, it would be foolhardy to attribute all dark tourism in Africa as simply African – or indeed all dark tourism as simply 'dark'. Different countries in Africa have specific 'painful pasts' and where this is the case I include them as belonging to pan-African dark tourism. I explore different contexts of dark tourism in Africa that might be conceptually framed with comparable dark tourism elsewhere, as well as exploring dark tourism that is peculiarly African. Ultimately, I set out an original taxonomy for the study of potential dark tourism in Africa, while highlighting some conceptual considerations that can theoretically frame future dark tourism scholarship.

Dark Tourism in Africa: Towards a Typology

In 2006, I published an oft-cited paper in dark tourism studies that offered a typology of death and macabre-related tourist sites, attractions, and exhibitions (Stone, 2006). Drawing upon ideas of conceptually 'shading' dark tourism products and supply, rather than tourist experiences and demand, the 'dark tourism spectrum model' constructed a comprehensive approach to an otherwise diverse and fragmented phenomenon (Stone, 2006). As a result, dark tourism was illustrated to be a rather 'fluid spectrum of intensity whereby particular sites may be conceivably *darker* [original emphasis] than others' (Stone, 2006, p. 146). Of course, I was not the first to conceptually 'shade' dark tourism. I drew inspiration from scholars including Strange and Kempa (2003) and their shades of penal heritage; Sharpley (2005) and his matrix model to incorporate demand into 'shades of darkness'; Dann's (2005) playful alliteration of dark tourism places; and, from Miles (2002), who suggested some dark tourism places are 'darker' than others because of their locational authenticity. Consequently, my dark tourism spectrum model and its supply characteristics, including issues of location, commercialism and commemoration, interpretation and authenticity, political ideology, chronology, infrastructure, and purposefulness, offered the first defined framework in which to locate diverse types of dark tourism. Subsequently, I proposed a classification of 'Seven Dark Suppliers' whereby different dark tourism contexts might be 'plotted' – albeit subjectively – along the spectrum of supply (also see Lacanienta et al., 2020; Raine, 2013; Wyatt, Leask, & Barron, 2020). In many ways, this visually appealing typology of dark tourism helped open scholarly scrutiny of difficult heritage and its consumption within visitor economies. Yet, the spectrum model of dark tourism has been criticised by some for failing to consider tourist experiences, as well as the semantic notion of 'darkness' in dark tourism (Ashworth & Isaac, 2015; Biran & Poria, 2012; Bowman & Pezzullo, 2010). Nevertheless, the 'construction of a firm and comprehensive typological foundation will lead not only to a better understanding of dark tourism supply, but also, and perhaps more importantly, to a better understanding of where to locate and explore dark tourism demand' (Stone, 2006, p. 146). At the very least, and regardless of its inherent limitations (as acknowledged in the original paper), the dark tourism spectrum model propelled critical discourse about difficult heritage(s) and touristification of our significant dead.

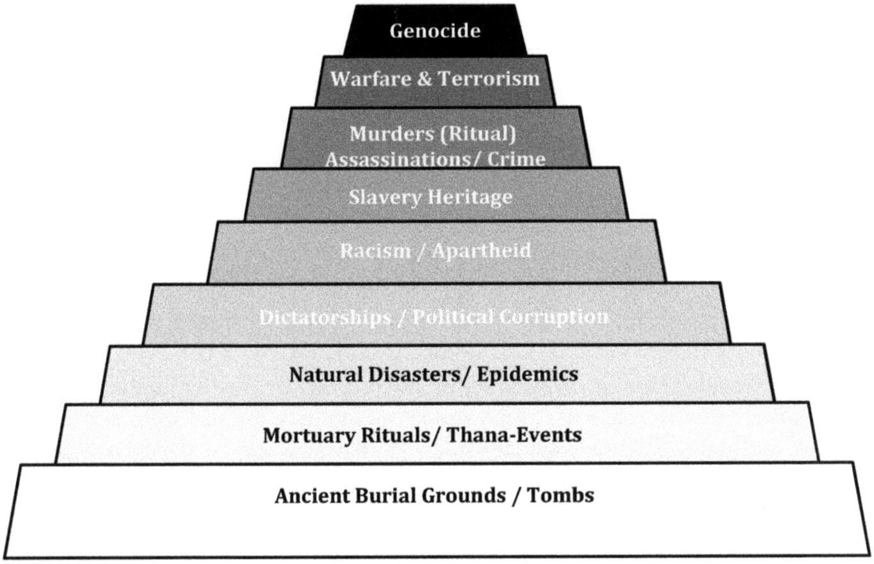

FIGURE 13.1 A taxonomy of African dark tourism themes.

At the crux of dark tourism as a concept is the (re)presentation of death or disaster. Notwithstanding semantic or linguistic complexities of the term 'dark' – dark tourism deals with the stuff of fatality – and that might be considered 'dark' across most cultures and societies (Stone, 2020b). Moreover, tourist experiences of various dark tourism sites remain at the core of dark tourism scholarship and fruitful research avenues are expected to continue (Stone, 2021). To that end, I have taken the shading aspects of my original dark tourism spectrum model and applied them here to a new typological model for dark tourism in Africa (Figure 13.1). Importantly, however, rather than applying judgmental parameters to define so-called 'darkness' as in the original spectrum – and without prejudice – I have taken different contexts that have perceived commonality. As a result, Figure 13.1 illustrates a broad taxonomy of potential dark tourism themes idiosyncratic to Africa, or that continental Africa has suffered – and continues to suffer. It is these specific themes that provide underpinning for potential dark tourism in Africa that I now outline (also see Werdler, 2012).

Genocide

In Namibia, which was known as German Southwest Africa under Germany's colonial occupation from 1884 to 1915, the 'forgotten genocide' of the early 20th century took place (Lemarchand, 2011). Between 1904 and 1907, after an indigenous Herero and Nama rebellion against the German seizure of land and cattle, the head of military administration – Lothar von Trotha – called for the extermination of the local populace. Up to 80,000 indigenous people were slaughtered, representing approximately 80% of the Herero people and 50% of the Nama people (ushmm.org, 2021). Through a systematic killing process carried out by German military forces – *Schutztruppe* – the intent of the slaughter was to rid the colony of people viewed as expendable and thus gain access to their land. Killings occurred through battles, starvation, and thirst in the Omaheke Desert, by sexual violence, and through forced labour, malnutrition, and medical experiments in concentration camps. This was the first genocide of the 20th century and was a prelude to the Holocaust in both methods employed and racial ideology. While the British were the first to utilise concentration camps during the Second Anglo-Boer War (1899–1902), the Germans *honed* their usage in what has been termed the 'continuity thesis' that links this holocaust in Africa with the Nazi Holocaust only three decades or so later (Bachmann, 2018; Baer, 2017). While the genocidal gaze of Germany dominated 20th-century geopolitics for the scale and horror of atrocities committed, it was only in 2021 that Germany officially recognised (with financial aid, not compensation) its colonial-era genocide in Africa (BBC News, 2021).

Africa is often associated with genocide. Despite the barbarisms of European colonial powers, including the German genocide in Namibia, post-colonial Africa has been the subject of accusations of genocide over the past few decades. Notwithstanding the official definition of genocide by the United Nations (1948), the number of such allegations relating to Africa has increased dramatically (Stapleton, 2017). Past and ongoing conflicts in countries such as Burundi, Ethiopia, Nigeria, the Democratic Republic of the Congo (DRC), Namibia, Rwanda, and Sudan have seen mass violence against certain populations. Thankfully, genocides in the strictest sense are rare but have indeed occurred. For instance, the 1994 genocide in Rwanda witnessed up to a million people slaughtered during a three-month period (Uvin, 2001). Victims were mostly Tutsis, but also included tens of thousands of Hutus who were killed as the country descended into the abyss. Racist mythology often advocates Africans belonging to 'tribes' that are inherently antagonistic towards one another and, therefore, engage in so-called 'tribal warfare' which cannot be easily rationalised. However, while the atrocities carried out in Rwanda were indeed genocide, according to Stapleton (2017), the many conflicts that plague post-colonial Africa have had logical explanations and, consequently, little of African warring has resulted in genocide.

Therefore, it is with caution and prudence that the language of genocide should be applied. Genocide memorialisation in Africa (as elsewhere) is open to contestation and, subsequently, to manipulation and perpetuation by dark tourism (Lischer, 2019). Hence, there needs to be an acknowledgement and understanding of how the emotive language and Western biases of genocide have been mobilised to describe episodes of tragic mass violence in Africa (Friedrich, Stone, & Rukesha, 2018). Indeed, as Stapleton (2017) points out, whilst there are a few instances where the international legal term 'genocide' has been applied appropriately in African history, there are many cases where it has not. Instead, the term has been utilised in a cynical manipulation to gain some political advantage

(Stapleton, 2017). Therefore, genocide accusations of post-colonial Africa often serve those who seek to maintain or gain hegemonic control, prolong wars and, thus, cause greater suffering and loss of life. Where actual genocides have occurred, a common history appears to be one of racial ideologies and hierarchies during colonialism, which when allied with psychological indoctrination, local geography and demography, religion, and/or socioeconomics (Dutton, 2007), the result has been tragic and appalling.

Warfare and Terrorism

The conventional wisdom is that African countries frequently go to war, but not with one another. Instead, they battle within their own internal borders. It is here that a high prevalence of civil wars, commonly attributed along ethnic lines, suggests that African rebel movements are almost always ethnically defined. Yet, contrary to widespread belief, the 'forever wars' of Africa are not due to the ethno-linguistic fragmentation of some of its states, but to elevated levels of poverty, failed political institutions, and economic dependence on natural resources (Elbadawi & Sambanis, 2000). In short, economic development miscarriages, deep political failures, and the lack of democracy – not tribalism or ethnic hatred – are the root causes of most of Africa's problems.

In 2020, at least 15 countries in Sub-Saharan Africa had active armed conflicts, including Burkina Faso, Burundi, Cameroon, the Central African Republic (CAR), Chad, DRC, Ethiopia, Kenya, Mali, Mozambique, Niger, Nigeria, Somalia, South Sudan, and Sudan (SIPRI, 2020). Almost all the conflicts are internationalised, including using state actors or through geopolitical proxies. Moreover, the transnational activities of violent Islamist groups, other armed groups, and criminal networks all add to conflict dynamics. As SIPRI (2020) reports, underlying forces of conflicts, as well as ethnic and religious tensions, are often manifested in lethal combinations of state weakness, corruption, lack of basic services, competition over natural resources (including food), acute inequality, and a sense of marginalisation. SIPRI (2020) goes on to suggest that two other cross-cutting issues will shape future regional security: the continuing internationalisation of counter-terrorism activities and the increasing impact of climate change (with water scarcity being a serious challenge). Additionally, since sub-state terrorism is endemic to Africa, future threats of atrocities are manifest in a complex mixture of sub-national and international terrorism (Cilliers, 2003; Elu & Price, 2015).

With the prevalence of armed conflict and terrorism in Africa being high, tourism is regarded as a phenomenon that requires peace to prosper. Moreover, tourism has been proposed as playing an interconnecting and facilitating role in the promotion of cross-cultural understanding amongst nations and peoples (Moufakkir & Kelly, 2010). Some commentators have even argued, optimistically, that tourism is a 'force for world peace' (D'Amore, 1988; Jafari, 1989; Salazar, 2006).

Regardless of sanguine notions of tourism acting for peace, the issue for African regions of perpetual conflict is that dark tourism is principally about 'recalling' conflicts for contemporary consumption. In other words, while war-related attractions might be considered the largest single category known within tourism (Smith, 1998), utilising military debacles where many people lose their lives raises unanswered questions about the relationship 'between cost and sacrifices of war and the response of the public in terms of tourism to war-related locations' (Butler & Suntikul, 2013, p. 1). Of course, utilising the detritus of war, whether in museums, monuments, or visitor attractions, is about cultural memory management and difficult heritage. As discussed later, the heritage of war is fraught with multiple interpretations and political ideologies and, consequently, is inherently 'dissonant' (Tunbridge & Ashworth, 1996). Nonetheless, the potential for dark tourism and African war/conflict-related tourism is both real and, regrettably, in abundance. The question is, of course, out of the many conflicts across Africa, 'whose battles' are remembered and why, and can African socio-political frameworks and economic infrastructure for tourism mobilities be secured and sustained?

(Ritual) Murders/Assassinations/Crime

The terrible affliction of murder crime is not unique to Africa. However, what is more prevalent is the act of ritual murders in many communities across the continent. Rugwiji and Masoga (2018) point out that ritual killing has become a major cause of concern for countries including Botswana, Nigeria, South Africa, Swaziland, Tanzania, Uganda, and Zimbabwe. Although ritual murders have been carried out elsewhere in the world, the mortiferous character of the killings in Africa often means that body parts, including genitals, are 'ritually' removed from the corpses, often when the victim is still alive. Ritual killings, or *muti* murders, are grounded in the supernatural, biblical doctrines, or chiefly rivalries. Consequently, body parts are used to 'make medicine' to politically or economically 'strengthen' those who commit the homicide (de Jong, 2015). However, while these murders are perceived in rational and utilitarian terms, the cultural dynamics of the slayings often go unheeded. As de Jong (2015) argues, ignoring the 'thick description' that can provide critical insights into the killings means that they persist in the present day. Whereas *muti* murders were once performed to strengthen a king or chief and, thus, the entire society over which he reigned, the notion of 'strengthening' has been omitted in motivational discourse dealing with these murders (de Jong, 2015). Therefore, rather than senseless brutality, the murders might be seen as 'meaningful violence' (de Jong, 2015). Crucially, therefore, national conversations and educative programmes are required to address issues of cultural ignorance and superstition, as well as necromantic tales of the supernatural and sorcery (Steyn, 2005; Vincent, 2008).

It is here that dark tourism might play a meaningful role in not only commemorating murder victims within African visitor economies, but also creating and mediating essential narratives as to the evolution, rationale, and consequences of these brutal killings. Rather than obliterating the murders as cultural 'black spots' in a post-modern landscape (after Rojek, 1993), the *muti* murder (victims) may be rectified and sanctified through a touristic process of remembrance (Foote, 2003). Remembrance of other significant murders within African contexts, including those against albinos (Rokosz, 2018), might also follow a similar pattern of collective memory and national conversation. While sexual homicide, family killings, vigilantism, terrorist and political assassinations, and serial murders in Africa (Bell, 2019; Labuschagne, 2018, 2021; Smith, 2019) may be the 'stuff of dark tourism', it is auratic and polysemic dark tourism experiences that could potentially open societal mediation for such perilous attacks (Apleni et al., 2020; Seaton, 2009).

Slavery Heritage

There is no universal framework in which a reconstruction of the history of slavery can be made. However, Africa as a whole and in part has suffered from the abominations of slavery throughout history, though this has never existed within a geopolitical vacuum. Instead, from antiquity to modern times, the stain of slavery has afflicted Africa in many forms. Undeniably, Africa is intimately allied with slavery – both as a source of slaves for ancient civilisations, the Islamic world, India, and the Americas – and, as a place where exploitative bondage was practised (Lovejoy, 2012). Broadly speaking, African slavery, which persisted in Africa well into the 20th century and where 'descendant-slavery' still exists today, expanded from about 1350 until the early 1900s (Rice, 2015). During this period, slavery became entrenched in the African political economy and was an international mercantile system that was commercialised, racialised, and inherited.

Much of slavery heritage today in Africa, certainly from a touristic perspective, is based on transatlantic slavery and the triangular slave trade between Africa (particularly West Africa), the Americas, and European colonial powers (most notably the British, Portuguese, Spanish, and French) (Teye & Timothy, 2004). Yet, while slavery heritage has suffered historical amnesia within scholarly agendas, it is only recently that public portrayals of transatlantic slavery legacies have ameliorated (Rice & Kardux, 2012). It is here that slavery heritage has not only re-entered Western public consciousness, but also has created critical conversations about slavery legacies, consequences, and diasporas (Dann & Seaton, 2001). Therefore, in Africa, opening slavery heritage discourse for mass (dark) tourism consumption will address the erasure of slavery memory for many. Through affect and emotion, dark tourism can reinvigorate cultural memory about the brutal impacts of slavery, as well as address the oft-forgotten African complicity in the slave trade (Rice, 2012). Indeed, while the role of the colonial powers in slavery is well documented, there is often silence on the role Africans played (Nwaubani, 2019). With roles of liaison brokers and dockland stevedores, and a system of slavery already in place before the Europeans arrived, Africans maintained a crucial link by being Black suppliers and facilitators of (their own) enslaved people to White colonial traders (Shenoy, 2019; Smith, 2009). Hence, with inclusive and holistic narratives, dark tourism can address the taboo of slavery with a sense of provocation, risk-taking, and outspokenness (Williams, 2002).

Racism/Apartheid

Racism is a scourge of prejudice, discrimination, and antagonism by individuals, communities, or institutions against people because of belonging to a specific racial or ethnic group. However, this is a simplified explanation of a complex issue where concepts of racism that have discrimination and racial superiority at their core are not always straightforward. Rather, racism is a fluid idea that changes over time and within new social contexts, yet false notions of racial difference have become embedded in the beliefs and behaviours of society. Of course, the concept of a 'racial group' which pertained from certain 19th- and 20th-century anthropological discourse has long been disproven. These theories were influenced by colonialism, and the desire to demonstrate that non-White people were inferior to White people to justify the imperialistic actions of Western (particularly European) nations.

Yet, Africa has suffered from many forms of racism for centuries. For instance, the ethnic pygmy populations in central Africa suffer from racialised discrimination from Bantu people, who see them as 'untouchables' (Survival International, 2021). In 2004, nationalists of the Republic of Côte d'Ivoire in the economic capital of Abidjan, rallied by the state media, beat and raped nationals of European and Lebanese descent. The Human Rights Watch organisation in 2001 reported that the Republic of Côte d'Ivoire government was guilty of fanning ethnic hatred for hegemonic ends. In Niger, in 2006, the government abandoned its plans to deport around 150,000 Diffa Arabs from eastern Niger to Chad. In Somalia, the Somali Bantus face significant stigmatisation due to physical differences and ancestry from the majority Cushitic. Marginalisation of the Somalia Bantu is based on ethno-racial factors, whereas other Somalian minorities, such as the Madhiban, face marginalisation primarily based on socially constructed caste classifications.

However, despite the many instances and multifaceted nature of racism in Africa, in South Africa, racism was legally institutionalised under a racial segregation system. From 1948 until the early 1990s, South Africa and South West Africa (now Namibia) operated 'apartheid' – an authoritarian political culture based on *baasskap* (boss-hood or boss-ship). It was here that the nation's minority White population politically, economically, and socially dominated Asians, 'Coloureds', and Black Africans. The dismantling of apartheid culminated in South Africa's general election on 27 April 1994 – the first-ever held with universal suffrage – and the inauguration of Nelson Mandela as the country's first Black president. However, the damaging economic and sociocultural legacy of apartheid continues to the present day, despite 27 April being declared an annual public holiday in South Africa called 'Freedom Day'.

Although racism continues to plague certain parts of Africa, dark tourism can help shine a potential light on its history and enduring impacts. With political support and willingness to address ethnic minority concerns (which seems unlikely at least in the short to medium term), dark tourism can open national debates and discourse on racial persecution and segregation. In South Africa, for example, dark tourism at the former penal colony of Robben Island, which incarcerated Nelson

Mandela, has become a global shrine of remembrance as South Africa comes to terms with its injustices and democratic transformation (Mangwane, Hermann, & Lenhard, 2019; Strange & Kempa, 2003). That said, however, dark tourism to Robben Island as a focal point of South Africa's negotiated transition from autocracy to democratic dispensation is often characterised by misconstruction and romanticism (Howell & Shearing, 2017). Indeed, the Island employs themes of triumph over oppression as narratives of nation-building which, in turn, may limit how visitors think about social (in)justice in contemporary South Africa (Wood, Berger, & Hasian, 2017).

Dictatorships/Political Corruption

Political and bureaucratic corruption is endemic in many African states. Indeed, corruption is often the *modus operandi* of governance where it is difficult to find an African nation that does not have a corrupt regime (Ebbe, 2005). Corruption is an illegal or immoral act with the intent to gain political or economic advantage inconsistent with official duty and the rights of others. However, before the colonisation of Africa by European imperialists in the 19th century, Africa was largely inhabited by stateless societies and kingdoms without formalised legislatures or complex administrative orders. In other words, pre-colonial governance in Africa was simple and homogenous and based upon religious aristocracies or gerontocracies (Ebbe, 2005). With the 'European Scramble for Africa' and subsequent colonisation of various ethnic groups (or tribes), an uncomfortable union ensued with amalgamations often achieved upon corrupting some tribal leaders to concede to others (Amundsen, 2019). Arguably, therefore, political corruption was imported into Sub-Saharan Africa. As the tribes became disintegrated, colonial 'masters' introduced a unitary system of government whereby a latent and uneasy peace was maintained.

After colonial administration in Africa, the struggle for leadership and control of wealth ensued. Those in control of economic and political power used all means to accumulate wealth to maintain hegemonic control, often resulting in despotism and dictatorships. As a result, political corruption became institutionalised in Africa by the nature, process, and structure of colonial governance – and the subsequent quest for hegemony by certain indigenous individuals and groups (Heidenheimer & Johnston, 2002). Undoubtedly, corruption at all political levels is complex and, realistically, eradicating corruption within African governance completely would take a socio-political tsunami to change both mindsets and policies. Nevertheless, a focus on the consequences of the painful heritage of political corruption and political autocracy could be a focus of a new museology in African institutions. Notwithstanding inherent opposition, this embracing (as well as problematic) museology in any African post-dictatorship would be driven by the visitor economy where discourses of corruption can be highlighted for contemporary consumption.

For instance, elsewhere, museums of Catalonia are examined by de Lluc Serra and Alcalde (2016) who evaluate not only the early cultural heritage impacts of the Francoist dictatorship in Spain, but also the role of the museum in a post-Franco world (also see Sutherland, 2014). Similarly, 'memorial museums' in Chile and Argentina are explored by Andermann (2012) who highlights a new mode of exhibitionary remembrance of traumatic political corruption. Andermann (2012) notes how museographical interpretation sidesteps 'objectivity' in favour of soliciting advice and invoking performative empathy from visitors. Other examples may be sought in eastern European museological representations of fascism and 'collaborative memory' (Apor, 2012; Jones, 2017), human rights and dictatorships in Latin America (Jelin & Herschberg, 2019), and contradictory memory management of political corruption in Cambodia and Italy (Violi, 2012). Indeed, the so-called 'transitional justice' museums can help persuade visitors to seek democracy and conciliatory policies, and to decrease support for repression, tyranny, and corruption (Balcells, Palanza, & Voytas, 2022).

Natural Disasters/Epidemics

Popular media portrayals of Africa are often as a continent in perpetual crisis. With afflictions of disease, environmental degradation, and famine, disasters occur when natural hazards and human vulnerability collide. Indeed, 'as far as natural hazards are concerned, many of them are as old as the continent and intrinsic to its ecology' (Loretti & Tegegn, 1996, p. 179). Notwithstanding the global COVID-19 pandemic from 2019 onwards, epidemic outbreaks often curse African countries, including diseases such as Ebola, Lassa fever, cholera, tuberculosis, AIDS and HIV, malaria, dysentery, yellow fever, and plague, as well as widespread measles. Throughout the 'Meningitis belt', hundreds of millions of people in twenty African countries are susceptible to recurrent epidemics (Loretti & Tegegn, 1996). These epidemics are often compounded by extreme weather storms, drought, cyclones, and earthquakes. Allied with poor disaster management, corrupt political institutions, and an over-reliance on external agencies and foreign aid programmes, natural catastrophes continue to cause significant destruction and loss of life (Aliyu, 2015).

With the consequences of global climate change yet to be fully realised, and with infectious disease epidemics emerging and re-emerging, natural hazards pose a major problem for healthcare and humanitarian coordination in Africa. However, the frequency at which these events occur means that the multiple causes and impacts of the disaster go unheeded. With one natural disaster following another, there has been limited appetite to commemorate what has been lost. Rather, the focus is on reconstruction. Therefore, tourism in post-natural-disaster destinations is fraught with difficulties. Apart from *voluntourism* in (post)disaster periods, while 'hoping' for a better future (Koleth, 2014), volunteer tourists may hinder rather than help host-residents (Rosenberg, 2018), yet tourism over the past two decades is vital to African economies. Indeed, tourism accounted for approximately 7% of Africa's GDP (Gross Domestic Product) in 2019 and contributed over $169 billion to its economy (Monnier, 2021). However, notwithstanding issues of remembering specific natural disasters through touristic memorialisation, it may not be dark tourism that is the panacea for Africa and its natural hazards. Rather, as COVID-19 has illustrated, the pandemic revealed an over-reliance on international tourism to Africa (Matiza, 2021). With a fall in international visitors to the continent during the pandemic, local and regional tourism flourished in the shadow of the world pandemic crisis. Moreover, as Monnier (2021) highlights, it is Africa's growing middle class, its soaring population of young travellers, and the recently launched African Continental Free Trade Area (AfCFTA), that are among the pillars for supporting the future growth of domestic and regional tourism – dark or otherwise.

Mortuary Rituals/Thana-Events

Mortuary rituals exist across societies and cultures where death is universal but dying is not. In other words, while death is a biological certainty, dying is a socio-cultural dynamic with varying end-of-life care systems, religious liturgy, and secular rituals that permit the dead to pass from this world (to the next). In the United Kingdom, for instance, the funeral is habitually a solemn affair and often based upon Victorian-era imaginations of funerary practice. However, global mortuary rituals of the Other exist and may appear exotic and enigmatic to the Western Self (Eaton, 2011). For example, the Iban populace of Sarawak, East Malaysia, have eschatological practices that focus on the notion of a single separable soul – the *semangat.* It is here that ghosts are invited from Sebayan – the Land of the Dead – and entertained through festivities. Consequently, the 'ghosts' are given goods which enable the *semangat* to 'live' self-sufficiently in Sebayan and, thus, the festival becomes a ceremony of final separation (Uchibori, 1978). Meanwhile, the Yű-lan-p'en is a Chinese Buddhist mortuary festival that synthesises ancestral sacrifices and the Chinese value of filiality with Indian notions of karma. It is within this envelope of mortuary festivity that 'China was made more Buddhist and Buddhism made more Chinese' (Teiser, 1986, p. 47).

Similarly, African mortuary rituals comprise numerous forces that have contributed to re-inventions of practices and beliefs that surround death, which are both modern yet rooted in historicity, folklore, and religions (Lee & Vaughan, 2020). For example, 'death celebration' became the most important cultural 'thana-event' throughout the Western Grassfields of Cameroon during the 20th century (Jindra, 2005). It is here that Christianity, with its doctrinal messages of personhood and an eternal afterlife morphed with the hierarchical structure of Grassfields society and ancestor worship. Ironically, the cultural appropriation of the mortuary festival has 'created a proliferation of ancestors for whom delayed mortuary rites such as *death celebrations* (original emphasis) are owed' (Jindra, 2005). Meanwhile, in Ghana, the mortuary institutions of the Lodagaa include funerary ceremonies, ideas of the afterlife, ancestor worship, and the exclusive system of property inheritance rights (Goody, 1962 [2004]). Comparably, in the highlands of Madagascar, secondary burials – or *famadihana* – are a funerary tradition of the Malagasy people. Dancing to live music and general festivity, the bereaved bring forth the bodies of their ancestral dead during a ceremony known as *the turning of the bones*. In doing so, they rewrap the corpses in fresh shrouds and rewrite the names of the deceased on the cloth in an act of perpetual remembrance (Larson, 2001).

Although there are many more mortuary festivals across continental Africa, some of these mortuary rituals and resultant festivals could be the touristic subject of cultural export. In other words, to maintain and conserve ancient ritualistic mortality events, African communities may wish to 'stage' their mortuary ceremonies for (dark) tourism consumption. In a similar vein, the indigenous Māori people of New Zealand and the South Pacific have used their traditional *haka* dance at *tangi* (funerals) on *marae* (Māori meeting grounds) where the dead are mourned and remembered. As an integral part of the Māori mourning process, it allows participants to express grief that a loved one has passed away, as well as being part of cultural memory and commemoration (Hoani, 2015). However, in so doing, the *haka* and its mortuary rituals are culturally exported to the mainstream. Consequently, Māori identity and spirituality may be used as culturally responsive learning interventions in addition to being part of the tourist gaze (after Urry, 2002; see also Whitinui, 2004). Of course, notwithstanding inherent issues of cultural authenticity and appropriation, African funerary rituals have the potential to display and conserve, promote, and culturally export creative approaches to death and the dead. Indeed, as in the case of the Mexican Day of the Dead mortuary ritual, a festivity that accompanies such thana-events can be symbolic and even be a form of national identity or cultural imperialism (Brandes, 1998).

Ancient Burial Grounds/Tombs

Africa is commonly considered the cradle of human modernity (Oppenheimer, 2004). As such, Africa is littered with ancient landscapes that provide cultural connections to our distant post-mortem past. Africa is also home to the earliest known human burial – that of a young child found in a cave in Kenya and named *Mtoto*, meaning 'the kid' in Swahili (Martinón-Torres, d'Errico, & Santos, 2021). Moreover, while ancient people have long departed, their olden funerary practices discovered by archaeology of burial grounds offer a glimpse into our ancestors' lives and deaths. Indeed, burials and disposal of the dead throughout history have resulted from fundamental human behaviour and are commonly distinctive and reflective of profound ideological beliefs (Tarlow & Nilsson-Stutz, 2013). Burials from the past allow us to observe how our ancestors treated the cadaver, how monuments and tombs were constructed, how bygone social structures were maintained, how grave goods and votive offerings made peace with the gods, as well as exploring human demographics, mobilities and migration, and health and nutrition. Of course, the mundane and inevitable event of death has become linked to complex processes of commemoration. Therefore, in African communities, as elsewhere, 'funerary evidence is often read as a sort of transcript of social complexity' (Gatto et al., 2019, p. 6). Consequently, the creation and enhancement of social memory through ancient funerary rituals and tombs appear to be the key stimulus behind mortuary practices in many bygone societies.

Thus, from the monumental pyramids of Giza to rock-cut tombs of the Valley of the Kings, Pharaonic Egyptian funerary traditions (including mummification) dominate Africa's earliest burial landscapes. Moreover, the Libyan Saharan region known as Fezzan is also notable for the range and complexity of its pre-Islamic funerary heritage (Stone & Stirling, 2007). Similarly, in Zimbabwe, during the building of the Kariba Dam on the Zambezi River in the 1960s, an ancient burial ground was revealed with tombs full of rich gems and funerary ornaments (Childress, 1989). Meanwhile, in the landscape of red sandstone in northwest Kenya, near the shores of Lake Turkana is a mysterious 5,000-year-old burial mould called Lothagam North Pillar which contains hundreds of Eastern Africa's early herders (Wei-Hass, 2018). While the full range of ancient African burial grounds is beyond the scope of this chapter, the first hominins emerged from 'Out of Africa' some 6–7 million years ago. With the earliest anatomically modern human remains discovered in Africa, including skulls at the Omo Kibish site in Ethiopia, the Jebel Irhoud site in Morocco, and the Florisbad skull representing the early Homo sapiens found in South Africa, our early ancestry is on display in Africa.

It is here that (dark) tourism can help conserve African ancient burial grounds through a new museology of interactive consumption and visitor experiences. Dark tourism can bring the ancient dead alive through new interpretive technologies that rely less on displaying skulls in glass cabinets but allowing us to intermingle with our ancestors through augmented realities or tactile virtual reality. Indeed, the futurology of dark tourism might assume the ancient dead as a starting point to bring mortuary customs and traditions to new technological horizons (Stone & Wright, Forthcoming). Notwithstanding issues of cultural conservation, archaeological integrity, and environmental sustainability, the ancient burial grounds of Africa are plentiful and, as such, might form a specific niche for dark tourism experiences.

Dark Tourism in Africa: Conceptual Considerations

Dark tourism is concerned with encountering places of fatality or acts of calamity that have perturbed the public consciousness. Moreover, dark tourism has domesticated death and exposes a cultural institution that mediates between the ordinary Self and the significant Other dead (Stone, 2020a). To that end, dark tourism in the so-called 'dark continent' of Africa is full of moral ambiguities, managerial dilemmas, and political conundrums. As I have outlined, potential dark tourism in Africa may be classified into a broad, but not exclusive pyramidal of mutual themes. Despite its increasing use to intellectually frame difficult heritage, dark tourism remains divisive as a concept as well as being ethically contentious in practice. Of course, not all noteworthy fatalities and significant disasters in Africa can be or even should be the focus of dark tourism. Importantly, it is where the lines of commemoration, remembrance, and commercialisation of 'heritage that hurts' blur, and with political and stakeholder goodwill, can dark tourism be 'packaged' for the visitor economy. However, as a global field of scholarly endeavour, dark tourism can traverse disciplinary borders, challenge social scientific subject gatekeepers, and engage directly with heritage-producers and tourist-consumers. Indeed, dark tourism as a post-disciplinary academic brand can connect difficult heritage with visitor experiences (Hartmann et al., 2018; Stone, 2013; Stone et al., 2018).

It is here that the study of dark tourism within African contexts may be underpinned by specific conceptual frameworks. Indeed, African contexts of dark tourism scholarship can act as an academic lens to scrutinise fundamental interrelationships of African contested and displaced heritage. One such field of study is *postcolonialism*, whereby the human, cultural and political consequences of imperialism may be interrogated within dark tourism sites and experiences. While the act of tourism and postcolonialism have been linked (see, for example, d'Hauteserre, 2004), the critical intersections between dark tourism and post-colonial scholarship are less well-made (Light, 2017). Indeed, within African contexts of genocide, slavery, racism, or political corruption, dark tourism can highlight tensions between the visitor economy, nationalism, and 'the reassertion of marginalised or suppressed histories' (Carrigan, 2014, p. 240). Indeed, Carrigan (2014) notes that intersections of dark tourism and postcolonialism need to recognise genealogies of colonialism and its influences on contemporary (re)presentations of dark sites. Moreover, Carrigan (2014) argues for a greater focus on the perspectives and voices of indigenous communities. Indeed, locating African dark tourism within postcolonial research may provide greater cultural and historical texture to the dark tourism literature and, subsequently, can offer a transformative approach to the ethics of dark tourism commoditisation. Such an approach would enhance our understanding of experiential and affective dimensions of dark tourism, including for tourists, site workers, as well as for local and national audiences. As Carrigan (2014, p. 238) suggests:

> If dark tourism sells – literally and academically – then it is possible to imagine a form of compatibility between the recovery of suppressed narratives in which post-colonial studies specializes, and the reinterpretation of many sites of touristic consumption.

Supressed narratives of traumascapes raise fundamental questions about the political and power dynamics of dark tourism which, in turn, lend to the concept of *dissonant heritage* (Tunbridge & Ashworth, 1996). In short, dissonant heritage represents discord or lack of agreement or coherence to a given heritage. Indeed, the memorialisation processes so inherent in dark tourism in general and dark tourism in Africa in particular, are fraught with dissonance. Within African dark tourism contexts, such as apartheid, slavery, war, or genocide, the notion of dissonance and interpretation has been investigated (Ashworth, 2004), yet the idea of *plural heritages* and community co-production of difficult heritage in Africa remains unexplored. Consequently, plural heritages are aligned with cultural and political imperatives that consider the plural values, meanings, affects, and relativities of dark tourism for vested communities and multiple stakeholders (Whitehead, Schofield, & Bozoğlu, 2021). Adopting such a conceptual approach within African dark tourism suggests a range of methods for locating and valorising alternative perspectives, while also exploring the inherent complexities of painful pasts in the present, as well as the ontology of displaced heritage (Roberts & Stone, 2014).

These alternative perspectives of a plural heritage approach can be allied to *meaning-making* and the tourist co-creation of meaning within the dark tourism process. Indeed, the 'emotional turn' within tourism studies has focussed on issues of emotion and affect (Buda, d'Hauteserre, & Johnston, 2014), whereby tourism is co-constructed by producers and consumers through negotiation and embodied performance (Chronis, 2005). Indeed, within African dark tourism contexts of racism, political corruption, or incidents of genocide, a co-creation approach to meaning-making would depend upon contingencies of the present, as well as the painful past. Consequently, traumascapes as socio-spatial encounters can be 'symbolically transformed and used by service providers and tourists alike to negotiate, define, and strengthen social values… when these values are most needed' (Chronis, 2005, p. 386). Moreover, where dark tourism and affect may frame African places of death and disaster (after Martini & Buda, 2020), dark tourism can contribute to multiple potentialities of *soft-power* and *place-making* (Clarke, Bull, & Deganutti, 2017; Lew, 2017). Notwithstanding issues of interpretation and commercial complexities, neoliberal place-making of dark tourism sites can add to issues of cultural diplomacy, Black empowerment, national identity, collective grief, hope, and transitional justice for African places of pain and shame (Goudie, Khan, & Kilian, 1999; Stone & Grebenar, 2021; White & Frew, 2013).

Similarly, conceptual frameworks that focus on *diaspora tourism* (Coles & Timothy, 2004; Li, McKercher, & Chan, 2020) may underpin dark tourism in Africa. Diasporic African communities exist across the world and much of dark tourism scholarship thus far has focussed on slavery heritage (e.g., Boateng, Okoe, & Hinson, 2018; Mensah, 2015; Otoo, Kim, & King, 2021). However, diaspora studies in other African dark tourism contexts could focus on other cultural heritage, such as mortuary rituals, whereby ancestral homelands, festivity, and funerary traditions form the attraction for the travelling diasporas (Leo & Jamal, 2013). Indeed, notions of roots-seeking and diasporic identity may also be allied to other disciplinary fields, particularly that of *thanatology*. Hence, the dominant Western-centric scholarship of dark tourism and the thanatological condition of contemporary society should be viewed from other sociocultural perspectives (Cohen, 2018; Speakman, 2019). To that end, mortuary festivals across Africa present an opportunity to examine not only differing cultural approaches to death, but also how the dark tourist gaze might be fixed on the dead.

Conclusion

While the social scientific academy has used dark tourism as a conceptual lens to scrutinise difficult cultural heritage, much of the research thus far has either originated in or focused on the Western world. However, as I demonstrated in this chapter, there are multiple potentialities for dark tourism across Africa as well as numerous post-disciplinary frameworks in which to locate dark tourism scholarship.

Thus, I have set out in this chapter several contexts and themes in which dark tourism in Africa might be categorised. In summary, I have provided an overview of these contexts with illustrative insights and, subsequently, constructed a comprehensive but not exclusive typology of dark tourism in Africa. Notwithstanding inherent limitations of subjectivity, the typological nature of dark tourism in Africa is as idiosyncratic as it is universal. Therefore, future dark tourism research within African contexts may be located across a broad pyramidal taxonomy and, in so doing, provide coherence to dark tourism as a subject field.

Importantly, however, conceptual underpinning to future dark tourism in Africa studies will strengthen and broaden post-disciplinary discourses and critical intersections of different subjects. I have outlined a few selective theoretical frameworks; there are others. Indeed, some of the conceptual considerations have already been used to worthy effect within dark tourism studies, but others are less so. Therefore, dark tourism in African contexts might be augmented by theoretically locating research in areas such as postcolonialism, dissonant heritage, plural heritages, cultural imperialism, meaning-making and affect, place-making, soft power, diasporas, thanatology, nationhood, and identity, as well as futurology. Of course, there are other fields that dark tourism in Africa may contribute to, including but not limited to, war and genocide studies, pilgrimages, performance and drama, existentialism, history, ethics and morality, consumerism and postmodernity, development and reconstruction, criminology and penal heritage, and education. Adopting grounded theoretical frameworks, dark tourism as a field of study can enlighten the challenges and consequences of (re)presenting death and calamity within African visitor economies. Ultimately, dark tourism in Africa is about finding out about life, living, and meaning through the sad demise of others – and to that end, 'the death' of the Other can often make the Self feel 'alive'.

References

Aliyu, A. (2015). Management of disasters and complex emergencies in Africa: The challenges and constraints. *Annals of African Medicine, 14*(3), 123–131.

Amundsen, I. (Ed.) (2019). *Political corruption in Africa: Extraction and power preservation.* Cheltenham: Edward Elgar.

Andermann, J. (2012). Showcasing dictatorship: Memory and the museum in Argentina and Chile. *Journal of Educational Media, Memory, and Society, 4*(2), 69–93.

Apleni, L., Mangwane, J., Maphanga, P.M., & Henama, U.S. (2020). The interface between dark tourism and terrorism in Africa: The case of Kenya and St Helena. In M.E. Korstanje & H. Seraphin (Eds.), *Tourism, terrorism and security: Tourism security, safety and post conflict destinations*, pp. 115–134. Bingley: Emerald.

Apor, P. (2012). Master narratives of contemporary history in Eastern European national museums. In D. Poulot, F. Bodenstein, & J. Guiral (Eds.), *Great narratives of the past: Traditions and revisions in national* museums, pp. 569–585. Linköping, Sweden: Linköping University Electronic Press.

Ashworth, G.J. (2004). Tourism and the heritage of atrocity: Managing the heritage of South African apartheid for entertainment. In T.V. Singh (Ed.), *New horizons in tourism: Strange experiences and stranger practices*, pp. 95–108. Wallingford: CABI.

Ashworth, G.J., & Isaac, R.K. (2015). Have we illuminated the dark? Shifting perspectives on 'dark' tourism. *Tourism Recreation Research, 40*(3), 316–325.

Bachmann, K. (2018). *Genocidal empires: German colonialism in Africa and the Third Reich*. Berlin: Peter Lang.

Baer, E. (2017). *The genocidal gaze: From German Southwest Africa to the Third Reich*. Detroit: Wayne State University Press.

Balcells, L., Palanza, V., & Voytas, E. (2022). Do transitional justice museums persuade visitors? Evidence from a field experiment. *The Journal of Politics, 84*(1), 496–510.

BBC News (2021). Germany officially recognises colonial-era Namibia genocide. *British Broadcasting Corporation*, 28 May, Available at https://www.bbc.co.uk/news/world-europe-57279008 [Accessed 22 July 2021].

Bell, L.N. (2019). Terrorist assassinations and societal unrest in Africa: A research brief. *Dynamics of Asymmetric Conflict, 12*(3), 242–256.

Biran, A., & Poria, Y. (2012). Reconceptualising dark tourism. In R. Sharpley & P.R. Stone (Eds.), *Contemporary tourist experience: Concepts and consequences*, pp. 59–70. London: Routledge.

Boateng, H., Okoe, A.F., & Hinson, R.E. (2018). Dark tourism: Exploring tourists' experiences at the Cape Coast Castle, Ghana. *Tourism Management Perspectives, 27*, 104–110.

Bowman, M.S., & Pezzullo, P.C. (2010). What's so 'dark' about 'dark tourism'? Death, tours, and performance. *Tourist Studies, 9*(3), 187–202.

Brandes, S. (1998). The Day of the Dead, Halloween, and the quest for Mexican national identity. *The Journal of American Folklore, 111*(442), 359–380.

Buda, D.M., d'Hauteserre, A.M., & Johnston, L. (2014). Feeling and tourism studies. *Annals of Tourism Research, 46*, 102–114.

Butler, R., & Suntikul, W. (2013). Tourism and war: An ill wind? In R. Butler & W. Suntikul (Eds.), *Tourism and war*, pp. 1–11. London: Routledge.

Carrigan, A. (2014). Dark tourism and postcolonial studies: Critical intersections. *Postcolonial Studies, 17*(3), 236–250.

Childress, D.H. (1989). *Lost cities and ancient mysteries of Africa and Arabia*. New York: Adventures Unlimited Press.

Chronis, A. (2005). Co-constructing heritage at the Gettysburg storyscape. *Annals of Tourism Research, 32*(2), 386–406.

Cilliers, J. (2003). Terrorism and Africa. *African Security Review, 12*(4), 91–103.

Clarke, D., Bull, A.C., & Deganutti, M. (2017). Soft power and dark heritage: Multiple potentialities. *International Journal of Cultural Policy, 23*(6), 660–674.

Cohen, E. (2018). Thanatourism: A comparative approach. In P.R. Stone, R. Hartmann, A. Seaton, R. Sharpley, & L. White (Eds.), *The Palgrave handbook of dark tourism studies*, pp. 157–171. London: Palgrave Macmillian.

Coles, T., & Timothy, D.J. (Eds.) (2004). *Tourism, diasporas and space*. London: Routledge.

D'Amore, L.J. (1988). Tourism: A vital force for peace? *Tourism Management, 9*(2), 151–154.

Dann, G. (2005). Children of the dark. In G.J. Ashworth & R. Hartmann (Eds.), *Horror and human tragedy revisited: The management of sites of atrocities for tourism*, pp. 233–252. New York: Cognizant.

Dann, G.M.S., & Seaton, T. (2001). Slavery, contested heritage and thanatourism. *International Journal of Hospitality & Tourism Administration, 2*(3–4), 1–29.

d'Hauteserre, A.M. (2004). Postcolonialism, colonialism, and tourism. In A.A. Lew, C.M. Hall, & A.M. Williams (Eds.), *A companion to tourism*, pp. 235–245. London: Blackwell.

de Jong, W. (2015). Makhosi a via (Chiefs commit ritual murder) – Why ritual murders in Southern Africa should be seen as meaningful violence (and not senseless). *Afrika Focus, 28*(2), 9–26.

de Lluc Serra, M., & Alcalde, G. (2016). Dictatorship and museums: The museums of Catalonia (Spain) in the early years of the Franco dictatorship (1939–47). *Museum History Journal, 9*(2), 205–214.

Dutton, D.G. (2007). *The psychology of genocide, massacres, and extreme violence*. Santa Barbara, CA: Praeger.

Eaton, K. (2011). Monthly lunar festivals in the mortuary realm: Historical patterns and symbolic motifs. *Journal of Near Eastern Studies, 70*(2), 229–245.

Ebbe, O.N.I. (2005). Political corruption in Africa. In R. Sarre, D.K. Das, & H.J. Albrecht (Eds.), *Policing corruption: International perspectives*, pp. 99–106. Lanham, MD: Lexington Books.

Elbadawi, E., & Sambanis, N. (2000). Why are there so many civil wars in Africa? Understanding and preventing violent conflict. *Journal of African Economies, 9*(3), 244–269.

Elu, J.U., & Price, G.N. (2015). The causes and consequences of terrorism in Africa. In C. Monga & J.Y. Lin (Eds.), *The Oxford handbook of Africa and economics: Volume 1 – contexts and concepts*, pp. 724–738. Oxford: Oxford University Press.

Foote, K.E. (2003). *Shadowed ground: America's landscapes of violence and tragedy*, 2nd Edn. Austin: University of Texas Press.

Friedrich, M., Stone, P.R., & Rukesha, P. (2018). Dark tourism, difficult heritage, and memorialisation: A case of the Rwandan genocide. In P.R. Stone, R. Hartmann, A. Seaton, R. Sharpley, & L. White (Eds.), *The Palgrave handbook of dark tourism studies*, pp. 261–289. London: Palgrave Macmillan.

Gatto, M.C., Mattingly, D.J., Ray. N., & Sterry, M. (Eds.) (2019). *Burials, migration and identity in the ancient Sahara and beyond*. Cambridge: Cambridge University Press.

Goody, J. (1962 [2004]). *Death, property and the ancestors: A study of the mortuary customs of the Lodagaa of West Africa*. London: Routledge.

Goudie, S.C., Khan, F., & Kilian, D. (1999). Transforming tourism: Black empowerment, heritage and identity beyond apartheid. *South African Geographical Journal, 81*(1), 22–31.

Hartmann, R., Lennon, J., Reynolds, D.P., Rice, A., Rosenbaum, A.T., & Stone, P.R. (2018). The history of dark tourism. *Journal of Tourism History, 10*(3), 269–295.

Heidenheimer, A.J., & Johnston, M. (Eds.) (2002). *Political corruption: Concepts and contexts*. New Brunswick, NJ: Transaction Publishers.

Hoani, S.H. (2015). Me Haka I te Haka a Tānerore? Māori 'post-war' culture and the place of Haka in commemoration at Gallipoli. *Australasian Canadian Studies Journal, 32*(1–2), 83–137.

Howell, S., & Shearing, C. (2017). Prisons, tourism, and symbolism: Reflecting (on) the past, present, and future of South Africa. In J. Wilson, S. Hodgkinson, J. Piché, & K. Walby (Eds.), *The Palgrave handbook of prison tourism*, pp. 277–294. London: Palgrave Macmillan.

Jafari, J. (1989). Tourism and peace. *Annals of Tourism Research, 16*(3), 439–443.

Jelin, E., & Herschberg, E. (Eds.) (2019). *Constructing democracy: Human rights, citizenship, and society in Latin America*. New York: Routledge.

Jindra, M. (2005). Christianity and the proliferation of ancestors: Changes in hierarchy and mortuary ritual in the Cameroon Grassfields. *Africa, 75*(3), 356–377.

Jones, S. (2017). Memory competition or memory collaboration? Politics, networks, and social actors in memories of dictatorship. In C. Kraenzle & M. Mayr (Eds.), *The changing place of Europe in global memory cultures*, pp. 63–86. London: Palgrave Macmillan.

Koleth, M. (2014). Hope in the dark: Geographies of volunteer and dark tourism in Cambodia. *Cultural Geographies, 21*(4), 681–694.

Labuschagne, G. (2018). Sexual homicide in South Africa. In J. Proulx, E. Beauregard, A.J. Carter, A. Mokros, R. Darjee, & J. James (Eds.), *Routledge international handbook of sexual homicide studies*, pp. 237–249. London: Routledge.

Labuschagne, G. (2021). An overview of serial murder and its investigation in South Africa. In H.C. Chan & S. Adjorlolo (Eds.), *Crime, mental health and the criminal justice system in Africa*, pp. 121–148. London: Palgrave Macmillan.

Lacanienta, A., Ellis, G., Hill, B.J., Freeman, P., & Jiang, J. (2020). Provocation and related subjective experiences along the dark tourism spectrum. *Journal of Heritage Tourism, 15*, 626–647.

Larson, P.M. (2001). Austronesian mortuary ritual in history: Transformations of secondary burial (*Famadihana*) in highland Madagascar. *Ethnohistory, 48*(1–2), 123–155.

Lee, R., & Vaughan, M. (2020). Death in Africa: A history c.1800 to present-day. In P.N. Stearns (Ed.), *The Routledge history of death since 1800*, pp. 320–335. London: Routledge.

Lemarchand, R. (2011). *Forgotten genocides: Oblivion, denial, and memory*. Philadelphia: University of Pennsylvania Press.

Leo, L., & Jamal, T. (2013). African Americans at sites of darkness: Roots-seeking, diasporic identities and place making. In L. White & E. Frew (Eds.), *Dark tourism and place identity: Managing and interpreting dark places*, pp. 28–45. London: Routledge.

Lew, A.A. (2017). Tourism planning and place making: Place-making or placemaking? *Tourism Geographies, 19*(3), 448–466.

Li, T.E., McKercher, B., & Chan, E.T.H. (2020). Towards a conceptual framework for diaspora tourism. *Current Issues in Tourism, 23*(17), 2109–2126.

Light, D. (2017). Progress in dark tourism and thanatourism research: An uneasy relationship with heritage tourism. *Tourism Management, 61*, 275–301.

Lischer, S. (2019). Narrating atrocity: Genocide memorials, dark tourism, and the politics of memory. *Review of International Studies, 45*, 805–827.

Loretti, A., & Tegegn, Y. (1996). Disasters in Africa: Old and new hazards and growing vulnerability. *World Health Statistics Quarterly, 49*(3–4), 179–184.

Lovejoy, P.E. (2012). *Transformations in slavery: A history of slavery in Africa*. Cambridge: Cambridge University Press.

Mangwane, J., Hermann, U.P., & Lenhard, A.L. (2019). Who visits the apartheid museum and why? An exploratory study of the motivations to visit a dark tourism site in South Africa. *International Journal of Culture, Tourism and Hospitality Research, 13*(3), 273–287.

Martini, A., & Buda, D.M. (2020). Dark tourism and affect: Framing places of death and disaster. *Current Issues in Tourism, 23*(6), 679–692.

Martinón-Torres, M., d'Errico, F., & Santos, E. (2021). Earliest known human burial in Africa. *Nature, 593*, 95–100.

Matiza, T. (2021). How to revive African tourism after the COVID-19 pandemic. *The Conversation*, July 13, Available from: https://theconversation.com/how-to-revive-african-tourism-after-the-covid-19-pandemic-163832 [Accessed: 18 Oct 2021].

Mensah, I. (2015). The roots tourism experience of diaspora Africans: A focus on the Cape Coast and Elmina Castles. *Journal of Heritage Tourism, 10*(3), 213–232.

Miles, W.F. (2002). Auschwitz: Museum interpretation and darker tourism. *Annals of Tourism Research, 29*(4), 1175–1178.

Monnier, O. (2021). A ticket to recovery: Reinventing Africa's tourism industry. *International Finance Corporation*, World Bank Group. Available from: https://www.ifc.org/wps/wcm/connect/news_ext_content/ifc_external_corporate_site/news+and+events/news/reinventing-africa-tourism [Accessed: 18 Oct 2021].

Moufakkir, O., & Kelly, I. (Eds.) (2010). *Tourism, progress and peace*. Wallingford: CABI.

Nwaubani, A.T. (2019). When the slave traders were African. *The Wall Street Journal*, online, Available from: https://www.wsj.com/articles/when-the-slave-traders-were-african-11568991595 [Accessed: 18 Oct 2021].

Oppenheimer, S. (2004). *The real Eve: Modern man's journey out of Africa*. New York: Carroll & Graf.

Otoo, F.E., Kim, S., & King, B. (2021). African diaspora tourism – How motivations shape experience. *Journal of Destination Marketing & Management, 20*, 100565.

Raine, R. (2013). A dark tourist spectrum. *International Journal of Culture, Tourism and Hospitality Research*, 7(3), 242–256.

Rice, A. (2012). *Creating memorials, building identities: The politics of memory in the Black Atlantic*. Liverpool: Liverpool University Press.

Rice, A. (2015). Shadows of the slave past: Memory, heritage and slavery. *Slavery and Abolition, 36*(4), 750–751.

Rice, A., & Kardux, J.C. (2012). Confronting the ghostly legacies of slavery: The politics of Black bodies, embodied memories and memorial landscapes. *Atlantic Studies, 9*(3), 245–272.

Roberts, C., & Stone, P.R. (2014). Dark tourism and dark heritage: Emergent themes, issues and consequences. In I. Convery, G. Corsane, & P. Davis (Eds.), *Displaced heritage: Dealing with disaster and suffering*, pp. 9–18. Woodbridge: Boydell & Brewer.

Rojek, C. (1993). *Ways of escape: Modern transformations in leisure and travel*. London: Palgrave Macmillan.

Rokosz, J. (2018). Unfortunately lucky: The persecution of albinos in East Africa. *The Ethnograph, 4*, 56–61.

Rosenberg, T. (2018). The business of voluntourism: Do Western do-gooders actually do harm. *The Guardian*, 13 September, Available from: https://www.theguardian.com/news/2018/sep/13/the-business-of-voluntourism-do-western-do-gooders-actually-do-harm [Accessed: 18 Oct 2021].

Rugwiji, T., & Masoga, M.A. (2018). A reflection on ritual murders in the biblical text from an African perspective. *Journal for Contextual Hermeneutics in Southern Africa, 117*(1), 1–13.

Salazar, N.B. (2006). Building a 'culture of peace' through tourism: Reflexive and analytical notes and queries. *Universitas Humanistica, 62*, 319–333.

Seaton, T. (2009). Purposeful otherness: Approaches to the management of thanatourism. In R. Sharpley & P.R. Stone (Eds.), *The darker side of travel*, pp. 75–108. Bristol: Channel View Publications.

Sharpley, R. (2005). Travels to the edge of darkness: Towards a typology of dark tourism. In C. Ryan, M. Aicken, & S. Page (Eds.), *Taking tourism to the limits: Issues, concepts and managerial perspectives*, pp. 217–228. Oxford: Elsevier.

Shenoy, R. (2019). 'Wilful amnesia': How Africans forgot – and remembered – their role in the slave trade. *The World*, online, Available from: https://www.pri.org/stories/2019-08-20/willful-amnesia-how-africans-forgot-and-remembered-their-role-slave-trade [Accessed: 18 Oct 2021].

SIPRI (2020). *Armed conflict and peace processes in sub-Saharan Africa*. SIPRI Yearbook 2020, Stockholm International Peace Research Institute, Available from https://www.sipri.org/yearbook/2020/07 [Accessed: 11 Nov 2021].

Smith, D. (2009). African chiefs urged to apologise for slave trade. *The Guardian*, 18 Nov, Available from: https://www.theguardian.com/world/2009/nov/18/africans-apologise-slave-trade [Accessed: 18 Oct 2021].

Smith, N.R. (2019). *Contradictions of democracy: Vigilantism and rights in post-apartheid South Africa*. Oxford: Oxford University Press.

Smith, V.L. (1998). War and tourism: An American ethnography. *Annals of Tourism Research, 25*(1), 202–227.

Speakman, M. (2019). Dark tourism consumption in Mexico City: A new perspective of the thanatological experience. *Journal of Tourism Analysis, 26*(2), 152–168.

Stapleton, T.J. (2017). *A history of genocide in Africa.* Santa Barbara, CA: Praeger.

Steyn, M. (2005). Muti murders form South Africa: A case report. *Forensic Science International, 151*(2/3), 279–287.

Stone, D.L., & Stirling, L. (2007). *Mortuary landscapes of North Africa.* Toronto: University of Toronto Press.

Stone, P.R. (2006). A dark tourism spectrum: Towards a typology of death and macabre related tourist sites, attractions and exhibitions. *Tourism, 54*(2), 145–160.

Stone, P.R. (2011). Dark tourism: Towards a new post-disciplinary research agenda. *International Journal of Tourism Anthropology, 1*(3/4), 318–332.

Stone, P.R. (2013). Dark tourism scholarship: A critical review. *International Journal of Culture, Tourism and Hospitality Research,* 7(3), 307–318.

Stone, P.R. (2020a). Dark tourism and 'spectacular death': Towards a conceptual framework. *Annals of Tourism Research, 83,* 102826.

Stone, P.R. (2020b). 'Following in the footsteps': Dark tourism as metempsychotic pilgrimage. In D.H. Olsen & M. Korstanje (Eds.), *Dark tourism and pilgrimage,* pp. xiii–xv. Wallingford: CABI.

Stone, P.R. (2021). The macabre world of dark tourism ft. Dr Philip Stone. *The Sense of Place Podcast,* Episode 18, Recorded 27 March, 52 mins. Available at https://www.senseofplacepod.com/podcast/episode/4966218c/ep-18-the-macabre-world-of-dark-tourism-ft-dr-philip-stone [Accessed: 21 July 2021].

Stone, P.R., & Grebenar, A. (2021). 'Making tragic places': Dark tourism, kitsch and the commodification of atrocity. *Journal of Tourism and Cultural Change.* https://doi.org/10.1080/14766825.2021.1960852

Stone, P.R., Hartmann, R., Seaton, A., Sharpley, R., & White, L. (Eds.) (2018). *The Palgrave handbook of dark tourism studies.* London: Palgrave Macmillan.

Stone, P.R., & Wright, D.W.M. (Eds.) (Forthcoming). *The future of dark tourism: Enlightening new horizons.* Bristol: Channel View Publications.

Strange, C., & Kempa, M. (2003). Shades of dark tourism: Alcatraz and Robben Island. *Annals of Tourism Research, 30*(2), 386–405.

Survival International (2021). *The 'Pygmies'.* Available from https://www.survivalinternational.org/tribes/pygmies/racism [Accessed: 4 Nov 2021].

Sutherland, C. (2014). 'Leaving and longing': Migration museums as nation-building sites. *Museum and Society, 12*(2), 118–131.

Tarlow, S., & Nilsson-Stutz, L. (Eds.) (2013). *The Oxford handbook of the archaeology of the dead.* Oxford: Oxford University Press.

Teiser, S.F. (1986). Ghosts and ancestors in Medieval Chinese religion: The Yű-Lan-P'en Festival as mortuary ritual. *History of Religions, 26*(1), 47–67.

Teye, V.B., & Timothy, D.J. (2004). The varied colours of slave heritage in West Africa: White American stakeholders. *Space and Culture,* 7(2), 145–155.

Tunbridge, J.E., & Ashworth, G.J. (1996). *Dissonant heritage: The management of the past as a resource in conflict.* London: Wiley.

Uchibori, M. (1978). *The leaving of this transient world: A study of Iban eschatology and mortuary practices.* Unpublished PhD thesis, Australian National University, Canberra, Australia.

United Nations (1948). *The Genocide Convention.* Office on Genocide Prevention and the Reasonability to Protect, United Nations, Available at https://www.un.org/en/genocideprevention/genocide-convention.shtml [Accessed: 22 July 2021].

Urry, J. (2002). *The tourist gaze,* 2nd edn. London: Sage.

ushmm.org (2021). Herero and Nama Genocide. *United States Holocaust Memorial Museum,* Available at https://www.ushmm.org/collections/bibliography/herero-and-nama-genocide [Accessed: 22 July 2021].

Uvin, P. (2001). Reading the Rwandan genocide. *International Studies Review, 3*(3), 75–99.
Vincent, L. (2008). New magic for new times: Muti murders in democratic South Africa. *Tribes and Tribals Special, 2*, 43–53.
Violi, P. (2012). Trauma site museums and politics of memory: Tuol Sleng, Villa Grimaldi and the Bologna Ustica Museum. *Theory, Culture & Society, 29*(1), 36–75.
Wei-Hass, M. (2018). Ancient riches discovered at mysterious burial monument. *National Geographic*, August 20, Available at https://www.nationalgeographic.com/science/article/news-ancient-pillar-monument-burial-mound-africa [Accessed: 9 Nov 2021].
Werdler, K. (2012). An introduction to dark tourism in Africa: Contested heritage or opportunity for a new proposition? *Mawazo: The Journal of the College of Humanities and Social Sciences, 11*(2), 103–114.
White, L., & Frew, E. (Eds.) (2013). *Dark tourism and place identity: Managing and interpreting dark places*. London: Routledge.
Whitehead, C., Schofield, T., & Bozoğlu, G. (2021). *Plural heritages and community co-production*. London: Routledge.
Whitinui, P. (2004). The indigenous factor: The role of kappa haka as a culturally responsive learning intervention. *Waikato Journal of Education, 10*, 85–98.
Williams, C. (2002). Beyond good and evil? The taboo in the contemporary museum: Strategies for negotiation and representation. *Open Museum Journal, 4*, Available from: http://amol.org.au/craft/omjournal/journal_index.asp [Accessed: 18 Oct 2021].
Wood, R., Berger, J., & Hasian, M. (2017). Public memory, digital media, and prison narratives at Robben Island. *Journal for Communication Studies, 1*, 173–195.
Wyatt, B., Leask, A., & Barron, P. (2020). Designing dark tourism experiences: An exploration of edutainment interpretation at lighter dark visitor attractions. *Journal of Heritage Tourism, 16*, 433–449.

14

COLONIALISM, POST-COLONIALISM, AND ITS HERITAGE IMPRINT

João Sarmento

Introduction

Colonialism is inescapable in any discussion of African heritage. Generally, we refer to African heritage as pre-colonial, colonial, and post-colonial, almost as if these were hermetic categories. We have internalized and naturalized these divisions, even though they are highly complex, fluid, and controversial. This chapter contributes to the ongoing destabilization of this ambiguity. It begins with a discussion of the indefinite geographical and heritage categories related to Africa, and the debate which follows is animated by three intersecting perspectives: the continuous looting of African heritage and its present consequences; the ways in which nation-building has de-Africanized and re-Africanized the past to serve political interests; and the present dominance of Western views in management, conservation and establishment of heritage sites on the continent. Numerous examples are used to illustrate the discussion. The aim is not to establish patterns, but to engage in a contrapuntal strategy that allows for provocative insights into larger processes, from local to global. While tourists, tourism, and cultural tourism are not repeatedly named throughout the chapter, their roles in ongoing processes of heritage-making are inevitable issues under analysis.

Destabilizing Time and Space in African Heritage

Colonialism derives from the Roman word '*colonia*', meaning 'settlement', which referred to Roman citizens who settled in other lands, retaining citizenship. While colonization assumed many forms in different places, the establishment of *colonias* always implied the domination of both land and indigenous people. As Loomba (1998) argues, everywhere the original inhabitants and the newcomers were implicated into the most complex and traumatic relationships

DOI: 10.4324/9781003153955-14

in human history. Whereas there are many differences between the terms colonialism and imperialism, they are often used interchangeably. Perhaps a consensual distinction is that imperialism is possible without direct colonial rule, without the formal takeover of a territory, as the economic, social, and cultural relations of dependency and control guarantee resources, captive labour, and markets. In the context of tourism and heritage, it is important to bear this fact in mind when we point to 'neo-imperialism' or 'neo-colonialism' practices and situations and observe 'imperialism nostalgia' as a sentiment and scenario for tourism production.

Since the age of colonialism is apparently over, the much-contested term 'post-colonialism', or 'postcolonialism' emerged, together with the dispute over the prefix 'post'. On the one hand, 'post' has a temporal dimension, which implies that it comes after the colonial. Yet, as there is a great temporal and contextual variation between the formal decolonization of say Brazil (1822) and Angola (1975), and the conditions under which decolonization occurred (occurs), it is difficult to use postcolonial in a single and unique sense. At the same time, the 'post' in post-colonialism, may represent that colonialism has been ideologically superseded by post-colonialism. Under this guise, the maintenance of the inequalities of the colonial rule in postcolonial times may lead to circumstances where nations are, at the same time, postcolonial (by law and history) and neo-colonial (by practice and exploitation). Tourism is often a key dimension of these ambiguities. It is also important to note that the use of post-colonialism for those who are on the economic margins of society or to those of the elites, and in the context of strong cultural and ethnical hybridity, is complex and far from consensual. Perhaps a useful way to think of postcolonialism is to understand it as a contestation of colonial rule and as a critical place from which to think of the legacies of colonialism, and the pervasiveness of colonial forms in the present day.

Ali A. Mazrui (1986), the renowned Kenyan scholar and postcolonial thinker, expanded, developed, and promoted the concept of Africa's triple heritage, after Blyden (1967) and Nkrumah (1970). His conceptualization, first related to the development of knowledge and the cultural forces that have given postcolonial Africa its present characteristics, encompasses indigenous values and ethos, Islam, and Western culture. African cultural identity and heritage are thus seen as eclectic, and composed of a multi-coloured interplay of these three realms: Indigenization, Islamification, and Westernization. In the context of heritage tourism studies, Teye (2009) used this tripartite approach to analyze heritage in Africa, dividing it into traditional or indigenous African heritage, Islamic heritage, and European colonial heritage. A postcolonial view of Africa's triple heritage, combining and retreating, complementing and opposing each other, can be a helpful departure point. But if an inelastic approach is taken, the ambiguity of dividing heritage into African, Islamic, and European, intermingling geographical and religious dimensions, and considering Islam as an external influence, for example, emerges. Christianity also had a particular geographical origin and expanded into territories or nations that internalized Christian heritage.

At first sight, Mazrui's (1986) tripartite model could suggest internal African heritage as only traditional or indigenous, emphasizing ideas of a primordial and 'authentic' African heritage, seen through an enduring pre-colonial African past, as in the case of archaeological rock art dating back over 10,000 years. Heritage of Punic or Roman origin, which went through a process of de-Africanization, should not only be understood under a process of conquest, colonization, and violence, in which the agency of conquerors is paramount, but also under the light of processes of resistance. According to Mazrui's approach, the 13th-century Rock-Hewn Churches in Lalibela, Ethiopia, are colonial Western heritage, and the Kasbah of Algiers is Islamic architecture. Yet both are imposed and colonial heritage, as the spread of Islam and Christianity is closely connected to the work of colonialists and the pacification of 'barbarian tribes and peoples'.

Mazrui (1986) not only defended that the Red Sea is an irrational colonial boundary imposed by Europe to divide Africa from the Arabian Peninsula, but also that 'Afrabia' is a notion or concept that could signal a decolonial reintegration of Africa and Arabia, to solve some identity issues and to bring Islam into an internal force of African heritage. On a different level, depth, and objective, the World Tourism Organization (UNWTO) and the United Nations Educational, Scientific and Cultural Organization (UNESCO), have their own malleable geographical divisions. For the UNWTO, Libya and Egypt are part of the Middle East; for UNESCO, Algeria, Egypt, Libya, Mauritania, Morocco, Sudan and Tunisia, and are included in the category of Arab States. Mazrui moves in a different direction, questioning even the northern boundary of Africa, almost proposing a re-Africanization of the Mediterranean northern shores. In fact, it is common to refer to Spanish or Portuguese colonial architecture (in Africa or Latin America), but unusual to point at African Islamic heritage in southern Spain and Portugal as colonial architecture. UNESCO (2021, n.p.) refers to the Acropolis in Athens and its monuments as 'universal symbols of the classical spirit and civilization'. Could we conceive them as indigenous or traditional European heritage? Are the Bantu civilizations of the Shona and the ruins of Great Zimbabwe more indigenous or traditional than the Greeks and the Acropolis in Athens? The ambiguity of the categorizations we impose on culture and heritage in different times and spaces, notably in Africa, can serve as a starting point to think of what colonial and postcolonial heritage in Africa means.

Hegel's (1975) 19th-century *Philosophy of History*, supported by the likes of Herodote, Robert Norris, Mungo Park, Kant, and Carl Ritter, constructed the idea that Africans had no history. In different ways, Eurocentric historiography based on a history of progress, technology, and expansion was nurtured by numerous intellectuals, politicians, artists, and the tourist imagination. They all contributed to the detachment of specific heritages from their African origins. Also, Hegel had a view of Africa divided into three parts. The first one was composed of the Mediterranean and northern Atlantic shores, where Carthage was once found and should be attached to Europe. In his view, the Maghreb was an extension of Spain. The second part comprised the Nile valley and was connected to Asia.

Lastly, 'Africa proper', or the lands south of the Sahara, which according to Hegel, had been cut off from the rest of the world and deleted from history and culture. This view has echoes in the exotic, primitive, timeless people, culture, and landscapes that the tourism industry helped to promote, since at least the 19th century.

Reid (2003) points to the fact that Ancient Egypt, as well as the neighbouring north-eastern African lands of Nubia and highland Ethiopia, went through a 'de-Africanization' process long ago. The pervasive disbelief among explorers, archaeologists, other intellectuals, and politicians that indigenous Africans could have built Great Zimbabwe, a large city and network of fortifications and monuments known since the early 16th century, continued well into the 20th century. The city, constructed between the 11th and 15th centuries, and a UNESCO World Heritage Site (WHS) since 1986, is testimony to the Bantu civilization of the Shona. Its existence was described on the basis of diffusion or migration mechanisms, rather than on the indigenous foundation of agency (Fontein, 2006). In the 1920s and 1930s, Great Zimbabwe's intrinsic African origin was only defended when supported by the argument that it was insufficiently developed to be European (Shepherd, 2002). This sentiment was true until recently, for any material signs of 'civilized society', such as the use of stone architecture, artistic achievements, and urbanism pointed to external or non-native initiatives (Finneran, 2013; Kusimba, 1999). Furthermore, as Mudimbe (1988) opines, it was common to say that Africans could produce 'nothing of value': Benin art was certainly a Portuguese creation, Yoruba statuary derived from the Egyptians, and Hausa and Buganda statecraft were creations of White invaders.

Looting African Heritage

From the mid-19th century to the mid-20th century, colonial museums, modelled closely after the European archetype—in architecture, design, and techniques of display—were built across Africa. They promoted European culture and trade and affirmed colonial domination. Often, sacred cultural objects were placed under the custody of foreign experts and the colonial state, and were permanently removed from the descendant communities where they belonged. At the same time, a vast range of ethnographic and anthropological museums flourished in the West, encompassing extensive collections of African material culture. In the context of Southwest Asia and North Africa, Timothy and Daher (2009) argue that most heritage sites and artefacts used as tourism resources are vestiges of the colonial and imperial periods. Thus, it is rather disconcerting that the best places for a tourist to engage with African art and heritage are museums in the so-called West. What is quite remarkable also is the meagre contemporary changes in these museums in this regard. As an illustration, the Africa Museum, also The Royal Museum for Central Africa, located in Tervuren, a quiet suburb of Brussels, Belgium, a key museum dedicated to the African continent and its people, reopened in late 2018 and still showcases 'traditional' African art in combination with stuffed animals such as lions and crocodiles.

Despite some efforts in providing narratives that contextualize African objects on display, such as in the Africa Museum, decolonial attempts in museums still embrace the presence of objects of suspicious provenience and accommodate the colonial violence and racist ideology which they represent. To a certain extent, tourism in London, Berlin, and Paris, among other cities, partially capitalizes on African heritage museums as no African city does. It is therefore inescapable, when discussing African heritage, to address the extension and consequences of European looting, especially from the late 19th century onwards. Important collections of African heritage have been assembled in European and North American museums and private collections, yet only an infinitesimal part is on public display. The vast majority of the collections are stacked in archives and dark basements, many even without any kind of cataloguing (Hicks, 2020). At the same time, many ethnographic museums in Africa are empty or struggle to consolidate their collections.

It is also ironic that it was precisely during the independence movements of the 1960s and 1970s that the illicit trafficking of antiquities boomed, accumulating today in colossal proportions (Hicks, 2020). As Shepherd (2002, p. 203) highlights, 'a trade that one associates with nineteenth century tomb robbers and the depredations of colonial 'collectors' shows no signs of abating, but in fact is on the increase'. The right of wanting heritage back is not the only reason for claims of repatriation. The ways in which objects are presented in Western museums, often purely for their aesthetic qualities, are totally decontextualized from their spiritual and cultural bonds to land and people.

Campaigns for the return of artefacts have been given an added impetus with the 2018 publication of the Felwine Sarr and Bénédicte Savoy *Rapport* (Sarr & Savoy, 2018), commissioned by the French president, Emmanuel Macron. The document calls for a detailed analysis of 'acquisition' circumstances, which may range from violent looting, colonial collectors, archaeological fieldwork, and purchases. While this report does not address the subject of private collections, Hicks (2020) points to changes in Europe's 'world culture' museums, with recent restitutions from Paris to Dakar and from Berlin and Stuttgart to Namibia. By proposing operational categories for restitution, Sarr and Savoy (2018) point directly to questions of consent and duress, which should or may have a future impact on the location, meaning, and role of African heritage globally and in the continent.

There are several interesting examples of new museums being built in Africa, under the light of restitution processes, foreign aid and development projects, and the continuous relationships and interests of former colonial countries in Africa. I will highlight two of them in the neighbouring countries of Nigeria and Benin. The Edo Museum of West African Art, in Benin City, Nigeria, is a project led by the Nigerian independent and non-profit Legacy Restoration Trust. The museum was designed by David Adjaye, a Tanzanian-born British architect of Ghanaian descent, and aims to gather looted objects from across the globe. The museum aims to create new opportunities to voice violence and to narrate the painful history of the invasion and destruction of Benin City by British troops in the late 19th century. On the site, British–Nigerian archaeological collaboration will develop an archaeological project up to 2026, for which 3 million euros have been secured from an anonymous European source. The construction of the museum is expected to start sometime before 2026, although no funds have yet been secured for building the museum. The British Museum, which holds 900 Beninese items in its collection, of which only 100 are on display, describes objects obtained in 1897 as 'looted'. Interestingly though, it only considers loaning material to the Edo Museum, on the basis that the British Museum is legally unable to deaccession these artefacts.

Construction of the Musée de l'Épopée des Amazones et des Rois du Dahomey, in Abomey, Benin, once the capital of the Kingdom of Dahomey, was scheduled to start in 2021. Supported by a 20 million-euro loan from the French Development Agency, it will exhibit artefacts taken by French troops over 125 years ago (1892) and previously kept in the Parisian Quai Branly Museum. This 'return' is partly the consequence of Emmanuel Macron's speech at the University of Ouagadougou, Burkina Faso, in November 2017, when the French president declared the end of 'Françafrique', that is, the French strategy of exerting military, political and commercial influence over its former African colonies. While Sarr and Savoy (2018) envision a juridical apparatus through a bilateral agreement of cultural cooperation with the formerly colonized countries, protectorates, and territories managed under French mandate, which would exempt the general principle of inalienability, the return of these objects is being made under a bill that was passed by the National Assembly in France in December 2020, which is far from Sarr and Savoy's recommendations.

Two vignettes serve to illustrate the economic, political, and geopolitical stakes that are at play at these news museums in Africa. The first one relates to Chinese soft power investment in culture in Africa. Recently, the National Agency for the Promotion of Heritage and the Development of Tourism of Benin has undertaken a joint venture with a Chinese company to invest in the redevelopment of cultural heritage property in the country. This includes the museums of Abomey, Portonovo, and the entire historic city of Ouidah. At the same time, the Museum of Black Civilizations, which opened in late 2018, in Dakar, Senegal (just two days before the re-opening of the Africa Museum in Tervuren, Belgium), is funded by China at a cost of 30 million dollars. Designed by Chinese architects, its shape is inspired by African traditional round huts, and its sheer size waits to be filled with African art.

The second vignette is about a five-year loan of a sword, which took place in late 2019, to this same Dakar museum. The temporary return of this symbolic object owned by the Musée des Invalides, Paris, allegedly belonged to El Hadj Omar Saidou Tall, the West African leader who founded the Toucouleur Empire in the early 18th century. The loan has nonetheless created some discomfort in neighbouring Mali, where the capital of this empire, encompassing much of what is now Guinea, Senegal, and Mali, was located. In late 2020, with the approval of the French National Assembly bill mentioned earlier, the sword was temporarily restituted to Senegal. Heritage and culture investments can be understood as acts of diplomacy and intersect with more pressing issues, such as the ever-present franc of the Financial Community of Africa (CFA, guaranteed by the French treasury), extractivism, and the military presence of France in the region. Significantly, at the same time, the sword was being handed over on loan, in December 2019, a significant arms contract was being signed between France and Senegal.

African material heritage also has a presence in public space across the West. Obelisks, for instance, are central pieces in Europeans' loathing of African heritage. Dating back to the time of Ramses II, an obelisk stolen in 1833 from Luxor, Egypt, after Napoleon's invasion in the 1798 campaigns, is one of the most photographed objects in the touristy Place de la Concorde, Paris. A 3,500-year-old Thutmose III-era obelisk belonging to Queen Cleopatra, embellishes Central Park, New York. The list is long, and Rome, London, and New York, among many other cities, sustain prestige and boost tourism with looted African heritage. A significant restitution of a public monument took place early this century, when the 4th century CE obelisk, which was stolen from Aksum, Ethiopia, returned home from Italy. The obelisk, or stela, was taken as a war trophy during the Italian fascist illegal aggression and annexation of Ethiopia (1936–1941), and from 1937 to 2003, was erected in front of the Ministry of Italian Africa (later FAO headquarters), embellishing an imperial public space in Rome. After a long and convoluted restitution process, which started with a Peace Treaty in 1947, the stela was disassembled and repatriated to Ethiopia in 2005. In 2008, it was re-erected in its original site and was included in the Aksum WHS, classified since 1980.

Aksum is key to the cultural tour of Ethiopian tourism, which also includes the other main Christian sites, Lalibela and Gondar (Finneran, 2013). Interestingly, the intricate return of the stela, which should have been the grounds for national celebrations and a further step towards decolonization, raised various identity contradictions. As Finneran (2013) explains, Aksum, once an essential part of the Ethiopian heritage metanarrative and the pride of Ethiopian civilization, is now perceived as a critical symbolic piece of a newly developing Tigrayan nationhood. Furthermore, Eritreans from the southern-central highlands, who historically and culturally have much in common with their neighbours from the Tigray region of northern Ethiopia, have also begun to reinterpret and claim their Aksumite heritage, creating tensions with non-Tigrayan Ethiopians. In fact, by late 2020, Aksum became the stage of various fights, involving the Ethiopian army, the Tigray Peoples' Liberation Front, and Eritrean forces, which continued into 2022 and had spread to other parts of the country.

Nation-Building and Heritage

Memoryscapes, understood as the sites occupied by public spaces such as squares, statues, avenues and urban layouts, museums, and other material objects, are in perpetual change, together with the transformation of political regimes and societal understandings. There is a great variety of historical pathways in the recapture of historical monuments and indigenous social and political practices into modern nationalism. By the mid-1930s, Aimé Césaire and Léopold Sédar Senghor's concept of negritude, reclaimed in the West and throughout the world, the recognition of Black African culture and heritage. After 1945, there was a re-Africanization of pan-Africanism (racial solidarity and the organization of the struggle against discrimination took a more militant form), and anticolonial movements, which led to the independence of 41 African countries in the 1950s and 1960s. This fact alone opened new avenues for development on the continent and renewed forms of conceiving and projecting African heritage. Most African states faced and still face the task of building unified nations out of many different ethnic and religious groups.

Strongly influenced by socialist conceptions of society, many leaders of the decolonizing nations of Africa—Amílcar Cabral, Leopold Sedar Senghor, Kwame Nkrumah, Sekou Toure, and Julius Nyerere, among others—imagined the new independent citizens as African prototypes of the 'new man', and favoured the idea of pre-colonial 'communalism'. Pan-African views looked for integrative views and writers, such as Franz Fanton, attempted to deconstruct the idea of an ahistorical Africa. If we can identify cases of heritage development of individual ethnic groups and localisms (see Ucko, 1994), the dominant ideas were related to the promotion of historically unified ethnic entities.

Various (violent) trajectories of demystification, destruction, appropriation, and protection of heritage can be found, often in contexts where national identity needed severe nurturing. They illustrate how political movements appropriated traditional arts and culture, adapting them to social and political needs (see McGovern (2013) and Sarró (2007) for the case of Guinea; Collier (2016) for Angola; and Bortolot (2013) for Mozambique). Rowlands (2007), for instance, explains how, formed out of the colonial ruins of French Western Sudan, the newly independent Mali from 1960, faced the difficult task of distancing itself from a French colonial image and nostalgia, while embracing the values of a modern state and nurturing national identity.

In the case of Egypt, the roots of nationalism were first grounded in a Pharaonic past, at the end of the 19th century (Winter, 2015). This idea, closely tied to independence movements against British rule, gradually developed from the 1910s, and the past glory of Egypt, materialized in the pyramids, was contrasted with existing poverty levels. Yet from 1953, independent Egypt embraced political discourses which were aligned with broader Arab nationalism (Hassan, 1998). At a larger scale, the scholar Cheikh Anta Diop (1991), and others thereafter, focused on claiming ancient Egypt for African scholarship and cultural history (see also Shepherd, 2002). Diop understood that Egypt is to Africa is what Greece is to Europe and China is to Asia (Asante, 2018). In fact, like Mazrui (1986), Diop conceptualized African heritage in three parts: a glorious past, a colonial experience (seen as centuries-long), and the future (a possible/desirable reincarnation of the first phase). Today, Egypt's Pharaonic past is used to project an international image of unity in an impoverished and unstable country, and to promote international tourism in the country. The recent glittering Pharaoh's Golden Parade in Cairo (3rd April 2021), when 22 mummies were moved from the Museum of Cairo to the about-to-open National Museum of Egyptian Civilization, illustrates the government's efforts along this path. Not only did the made-for-TV lavish and multimillion-dollar parade contrast with the 5-km derelict working-class neighbourhoods, along which the mummies were exhibited, but as Hassan (1998, p. 212) argues, this Pharaonic past is hardly used in contemporary celebrations or festivities, and is not 'an integral or a predominant element of the materiality of Egyptian life'. Except for school visits, residents do not visit the pyramids or the Egyptian museum, as it is Islamic heritage that is an integral part of daily life.

I wish to close this section referring to the capital of Eritrea. In 2017, the modernist city of Asmara was inscribed on UNESCO's World Heritage List (WHL). Just 185 km north of Aksum, modernist Asmara was built during the Italian occupation (1890–1941), especially during the 1930s. As a dual city, with a disciplinary design and fascist master plan dictating segregation, it remained mostly unchanged during the British protectorate period (1941–1950) and was miraculously preserved during the war with Ethiopia (1961–1991). In 2001, the Cultural Assets Rehabilitation Project, financed by a loan from the World Bank, and led by architect Naigzy Gebremedhin, started to survey heritage and plan preservation, with an eye on promoting the city for future investment and tourism. Furthermore, the World Bank, while clearly supporting economic development through cultural heritage, participated in carving Eritrea's identity. In that same year, the International Council on Monuments and Sites (ICOMOS), published a report that urged the inventorying and protection of modern heritage in Africa, labelling it a 'shared colonial heritage at risk, and a category of heritage that combines some of the thematic and regional shortcomings of the UNESCO WHL (ICOMOS 2001, n.p.). Moreover, it noted that one of the threats to 'Shared Colonial Heritage', is the 'Emerging nationalism and the need to establish an independent identity has often encouraged former Colonies to reach back to their Indigenous traditions at the expense of the remaining Colonial-period architecture and infrastructure' (ICOMOS 2001, n.p.). The domestic aim of shaping Asmara's international image converged with outsiders' colonial nostalgia, and alarmingly, as Fuller (2011) argues, the delineation of what is valuable in Asmara excludes the same area Italian colonizers excluded under the rubric of native quarters. Interestingly though, as Fuller (2011) also notes, Italian rule and colonial architecture help Eritreans build a sense of separateness from Ethiopians, and to claim long-standing cosmopolitanism that provides a further distinction from their neighbours. Still, as Rausch (2021) explains, heritage has been intricately caught in Eritrea's assertion of a national politics of self-reliance and sovereignty, and in transnational appeals to post-conflict reconstruction, nation-building, economic development, and poverty reduction.

African Heritage and the Colonial Present

Gregory (2004, p. 7) argues that 'the capacities that inhere within the colonial past are routinely reaffirmed and reactivated in the colonial present'. To a large degree, many of the contemporary measures to protect heritage in Africa were put into place during colonialism: the organization of archaeological findings, the establishment of heritage institutions, including museums, the introduction of legal frameworks, and research. Ideas of monumentality, ruins, relics, and antiquities were also brought from imperial centres. Furthermore, tourism was promoted vis-à-vis heritage development. Very often people living at heritage sites were chased away, detached from their cultural inheritances (Ndoro & Wijesuriya, 2015), and sacred places were taken away from the custody of African communities to that of the colonial government. Significantly, the conservation values which were imposed on the African colonies, that is, Western conservation values that emphasize the idea of materiality and authenticity, still dominate today (Chirikure et al., 2016; Makuvaza & Makuvaza, 2013; Shepherd, 2002). At the same time, colonial narratives and interpretation in many museums and heritage sites often celebrate empire, highlighting the remarkable work of Western architects and engineers, failing to engage in the dense spaces of struggle, past and present, and mute the story of colonization (Sarmento, 2011).

It is important to note, however, that before colonialism there were systems of heritage management and conservation. Often, elders, chiefs, and other custodians had the responsibility of looking after heritage resources, and there was a deep engagement of local communities in conservation, spiritually, materially, or both. The masons of Djeneé, Mali, have been building and maintaining houses for centuries, and also performing particular spiritual functions of protection, through foundation deposits and sacrifices (Rowlands, 2007). Furthermore, as there was a great variety of colonial powers and administrative systems, pre-existing understandings and relations to heritage were not, in many situations, completely erased, even if legal systems largely made them invisible (Baillie & Sørensen, 2021).

In the case of Great Zimbabwe and Khami, Pwiti (1996) has shown how the apparent neglect of monuments by locals, seen through the eyes of Westerners in the vegetation-covered, crumbling monuments, has been misunderstood. Local custodians, such as spirit mediums, look after the spiritual values of ruins and focus far less on their material dimensions. Also, Chirikure, Mukwende, and Taruvinga (2016) note that nearby traditional custodians have condemned Western-centric conservation initiatives that seek to promote tourism at these sites. Tourism implies public accessibility, which under traditional views, is not compatible with the ways in which these sites should be protected. By contrast, Ndoro (2001) notes the conflicting views between UNESCO's World Heritage Committee, which argues that the archaeological remains and architectural features of the Great Zimbabwe National Monument require protection against use, and the local communities, who understand that the site possesses great spiritual value. Therefore, restrictions on access to, and use of, the site is incompatible with local heritage values.

Ndoro and Wijesuriya (2015) have argued that colonial heritage management systems gained a powerful ally in international and global institutions such as

UNESCO and ICOMOS. While progress has been made in recent years, there are sustained critiques of the World Heritage system to postcolonial interventions against what is seen to be a traditionally European-dominated body (Meskell, 2013; Ndoro, 2001). Often, conservation and management in heritage places are framed by universalist approaches, established during colonialism, and perpetuated in post-colonial times, and there is an inadequate envelopment of local communities (Makuvaza & Makuvaza, 2013; Ndoro & Wijesuriya, 2015). In fact, it is Western ontologies and epistemologies that continue to direct heritage management and scholarship throughout the world. UNESCO typifies heritage as aesthetical, monumental, mostly physical, static, national, and expert-driven, and the values and narratives that support listing are generally developed by archaeologists and architects without consultation with the local communities.

In 2021, the WHL included 145 sites in Africa[1] of a total of 1,121 sites worldwide. Of these, 90 are cultural properties, and 23 are part of the endangered list (roughly half of the worldwide properties under this category). While 34 cultural properties are located in only five countries, all in northern Africa—Morocco (9), Algeria (7), Tunisia (7), Libya (5), and Egypt (6)—13 countries are not listed, and 7 others have no cultural properties listed. While it is important to underline that the process of listing is totally dependent on states' willingness to propose certain cultural heritage for inscription, in most African countries this process is clearly dominated by third-party countries, which provide the necessary technical expertise and resources to advance the nomination process (Meskell, 2013; Rowlands, 2007; Scholze, 2008).

At the same time, as Rowlands (2007, p. 129) argues, 'choosing what enters a World Heritage list turns out to be a highly subjective matter resting on claims about the objectification of "human value" in monumental art'. The two first African sites to be inscribed on the WHL in 1978 reflected European interests: the 13th-century 'New Jerusalem' Christian churches at Lalibela, Ethiopia, and the slave trade island of Gorée, Senegal. Indicating how heritage-making is volatile, politicized, and frequently subject to omissions and effacement, is the fact that the current list does not truly reflect cultural diversity, and there is a bias in favour of some regions. Sub-Saharan Africa is underrepresented (Abungu, 2016), having roughly 6% of the world's cultural properties (52 out of 869).

Furthermore, as Gottlieb (2020) argues, certain types of cultural property are disproportionately represented on the WHL. It is clear that monumental architecture from Egypt and Ethiopia, as well as sites that refer to the culture of invasion of Romans, European colonial rule, the introduction of Islam, and the slave trade, dominate when compared to (other) indigenous African cultures. With the launch of the UNESCO Convention for the Safeguarding of the Intangible Cultural Heritage (2003), several nominations merged intangible and tangible values, such as the Sacred Mijikenda Kaya Forests in Kenya (2008), the Konso Cultural Landscapes in Ethiopia (2011), the Bassari Country Cultural Landscapes in Senegal (2012), and the ǂKhomani Cultural Landscape, in South Africa (2017). In fact, during the past quarter century, UNESCO has attempted to embrace more inclusive models of heritage, acknowledging the importance of cultural landscapes, attachment to place, and diversity in heritage.

UNESCO also addressed the under-representation of properties dating to periods of early human history (Deacon, 2014). Presently, there are only 13 properties in Africa which emphasize human evolution and/or rock art. Furthermore, in some of these, hunter-gatherer subsistence over hundreds of thousands of years, and sometimes rock art as well, were not cited as a primary value (Deacon, 2014). Two sites in Africa provide very different narratives about humankind and our common human heritage; both are major tourist attractions. The first one is the National Museum of Ethiopia in Addis Ababa, where the skeletal remains of the earliest known hominid, *Australopithecus afarensis* are housed. Known to Ethiopians as 'Denqunash' ('You are fantastic/marvellous') and internationally as 'Lucy', the remains were found by an American anthropologist in 1974, in the Lower Valley of the Awash, Ethiopia, a WHS since 1980. As mentioned earlier in relation to Aksum, other later periods of Ethiopian pre-history and history are open to the emergence of contested, conflicting, and disputed heritage meanings and ownership. Differently, Denqunash, as Finneran (2013) argues, offers a starting point for an Ethiopian cultural meta-narrative, enveloped in pride and triumphalism. Importantly here, Denqunash is key to a pan-Ethiopian heritage meta-narrative, 'neither claimed as an Amhara, an Oromo or a Harari: she is resolutely Ethiopian' (Finneran, 2013, p. 49).

The other site is the Maropeng visitor centre, located 55 km northwest of Johannesburg, South Africa. Maropeng provides an entertaining exhibition related to the 15 fossil hominin sites spread over three of South Africa's provinces, of which Sterkfontein Caves, is the only one open to the public (WHS since 1999, extended in 2005). Known as the Cradle of Humankind World Heritage Site, visitors are greeted with the slogan 'welcome home'. The narrative, constructed for the tourist, emphasizes the common heritage of humanity. While it displays some authentic archaeological, palaeontological, and palaeoanthropological artefacts, on a rotating basis, the core of the site rests in interactive and alluring displays (McGhie, 2009). Maropeng is built like a mound, representing a tumulus, which points more to European Neolithic than to Africa, where these structures are rare, and positioning visitors in a different time period than the one represented at the site (McGhie, 2009). While apparently Lucy is consensual heritage in Ethiopia, both internally and internationally, Maropeng is a compelling tourism experience, a convenient weekend getaway whose narrative echoes the 'Out-of-Africa' origins story. The site nests perfectly in the post-apartheid South African fascination for heritage tourism development at the same time as it caters to Western cultural consumption of African heritage. Alarmingly though, as Scott (2007, p. 21) notes in her study on the presentation of hominin evolution in museums, is the way Europeans accept that 'Africa serves as the cradle of humanity, while Europe serves as the evolutionary finishing school'. Furthermore, findings of research into young South Africans' knowledge and understanding of 'their' national histories illustrates how histories written by White students more often included human origins, the Cradle of Humankind underlining a common past at the expense of the more recent period of apartheid (Angier, 2017). By contrast, Black students focused their stories in the 20th century, especially on the formation of the African National Congress, and on particular events, such as the Sharpeville massacre (1960), the Soweto uprisings (1976), and the first democratic election (1994).

Conclusion

The ambiguities discussed above support the rejection of a reductive approach that divides African heritage meta-narratives into three broad timescales: pre-colonial, colonial, and postcolonial. There is an increasing prevalence of forms of subjectivity that are more fluid, mobile, and flexible. Heritage has been de-Africanized and re-Africanized, at various scales, according to specific political projects and identity aspirations. Complex realities on the ground imply that global, national, and local claims to, and understandings of, heritage, overlap, and often collide with 'outstanding universal value'.

In many of the multi-ethnic countries of Africa, what constitutes 'national heritage' is not consensual. While UNESCO works on the presupposition of national cultures, it is often particular ethnic political hegemonies that make decisions and attempt to forge collective memory. There are many examples throughout the continent of struggles to redefine heritage in postcolonial times, allowing for alternative narratives, but also western conceptualizations of heritage that prevail and prevent community participation and other modes of interpreting and living heritage sites.

Recently, new flagship museums have been inaugurated in various African countries. While they do not account for the thousands of museums that face severe financial constraints to open each day, difficulties in maintaining collections, and in narrating and reinterpreting in other than colonial ways, they signal some changes. On the one hand, restitution of collections may provide new impetuses in various places. The challenge at play is the continuous interpretation and reinterpretation of heritage, according to particular, local, and national circumstantial political moments, and their insertion in a global political economy. On the other hand, breaks with the European archetype of colonial museums and techniques of display may open new avenues. Land dispossession, structural racial continuity between the colonial and postcolonial, modern slavery, and global migration are just some themes that could be addressed. This involves fluidity and flexibility, in the engagement with the past and with the present. It is in the interstices of the colonial past, the present needs, the restitution of heritage, the agendas of UNESCO and other international organizations, donors, funding agencies, and international states that African communities must reclaim and use their heritage.

Note

1 UNESCO includes several African countries in the category of Arab States.

References

Abungu, G.O. (2016). UNESCO, the World Heritage Convention and Africa: The practice and practitioners. In W. Logan, M.N. Craith, & U. Kockel (Eds.), *A companion to heritage studies*, pp. 373–391. Chichester: Wiley-Blackwell.

Angier, K. (2017). In search of historical consciousness: An investigation into young South Africans' knowledge and understanding of 'their' national histories. *London Review of Education, 15*(2), 155–173.

Asante, M.K. (2018). Meeting Cheikh Anta Diop on the road to African resurgence. *International Journal of African Renaissance Studies—Multi-, Inter- and Transdisciplinarity, 13*(1), 4–15.

Baillie, B., & Sørensen, M.L.S. (2021). Heritage challenges in Africa: Contestations and expectations. In B. Baillie & M.L.S. Sørensen (Eds.), *African heritage challenges: Communities and sustainable development*, pp. 1–43. Singapore: Palgrave Macmillan.

Blyden, E. (1967). *Christianity, Islam and the Negro race*. Edinburgh: University of Edinburgh Press.

Bortolot, A. (2013). Artesãos da Nossa Pátria: Makonde blackwood sculptors, cooperatives and the art of socialist revolution in postcolonial Mozambique. In S.L. Kasfir & T. Förster (Eds.), *African art and agency in the workshop*, pp. 253–273. Bloomington: Indiana University Press.

Chirikure, S., Mukwende, T., & Taruvinga, P. (2016). Post-colonial heritage conservation in Africa: Perspectives from drystone wall restorations at Khami World Heritage site, Zimbabwe. *International Journal of Heritage Studies, 22*(2), 165–178.

Collier, D. (2016). *Repainting the walls of Lunda: Information colonialism and Angolan art*. Minneapolis: University of Minnesota Press.

Deacon, J. (2014). An overview of the history of the nomination of cultural world heritage sites in Africa. In S. Makuvaza (Ed.), *The management of cultural World Heritage Sites and development in Africa*, pp.1–16. New York: Springer.

Diop, C.A. (1991). *Civilization or barbarism*. New York: Lawrence Hill Books.

Finneran, N. (2013). Lucy to Lalibela: Heritage and identity in Ethiopia in the twenty-first century. *International Journal of Heritage Studies, 19*(1), 41–61.

Fontein, J. (2006). *The silence of Great Zimbabwe: Contested landscapes and the power of heritage*. London: UCL Press.

Fuller, M. (2011). Italy's colonial futures: Colonial inertia and postcolonial capital in Asmara. *California Italian Studies, 2*(1), 1–18.

Gottlieb, Y. (2020). Attacks against cultural heritage as a crime against humanity. *Case Western Reserve Journal of International Law, 52*(1), 287–330.

Gregory, D. (2004). *The colonial present: Afghanistan, Palestine, Iraq*. Malden, MA: Blackwell.

Hassan, F. (1998). Memorabilia, archaeological materiality and national identity in Egypt. In L. Meskell (Ed.), *Archaeology under fire: Nationalism, politics and heritage in the eastern Mediterranean and Middle* East, pp. 200–216. London: Routledge.

Hegel, G.W.F. (1975). *Lectures on the philosophy of world history*. Cambridge: Cambridge University Press.

Hicks, D. (2020). *The Brutish museums: The Benin Bronzes, colonial violence and cultural restitution*. London: Pluto Press.

ICOMOS (2001). Heritage at risk 2001–2002: Shared colonial heritage. Online at: http://www.international.icomos.org/risk/2001/colonial2001.htm.

Kusimba, C.M. (1999). *The rise and fall of Swahili states*. Walnut Creek, CA: AltaMira Press.

Loomba, A. (1998). *Colonialism/postcolonialism*. London: Routledge.

Makuvaza, S., & Makuvaza, V. (2013). The challenges of managing an archaeological heritage site in a declining economy: The case of Khami World Heritage Site in Zimbabwe. *Conservation and Management of Archaeological Sites, 15*(3–4), 281–297.

Mazrui, A. (1986). *The Africans: A triple heritage*. London: BBC Publications.

McGhie, L-M. (2009). Archaeology and authenticity in South African heritage locales and public spaces. *Public Archaeology, 8*(4), 351–372.

McGovern, M. (2013). *Unmasking the state: Making Guinea modern*. Chicago: University of Chicago Press.

Meskell, L.M. (2013). UNESCO's World Heritage Convention at forty: Challenging the economic and political order of international heritage conservation. *Current Anthropology, 54*(4), 483–494.

Mudimbe, V.Y. (1988). *The invention of Africa: Gnosis, philosophy and the order of knowledge*. Bloomington: Indiana University Press.

Ndoro, W. (2001). *Your monument, our shrine: The preservation of Great Zimbabwe*. Uppsala: Societas Archaeologica Upsaliensis.

Ndoro, W., & Wijesuriya, G. (2015). Heritage management and conservation: From colonization to globalization. In L. Meskell (Ed.), *Global heritage: A reader*, pp. 131–149. New York: Wiley-Blackwell.

Nkrumah, K. (1970). *Conscientism: Philosophy and the ideology for decolonization*. New York: Monthly Review Press.

Pwiti, G. (1996). Let the ancestors rest in peace? New challenges for cultural heritage management in Zimbabwe. *Conservation and Management of Archaeological Sites, 1*(3), 151–160.

Rausch, C. (2021). Modern nostalgias for sovereignty and security: Preserving cultural heritage for development in Eritrea. In B. Baillie & M.L.S. Sørensen (Eds.), *African heritage challenges: Communities and sustainable development*, pp. 297–329. Singapore: Palgrave Macmillan.

Reid, D. (2003). Introduction – locating ancient Egypt in Africa: Modern theories, past realities. In D. O'Connor & D. Reid (Eds.), *Ancient Egypt in Africa*, pp.1–22. London: UCL Press.

Rowlands, M. (2007). Entangled memories and parallel heritages in Mali. In F. de Jong & M. Rowlands (Eds.), *Reclaiming heritage: Alternative imaginaries of memory in West Africa*, pp.127–144. Walnut Creek, CA: Left Coast Press.

Sarmento, J. (2011). *Fortifications, post-colonialism and power: Ruins and imperial legacies*. Farnham: Ashgate.

Sarr, F., & Savoy, B. (2018). *The restitution of African cultural heritage: Toward a new relational ethics*. Paris: Ministère de la Culture.

Sarró, R. (2007). Demystified memories: The politics of heritage in post-socialist Guinea. In F. de Jong & M. Rowlands (Eds.), *Reclaiming heritage: Alternative imaginaries of memory in West Africa*, pp. 215–229. Walnut Creek, CA: Left Coast Press.

Scholze, M. (2008). Arrested heritage: The politics of inscription into the UNESCO World Heritage List: The case of Agadez in Niger. *Journal of Material Culture, 13*(2), 215–231.

Scott, M. (2007). *Rethinking evolution in the museum: Envisioning African origins*. London: Routledge.

Shepherd, N. (2002). The politics of archaeology in Africa. *Annual Review of Anthropology, 31*(1). 189–209.

Teye, V.B. (2009). Tourism and Africa's tripartite cultural past. In D.J. Timothy & G.P. Nyaupane (Eds.), *Cultural heritage and tourism in the developing world: A regional perspective*, pp. 165–185. London: Routledge.

Timothy, D.J., & Daher, R.F. (2009). Heritage tourism in Southwest Asia and North Africa. In D.J. Timothy & G.P. Nyaupane (Eds.), *Cultural heritage and tourism in the developing world: A regional perspective*, pp.146–163. London: Routledge.

Ucko, P. (1994). Museums and sites: Cultures of the past within education—Zimbabwe, some ten years on. In P.G. Stone & B.L. Molyneaux (Eds.), *The presented past: Heritage, museums and education*, pp. 237–282. London: Routledge.

UNESCO (2021). *Acropolis, Athens*. Online at: https://whc.unesco.org/en/list/404/.

Winter, T. (2015). Heritage and nationalism: An unbreachable couple? In E. Waterton & S. Watson (Eds.), *The Palgrave handbook of contemporary heritage research*, pp. 331–345. London: Palgrave Macmillan.

15

AFRICAN ARCHAEOLOGY AND TOURISM

Liza Gijanto

Introduction

At the most basic level, archaeotourism involves passive or purposive visitation at archaeological sites, including ongoing digs, and engagement with archaeological artifacts (Little & McManamon, 2005), including the development of programming for families and students. Among serious archaeotourists, however, it is a form of tourism in which the visitor seeks a unique, tailored experience that they view as sustainable or beneficial to a local community (Weaver, 2015). The partnership between tourism and archaeologists and other heritage professionals is often fraught but can yield positive benefits for site preservation when done in a responsible, informed manner (see Timothy & Tahan, 2020 for a detailed discussion). Archaeologists approach tourism collaboration hoping to avoid an overly commodified product, site destruction, and misappropriation of the past for political and economic gain (Gillot, 2020; Rowan & Baram, 2004). Ideally, tourist visitation should raise awareness of archaeological resources among the general public and spur a passion for conservation and preservation (Baram, 2008). The development of archaeological sites for tourist consumption occurs in multiple ways, including viewing active excavations, touring heritage sites with archaeological components (excavated or not), participating in volunteer activities at heritage sites, and viewing artifacts in a museum display. Much archaeology-based tourism falls under the guise of public archaeology, a recognized 'sub-sub-field' that began in the United States in the 1990s (Jameson, 1997) and expanded to other regions in the early 2000s (Okamura & Matsuda, 2011). Such projects, including many now occurring in parts of Africa, often intersect with community-based inquiries with efforts to involve multiple stakeholders in the development and implementation of research (Jopela & Fredrikson, 2015; Posnansky, 2017; Schmidt, 2014).

DOI: 10.4324/9781003153955-15

Although public archaeology is becoming more common in Africa, archaeotourism has yet to be fully developed in the region, despite Africa being rich in archaeological remains. The African archaeological record includes the remains of our earliest hominin and human ancestors, evidence for initial experimentation in animal and plant domestication, the human use of natural resources (e.g., mining and metallurgy), and the rise of cities and empires. The archaeological sites representing these pivotal moments in history span the continent. For example, eastern and southern Africa have yielded numerous hominin species, revealing much about the origins of humanity. Likewise, portions of northern Africa provide evidence of early plant and animal domestication. Archaeological sites spanning the last 10,000–20,000 years have allowed archaeologists and historians to trace the technological, social, and economic trajectories of the continent including regional involvement in oceanic and global trade networks. Yet, despite this rich record, and with the obvious exception of Egypt, Tunisia, and Morocco, much of Africa is not a major destination for archaeotourism, compared to Asia, Latin America, Europe, and the Middle East. This is partially due to the underwhelming nature of many sites to the non-specialist because the sites often lack recognizable above-ground components and are generally inaccessible in the tourism marketplace. However, even the archaeological sites that do contain standing ruins have not been fully developed for tourism due to a general lack of funds and staff, an underdeveloped infrastructure, and a colonial history that sometimes ignored or downplayed the native heritage of the continent.

This chapter is offered as an initial assessment of the incorporation of archaeology into tourism development in Africa and not an exhaustive overview. The bulk of the archaeological work conducted throughout most of Africa is done by foreign researchers affiliated with universities and museums in Europe and the United States, which also supply the majority of funding for the excavations as well. Though many are aware of the need to involve local partners and attempt to ensure local communities see some benefits of archaeology-based tourism, this is not always attainable (Chirikure & Pwiti, 2008, Gavua, 2006; McIntosh, 2017). The regions where archaeological sites have been incorporated into heritage tourism have seen varying degrees of success, which are reviewed herein. Finally, it is not possible to review work conducted in every African nation; as a result, I focus on those where tourism sites include some archaeological component, even if archaeology is not the focal point of the locale.

Heritage Tourism and Archaeology in Africa

Tourism has long been hailed as a key tool in the economic development of Africa, as well as a means to highlight cultural heritage (Adeola et al., 2017; Mkono, 2019; Novelli, 2016). Central to many countries' efforts to appeal to foreign travelers, primarily those from Europe, has been the promotion of cultural heritage. Bound up in this are often archaeological resources, often unmitigated. Archaeological sites are often listed as assets that determine a country's competitiveness in the broader tourism marketplace, but the economic impact of these locales is dependent on the overall tourism infrastructure, which is in turn limited by the country's accessibility, general infrastructure, and other measures of development (Krstic et al., 2016, p. 4). In addition to in-country development, the locales must be accessible to international tourists (Adeola et al., 2017). This is a major obstacle for archaeotourism on much of the continent.

To varying degrees, archaeological material has found its way into different local, regional, and national museums in most countries of Africa. While national or local museums are sometimes incorporated into tourist visits, the stronger draw for most tourists is standing structures, megalithic monuments, or other historic sites that often include archaeological components. Heritage tourism in much of Africa is centered on UNESCO World Heritage Sites (WHSs) and areas as discussed elsewhere in this volume. Though there are fewer WHSs in Africa than in Europe, where the list began, an initiative launched in 2010 to include more archaeological sites regardless of the presence of remaining built heritage was undertaken with an eye on Africa (Willems & Comer 2011, p. 163). Since this initiative, there has been a steady rise in nominations. Regardless of the designation, several of these sites were available for heritage tourism prior to being conferred WHS status such as James Island in The Gambia and Djenné-Djenno in Mali. Thus, the inscription of these and other sites on the list was meant to raise international awareness and access to funding for preservation and tourism development, as no support is given by UNESCO once a site is designated. This is particularly pertinent for heritage locales with archaeological components that require mitigation and long-term management to prevent damage by tourists and the continuing threat of looting (Ashley & Bouakaze-Khan, 2011). Due to these and other circumstances challenging heritage protection, such as extreme weather, tourism development around these sites often goes unchecked and unmanaged. Portions of areas with standing monuments are often overhauled for current use as restaurants or shops with no oversight, while others are subject to looting by locals to sell artifacts to tourists or local markets. This is seen in the Galamsy gold 'mining' activities in Ghana whereby local men dig shafts into archaeological sites to find gold objects to sell either as-is or melted down. These activities not only destroy archaeological resources, but have also led to over-silting of many local waterways (Reid & Amartey, 2019). Because of such activities, the promise of local development and poverty alleviation is often unrealized (Breen, 2007).

While most African countries are home to colleges and universities that offer undergraduate degrees in archaeology, few can support graduate programs. The reliance on foreign-trained professionals and, at times, a lack of professional opportunities at home have resulted in many underdeveloped heritage sectors, to say nothing of archaeological resource management. As such, archaeology has not permeated the public consciousness in most African states. Likewise, volunteer archaeology is limited and often done under the guise of archaeological field schools that primarily enroll foreign students for course credit in American or European colleges and universities. An exception to this are programs such as the Central Region

Archaeological Project in Ghana directed by Dr. Christopher DeCorse from the late 1990s into the early 2000s that was partially funded by Earthwatch, allowing American tourists to participate in excavations[1] as a form of archaeology volunteer tourism. However, volunteer tourism in archaeology is rare in Africa, compared to other world regions (Timothy, 2020). The overall lack of government oversight in most nations, a dearth of professional opportunities, and poorly funded museums have hindered the overall assessment and management of historical resources throughout most of the continent (Breen & Rhodes, 2010, p. 26).

Archaeotourism and Regional Variability

Given the disparate nature of the African continent including available resources, ongoing conflicts, and varying degrees of local interest in heritage and archaeological development, there is no one-size-fits-all approach to how, or even if, archaeological resources should be harnessed for tourism. This section presents a general, descriptive overview of what is being done across Sub-Saharan Africa. This section is not meant to be exhaustive but rather a survey of how archaeological resources have been mobilized in the major regions of Sub-Saharan Africa, highlighting key examples where possible. It should be noted that, with a few exceptions, it is the more politically stable, developed (as perceived by international travelers) countries that have made the greatest strides in incorporating archaeology into heritage tourism.

West Africa

Archaeological research in West Africa has a long history and includes long-term investigations of sites associated with the transatlantic slave trade (DeCorse, 2001b; Gijanto, 2017; Kelly, 1997; Monroe, 2007; Ogundiran, 2007). Because of this, much of West Africa's tourism involving archaeology is directly related to African diaspora tourism. Diaspora tourism, often defined as a subset of heritage tourism, is also described as a pilgrimage, or homecoming, by many tourists of African descent from the Americas and Europe who visit the slave forts and castles along Africa's west coast (Mensah, 2015; Reed, 2014; Schramm, 2010), as well as by those selling the destinations (Bruner, 1996; Gebauer & Umscheid, 2021).

The coastal slave forts are relatively easy to access from Accra and other cities in Ghana and are the most visited heritage sites in Ghana, often eclipsing nature-based tourism and other heritage tourisms in other parts of the country. Even though the tourism economy of Ghana is well-developed, archaeology and heritage tourism sites outside the coastal regions have received little attention outside of local attempts to attract tourists. Rather, Ghana has gone all in on promoting African diaspora tourism, which focuses on the region's slavery heritage. Senegal and The Gambia are following distantly in terms of diasporic market share. Recently, Benin and Nigeria have made serious strides in developing diaspora- and slavery-based heritage tourism.

Diaspora tourism began in Ghana focusing on Elmina and Cape Coast castles (slave trading forts). While archaeology was never explicitly part of the tourist draw at these sites, extensive excavations have been carried out in the town that grew up around Elmina (DeCorse, 2001a) and some of the materials from these activities are on display at each locale. Of particular note are the remains of two persons of African descent that were found during archaeological excavations in the Americas and which were repatriated to Ghana—a woman from Seville Plantation in Jamaica and a man from the American northeast. While it is unknown if these individuals originated in Ghana, they have nevertheless been reinterred symbolically in the castle courtyard and are part of standard tours (Singleton, 2006, p. 249). In addition to these individuals, two more sets of remains from New York City's African Burial Ground have been reinterred in Ghana following a ceremony at Cape Coast Castle marking their return to the homeland via the Door of No Return. These two are buried at Assin Manso along the former slave route and are conveniently located near Cape Coast allowing easy tourist access (Osei-Tutu, 2014, p. 332). Even though these are not 'African archaeological remains' in the strictest sense, they are archaeological resources that have been repatriated to Ghana as part of the slavery narrative of West Africa and to add authenticity to the tourism circuit.

The Gambia and Senegal are home to several UNESCO sites and numerous archaeological resources that have been developed for tourism. For example, the Stone Circles of Senegambia is comprised of sites in both countries. African diaspora tourism, commonly referred to as 'roots tourism' in these two countries also includes visits to archaeological sites and materials from excavations in museums. However, like Ghana, archaeology itself is not the main attraction. Archaeological excavations carried out in the 1980s and 1990s at Wassu and other stone circle sites contributed to their interpretation prior to their inscription on the World Heritage List. The findings of the work are incorporated into site interpretation and small museum exhibits; however, currently, there is no active archaeological project. Likewise, archaeological investigations have occurred sporadically at James Island (now known as Kunta Kinteh Island), but not as part of a formal tourism project (Gijanto, 2009). All archaeological work on the island is conducted by foreign researchers because the National Centre for Arts and Culture does not have a staff archaeologist and The Gambia's education system lacks archaeology training anywhere in the country.

In Senegal, work at Goree Island, including the fort, was done in part to support tourism development. Since the initial preservation efforts, the small island has become a major attraction in Senegal. The perceived success of Goree in incorporating archaeological and architectural heritage is marred by the large number of stakeholders involved on the island ranging from the IFAN and other government agencies that manage the historical resources, local property and business owners, and diaspora tourists (Thiaw & Wait, 2018). Largely under the direction of Dr Ibrahima Thiaw, long-term archaeological and historical investigations of Goree Island including archaeological field schools in partnership with Rice University were carried out on Goree Island in the early 2000s. Much of the excavations were open to the public, including tourists, but the overall findings of a rich, multi-ethnic community have not made it into the overall heritage discourse that is focused on the Atlantic slave trade catering to the African diaspora tourist market (Thiaw, 2011).

The exception in the region is Nigeria. Like Ghana, there are many home-trained archaeologists with four universities offering programs in archaeology, but the heritage sector is not dominated by African diaspora tourism as it is in some other West African countries. The advancement of the field locally also includes a strong push by archaeologists to sustainably incorporate archaeological resources within the tourism sector (Gubam & Nominshan, 2020). Several archaeology sites of note have been excavated in the last several decades in Nigeria, including the Nok and Ile-Ife dig sites (Babalola et al., 2020; Franke, 2016). The materials from these excavations are dispersed, and the sites are neither listed for visitation by the Nigerian Tourism Board nor are the associated villages being developed for tourism (Gubam & Nominshan, 2020). There is an awareness of the potential for archaeotourism as both an engine for economic growth and site preservation in Nigeria. Odum and Oguamanam (2020) make the case for sites in Igboland as an archaeotourism destination boasting a large number of sites that could be under active excavation. The push for the growth of tourism includes advocacy by local archaeologists to incorporate sites and monuments when possible. This includes physical site visits, as well as the use of virtual reality to engage a broader audience to take an interest in the country, and its heritage, and perhaps take the next step to visit in person (Chigozie & Ukekwe, 2020).

Finally, there is potential in Liberia and Sierra Leone, and archaeological excavations were beginning to be undertaken again prior to the COVID-19 pandemic (DeCorse, 2015; Reilly et al., 2019). However, relatively recent Ebola outbreaks coupled with slow recovery from decades-long civil wars have seriously crippled infrastructure in these countries and drained the already limited resources devoted to the heritage sector.

East Africa

East Africa has the most extensive archaeological record in the world, ranging from our earliest hominin ancestors, including the first anatomically modern humans, up until the recent past. This record provides the potential for a range of archaeological sites to be incorporated into tourism schemes and itineraries. In fact, other than Nigeria and South Africa, this part of the continent has the greatest push for archaeotourism development by professional archaeologists.

Kenya, Ethiopia, and Tanzania have some of the most important hominin sites in terms of the development of the human genus. However, many of these are inaccessible to average tourists, and the closest one can get to an archaeological site is as a student in one of a handful of archaeological field schools. For example, in 2012, George Washington University took over the Koobi Fora Field School in Kenya from Rutgers University, maintaining the partnership with the National Museum of Kenya. Students enrolled in this program to work at dig sites on the eastern shore of Lake Turkana. Although the primary focus was educational, many consider studying abroad a form of tourism, and students do participate in other typical tourist activities. From a broader industry perspective, Olduvai Gorge and Laetoli are incorporated into many safari tours of Tanzania and the visits include stops at museums containing some materials and information from decades-long investigations.

Other areas of East Africa have sought to incorporate archaeological resources into tourism programs but have been relatively unsuccessful due to local conflicts, as is the case for Sudan and Ethiopia. Alternatively, Zanzibar (Tanzania) has experienced moderate success due to the high density of historic architecture, though these buildings are under constant threat of environmental degradation and dangers derived from tourism development (Chami, 2018).

Southern Africa

In southern Africa, heritage sites containing archaeological components that have received the greatest attention in terms of investment and integration with local and national tourism policies and plans are found in South Africa, Botswana, and Zimbabwe. Prior to the onset of the COVID-19 pandemic in 2020, South Africa and Zimbabwe consistently ranked among the top five African nations for international arrivals (Adeola et al., 2017). In 2019, South Africa (10.23 million) held its position at number two with Zimbabwe (2.29 million) fourth and Mauritius (1.38 million) rounding out the top five for the first time in the last decade (Statista, 2021). It is telling that the rankings are consistently dominated by countries in northern and southern Africa. It is also notable that the gap in visitation between these is so large. On the surface, adequate infrastructure, level of development, perceived safety, and the general awareness of a country as a tourist destination are the obvious factors reflected in the visitation numbers. The opposite is true with regard to non-visitation. Another influential factor is that not all international visitors travel for leisure tourism, as business travelers are also counted in these numbers. Nonetheless, many of these non-leisure tourists do take part in traditional tourist activities, which pair well with business travel, such as South Africa's heritage sites that include some of the earliest and most complete hominid sites, rock art, battlefields, and British or Dutch colonial monuments.

Despite the presence of these resources, their full potential to attract tourists is rarely achieved. A clear example of this is seen in the management and promotion of rock art sites in the region. Like the Stone Circle UNESCO WHS in Senegambia, the Maloti-Drakensberg Park WHS spans two countries and is a joint effort at conservation and tourism promotion. However, unlike the example of Senegal and The Gambia, this is a designated park with clear boundaries that was also designated a WHS based on its natural resources and the rock art associated with the San within its boundaries (Laue et al., 2018). Despite this dual designation by UNESCO (i.e., natural and cultural), the local partnership between Lesotho and South Africa highlights the natural resources, specifically the need to protect these for those who rely on them for their survival. This designation was received in 2000 for South Africa and expanded to include Lesotho in 2013. This park offers a cautionary example of the desire to gain WHS inscription without adequate resources or plans in place to manage visitation and protection. The relatively recent desire to democratize the list of WHSs by upping African representation (see Willems & Comer, 2011) has left many states prematurely nominating or inscribing sites without proper planning to ensure the intended benefits are realized, including increased visitation and revenue (Meskell, 2012, pp. 149–150).

The designation of the transfrontier park based equally on natural and cultural criteria has not translated into equal funding or attention to both types of resources and has resulted in a lack of promotion of rock art (Laue et al., 2018). The 2013 expansion to include Lesotho did not include a detailed plan to manage additional rock art sites or to include long-term archaeological or conservation management in Lesotho. This in turn is a chronic problem in many small African states that lack the infrastructure, training, and professional opportunities for local professionals, and are thus reliant on foreign researchers to conduct archaeological research and carry out site management. Much of the training available for local communities is condensed into a few days or weeks-long workshops (Cain, 2009). On the South African side, the primary management board for the park does not include heritage experts to the ongoing detriment of the archaeological resources of which as much as 77% are currently damaged or threatened (Mazel, 2018, p. 132). The critique of the area's resource management extends to the museums that include removed rock art and disparate interpretations of these alongside de-contextualized archaeological materials from rock shelters.

This example highlights the uphill battle South Africa and other nations with rock art resources to compete with more organized and better-funded rock art sites in Europe where ongoing or preserved archaeological excavations contribute to the tourist experience (Dans & González, 2019). Balanced site management is key to promoting sustainable visitation that includes a mix of management/conservation where specific sites are set aside for visitation and interpretation. However, even though 23 of the known 600 sites in the South African portion of the park are set up for visitation, very few tourists visit to take advantage of the heritage opportunity there (Duval & Smith, 2014).

Botswana boasts similar rock shelter sites including the Tsodilo Hills WHS. The Monuments and Relics Act of 2001 makes a clear distinction between a monument (post-June 1, 1902) and relics (pre-June 1, 1902) (Brown, 2020a, p. 19). This determines the type of designation and attention the site is given. Much of the relic sites include moveable heritage (i.e., artifacts and other objects) that are threatened by tourist activities. The archaeological sites that have received the most national attention in terms of tourism development are the rock shelters—including paintings—in the Tsodilo Hills, though these are not the only ones within the country that local communities deem important or significant (Brown, 2020b). These sites are largely part of standard heritage tourism visits and are incorporated into Botswana's tourism policies as such but have not been developed as archaeotourism attractions in their own right the way those in South Africa have been developed.

Zimbabwe is home to perhaps the most well-known African WHS: Great Zimbabwe. The country has a long history of site preservation beginning under the colonial government in 1902 and the active promotion of archaeological sites to tourists (primarily White) in the 1940s (Nhamo, 2016). However, internal conflict and economic struggles have hindered growth in tourism and heritage preservation in recent decades.

Central Africa

Archaeology has a long history in Central Africa (see Denbow, 2013), yet it has yet to be harnessed for tourism. Much of the work in countries such as Chad and Cameroon, is done under the guise of cultural resource management related to oil and other natural resource development, rather than for investment in heritage resources and tourism (Lavachery et al., 2012; MacEachern, 2010). In fact, one of the only countries in the region actively building its heritage tourism sector to include archaeological resources is Rwanda. In recent years, Rwanda has begun developing heritage tourism, including localities with archaeological components. Archaeology in general is complicated by the post-genocide political climate (Giblin, 2012), with most of the focus of heritage tourism in the post-genocide era has been on 'gorillas and genocide' (di Domenico & di Domenico, 2009). Despite a recognition that archaeological resources can be harnessed for economic development, in cases where significant archaeological remains are present, such as Musanze Caves, many secondary assets have been destroyed to provide access to larger heritage sites that are deemed more valuable (Giblin et al., 2017). Thus, the emphasis has been on developing larger spaces and more 'important' artifacts and localities rather than movable resources or engaging in archaeotourism specifically.

Conclusion: Is There a Future for Archaeotourism in Africa?

Large-scale archaeological digs and public archaeology programs as conceived in the Global North are far from accessible or attainable throughout much of Sub-Saharan Africa, and those that do exist rely overwhelmingly on international partnerships. This is largely a result of inadequate infrastructure, financial resources, secure environments in some regions, and trained local professionals in archaeology and other areas in the cultural heritage sector. In many countries, there is also a lack of political will to effect legal and policy changes that might help promote tourism and protect archaeological resources simultaneously.

The lack of emphasis on archaeological remains and other heritage assets is compounded by many countries' overwhelming tourism development focus on urban and coastal areas, where large hotels and resorts, such as those in Ghana, Senegal, The Gambia, Kenya, and South Africa, emphasize urban tourism and beach vacations. Likewise, several countries in eastern and southern Africa focus almost exclusively on fauna-based safari tourism. Thus, at present, although it is more prevalent in North Africa, archaeotourism on its own is an unrealistic goal in Sub-Saharan Africa in the short term.

Nonetheless, there is significant potential for archaeology-based tourism throughout the continent, especially if paired with other types of tourism and holiday experiences, such as safaris in Tanzania, beach-based tourism in The Gambia, or urban tourism in South Africa. For instance, incorporating visits at Olduvai Gorge into Tanzanian safari itineraries can help build awareness of the region's magnificent archaeological record. This awareness, in turn, could be harnessed to push for the development of archaeological management systems, including countrywide inventories and assessment programs that would create positions for professionals, enhance tourist visitation, and help protect delicate cultural resources. While Africanist archaeologists have been pushing for this since the 1990s, little progress has been made (Breen & Rhodes, 2010).

From the perspective of the entire continent, Africa remains the least visited region in the world, yet its material culture (and living culture), including archaeological assets, are valuable and important not only to Africans but also to the world community at large. There is significant latent potential to develop a robust archaeotourism sector, just as many destinations in Latin America, Europe, and Asia have done, but it will require a strong political will among Africa's governments, industry leaders, and heritage managers to collaborate, train archaeologists and heritage managers, sustain sites, and create opportunities for tourism development that have heretofore not been tapped except by a small handful of countries such as Tunisia, Egypt, Ethiopia, and Zimbabwe.

Note

1 In 2000, Earthwatch volunteers participated in archaeological excavations as part of the Central Region Project (Ghana) at numerous sites that were served as the basis for a number of doctoral dissertations for Syracuse University graduate students (DeCorse et al., 2000).

References

Adeola, O., Boso, N., & Olaniyi, E. (2017). Drivers of international tourism demand in Africa. *Business Economics*, *53*(1), 25–36.

Ashley, C., & Bouakaze-Khan, D. (2011). Conservation and management of archaeological sites in Sub-Saharan Africa. *Conservation and Management of Archaeological Sites*, *13*(2–3), 95–102.

Babalola, A.B., Ogunfolakan, A.B., & Rehren, T. (2020). Semi-finished glass from Ile-Ife, Nigeria: Implications for the archaeology of glass in sub-Saharan Africa. *Antiquity, 94*(375), e16.

Baram, U. (2008). Tourism and archaeology. In D.M. Pearsall (Ed.), *Encyclopedia of archaeology*, pp. 2131–2134. Amsterdam: Elsevier.

Breen, C. (2007). Advocacy, international development and World Heritage Sites in Sub-Saharan Africa. *World Archaeology, 39*(3), 355–370.

Breen, C., & Rhodes, D. (2010). *Archaeology and international development in Africa*. London: Gerald Duckworth and Co.

Brown, B. (2020a). Cultural heritage: Concepts, places and issues. In B. Brown, L.L. Mokgalo, & T. Cipfuva (Eds.), *Botswana cultural heritage and development: A handbook of theory and practice*, pp. 16–57. Newcastle upon Tyne: Cambridge Scholars.

Brown, B. (2020b). The politics of cultural heritage and its packaging for tourism consumption in Botswana. In B. Brown, L.L. Mokgalo, & T. Cipfuva (Eds.), *Botswana cultural heritage and development: A handbook of theory and practice*, pp. 58–97. Newcastle upon Tyne: Cambridge Scholars Publishing.

Bruner, E.M. (1996). Tourism in Ghana: The representation of slavery and the return of the black diaspora. *American Anthropologist, 98*(2), 290–304.

Cain, C. (2009). Cultural heritage survey of Lesotho for the Maloti-Drakensberg Transfrontier Project, 3005–2006: Palaeontology, archaeology, history and heritage management. *The South African Archaeological Bulletin, 64*(189), 33–44.

Chami, M.F. (2018). Tourist touching loyalty in cultural heritage sites in Tanzania: A case of Zanzibar Stone Town, Zanzibar Island. *International Journal of Economics, Business Management Research, 2*(4), 306–316.

Chigozie, O.J., & Ukekwe, C. (2020). Tourism and virtual reality (VR) in developing nations. *African Journal of Hospitality, Tourism and Leisure, 9*, 1–16.

Chirikure, S., & Pwiti, G. (2008). Community involvement in archaeology and cultural heritage management: An assessment from case studies in Southern Africa and elsewhere. *Cultural Anthropology, 49*(3), 467–485.

Dans, E.P., & González, P.A. (2019). Sustainable tourism and social value at World Heritage Sites: Towards a conservation plan for Altamira, Spain. *Annals of Tourism Research, 74*, 68–80.

DeCorse, C.R. (2001a). *An archaeology of Elmina: Africans and Europeans on the Gold Coast, 1400–1900*. Washington, DC: Smithsonian Institution Press.

DeCorse, C.R. (Ed.) (2001b). *West Africa during the Atlantic slave trade: Archaeological perspectives*. London: Leicester University Press.

DeCorse, C.R. (2015). Sierra Leone in the Atlantic world: Concepts, contours, and exchange. *Atlantic Studies, 12*(3), 296–316.

DeCorse, C.R., Carr, E., Chouin, G., Cook, G., & Spiers, S. (2000). Central Region Project, coastal Ghana, Perspectives 2000. *Nyame Akuma, 53*, 6–11.

Denbow, J. (2013). *The archaeology and ethnography of Central Africa*. Cambridge: Cambridge University Press.

di Domenico, C., & di Domenico, M. (2009). Tourism in Rwanda: The challenges of managing and presenting a sensitive heritage. In H. Andrews, P. Hackett, M. Meadows, & M. Selby (Eds.), *Leisure and Tourism: International Perspectives on Cultural Practice*, pp. 3–15. Liverpool: Leisure Studies Association.

Duval, M., & Smith, B. (2014). Seeking sustainable rock art tourism: The example of Maloti-Drakensberg Park World Heritage Site. *The South African Archaeological Bulletin, 69*(199), 34–48.

Franke, G. (2016). A chronology of the central Nigerian Nok culture–1500 BC to the beginning of the common era. *Journal of African Archaeology, 14*(3), 257–289.

Gavua, K. (2006). Rethinking African archaeology from inside-out. Paper preentated at the *18th biennial meeting of the Society for Africanist Archaeology, Calgery*, Alberta, Canada.

Gebauer, M. & Umscheid, M. (2021). Roots tourism and the year of return campaign in Ghana: Moving belonging beyond the history of slavery. In J. Saarinen & J.M. Rogerson (Eds.), *Tourism, change and the Global South*, pp. 123–134. London: Routledge.

Giblin, J.D. (2012). Decolonial challenges and post-genocide archaeological politics in Rwanda. *Public Archaeology, 11*(3), 123–143.

Giblin, J.D., Mugabowagahunde, M., & Ntagwabira, A. (2017). International heritage tourism in Rwanda: Paving over the past at the Musanze Caves. *Conservation and Management of Archaeological Sites, 19*, 126–140.

Gijanto, L. (2009). Stabilization and tourism at the Gambia River's Atlantic trade sites: The James Island Conservation and Survey Project. *African Diaspora Archaeology Newsletter, 12*(3), 1–25.

Gijanto, L. (2017). *The life of trade: Events and happenings in the Niumi's Atlantic Center.* New York: Routledge.

Gillot, L. (2020). Archaeologists and tourism: Symbiosis or contestation? In D.J. Timothy & L.G. Tahan (Eds.), *Archaeology and tourism: Touring the past*, pp. 26–40. Bristol: Channel View Publications.

Gubam, D., & Nominshan, T. (2020). Archaeology and tourism in Nigeria: An overview. *Journal of Tourism and Heritage Studies, 9*(1), 90–101.

Jameson, J. (Ed.) (1997). *Presenting archaeology to the public.* Walnut Creek, CA: Altamira Press.

Jopela, A., & Fredrikson, P.D. (2015). Public archaeology, knowledge meetings and heritage ethics in Southern Africa: An approach from Mozambique. *World Archaeology, 47*(2), 261–284.

Kelly, K.G. (1997). The archaeology of African-European interaction: Investigating the social roles of trade, traders, and the use of space in the seventeenth and eighteenth century Hueda Kingdom, Republic of Benin. *World Archaeology, 28*(3), 77–95.

Krstic, B., Jovanovic, S., Jankovic-Milic, V., & Stanisic, T. (2016). Examination of travel and tourism competiveness contribution to national economy competiveness of Sub-Saharan Africa countries. *Development Southern Africa, 33*(4), 470–485.

Laue, G., Challis, S., & Mullen, A. (2018). Concerning heritage: Lessons from rock art management in the Maloti-Drakensberg Park. In S. Makuvaza (Ed.), *Aspects of management planning for cultural World Heritage Sites: Principles, approaces and practices*, pp. 119–130. New York: Springer.

Lavachery, P., MacEachern, S.A., Bouiman, T., & Mindzie, C.M. (2012). *Komé-Kribi: Rescue archaeology along the Chad-Cameroon oil pipeline, 1999–2004.* Frankfurt: Africa Magna Verlag.

Little, B., & McManamon, F.P. (2005). Archaeology and tourism in and around America's National Parks. *The SAA Archaeological Record, 5*(3), 12–14.

MacEachern, S.A. (2010). Seeing like an oil company's CHM programme: Exxon and archaeology on the Chad Export Project. *Journal of Social Archaeology, 10*(3), 347–366.

Mazel, A.D. (2018). Managing the rock art of the uKhahlamba-Drakensberg: Progress, blind spots and challenges. In S. Makuvaza (Ed.), *Aspects of management planning for cultural World Heritage Sites: Principles, approaches and practices*, pp. 131–145. Cham, Switzerland: Springer.

McIntosh, S.K. (2017). Archaeology in Africa: Who or what sets the agenda? In A.L. Smith, E. Cornelissen, O.P. Gosselain, & S.A. MacEachern (Eds.), *Field manual for African archaeology*, pp. 15–17. Tervuran: Royal Museum for Central Africa.

Mensah, I. (2015). The roots of tourism experiece of diaspora Africans: A focus on the Cape Coast and Elmina Castles. *Journal of Heritage Tourism, 10*(3), 213–232.
Meskell, L. (2012). The rush to inscribe: Reflections on the 35th session of the World Heritage Committee, UNESCO Paris, 2011. *Journal of Field Archaeology, 37*(2), 145–151.
Mkono, M. (Ed.) (2019). *Positive tourism in Africa*. London: Routledge.
Monroe, J.C. (2007). Dahomey and the Atlantic Slave Trade: Archaeology and political order in the Bight of Benin. In A. Ogundiran & T. Falola (Eds.), *Archaeology of Atlantic Africa and the African diaspora*, pp. 100–120. Bloomington: Indiana University Press.
Nhamo, A. (2016). Archaeology and tourism: Lessons from the Zimbabwean experience. In C. Walker & N. Carr (Eds.), *Tourism and archaeology: Sustainable meeting grounds*, pp. 211–222. Routledge.
Novelli, M. (2016). *Tourism and development in Sub-Saharan Africa: Current issues and local realities*. London: Routledge.
Odum, C.J., & Oguamanam, C.C. (2020). Archaeotourism and archaeological heritaae in Igboland, Nigeria. *Africa Journal of Hospitality, Tourism and Leisure, 9*(4), 639–654.
Ogundiran, A. (2007). Living in the shadow of the Atlantic world: History and material life in a Yoruba-Edo hinterland, ca. 1600–1750. In A. Ogundiran & T. Falola (Eds.), *Archaeology of Atlantic Africa and the African diaspora*, pp. 77–99. Bloomington: Indiana University Press.
Okamura, K., & Matsuda, A. (Eds.) (2011). *New perspectives in global public archaeology*. New York: Springer.
Osei-Tutu, B. (2014). Cape Coast Castle and rituals of memory. In A. Ogundiran & P. Saunders (Eds.), *Materialities of ritual in the black Atlantic*, pp. 317–338. Bloomington: Indiana University Press.
Posnansky, M. (2017). Archaeology and the local community in Africa: A retrospective. *Journal of Community Archaeology and Heritage, 4*(2), 77–84.
Reed, A. (2014). *Pilgramage tourism of diaspora Africans to Ghana*. New York: Routledge.
Reid, S., & Amartey, S. (2019). Archaeology, GIS, and remote sensing near the Pra River, southern Ghana. Central River Project Update 2015–2017. *Nyame Akuma, 91*, 15–22.
Reilly, M., Banton, C., Stevens, C., & Gijanto, L. (2019). Back-to-Africa archaeology and heritage in Liberia. *Nyame Akuma, 92*, 4–11.
Rowan, Y., & Baram, U. (Eds.) (2004). *Marketing heritage: Archaeology and the consumption of the past*. Walnut Creek, CA: Altamira Press.
Schmidt, P. (2014). Rediscovering community archaeology in Africa and reframing its practice. *Journal of Community Archaeology and Heritage, 1*(1), 37–55.
Schramm, K. (2010). *African homecoming: Pan-African ideology and contested heritage*. Walnut Creek, CA: Left Coast Press.
Singleton, T. (2006). African diaspora archaeology in dialogue. In K.A. Yelvington (Ed.), *Afro-Atlantic dialogues: Anthropology in the diaspora*, pp. 249–287. Santa Fe: School of American Research Press.
Statista (2021). Selected African countries with the largest number of international tourist arrivlas in 2019. Online: https://www.statista.com/statistics/261740/countries-in-africa-ranked-by-international-tourist-arrivals/
Thiaw, I. (2011). Digging on contested grounds: Archaeology and the commemoration of slavery on Gorée Island, Senegal. In K. Okamura & A. Matsuda (Eds.), *New perspectives in global public archaeology*, pp. 127–138. New York: Springer.
Thiaw, I., & Wait, G. (2018). Presenting archaeology and heritage at a UNESCO World Heritage Site: Goree Island, Senegal. *Advances in Archaeological Practice, 6*(3), 238–247.

Timothy, D.J. (2020). Archaeological heritage and volunteer tourism. In D.J. Timothy & L.G. Tahan (Eds.), *Archaeology and tourism: Touring the past*, pp. 87–105. Bristol: Channel View Publications.

Timothy, D.J., & Tahan, L.G. (Eds.) (2020). *Archaeology and tourism: Touring the past*. Bristol: Channel View Publications.

Weaver, D. (2015). Enlightened mass tourism as a 'third generation' aspiration for the twenty-first century. In M. Hughes, D. Weaver, & C. Pforr (Eds.), *The practice of sustainable tourism: Resolving the paradox*, pp. 11–23. London: Routledge.

Willems, W.J., & Comer, D. (2011). Africa, archaeology, and Word Heritage. *Conservation and Management of Archaeological Sites, 13*(2–3), 160–173.

16

NATURAL HERITAGE

Wildlife and Nature Preserves in the African Tourism Landscape

Chiedza Ngonidzashe Mutanga and Edson Gandiwa

Introduction

Protected areas (PAs), for example, national parks and nature reserves, are widely acknowledged as a key component of biodiversity conservation strategies because they serve as refuges for wildlife that would otherwise perish in altered landscapes and seascapes, as well as providing space for natural evolution and future ecological restoration (Dudley & Parish, 2006). Article 2 of the Convention on Biological Diversity defines a protected area as 'a geographically delimited region that is designated, regulated, and managed to fulfill certain conservation objectives' (CBD, 2011). As a result, wildlife and nature preserves are a type of protected area that is important for biodiversity conservation and tourism in many African countries (Brito, 2020; Forje, Tchamba, & Eno-Nku, 2021; Mutanga, 2017; Twining-Ward et al., 2018).

Nature-based tourism mainly depends on natural resources in a moderately undeveloped state, including scenery, water features, vegetation, and wildlife (Job, Becken, & Lane, 2017). Many African countries have benefited economically from considerable growth in nature-based tourism, which has had favourable effects on local and national economies (UNWTO, 2018). Nature-based tourism provides long-term economic benefits while also contributing to environmental conservation (Naidoo et al., 2016). As a result, wildlife and nature preserves are seen to be significant tourist attractions and substantial foreign currency earners that contribute immensely to the tourism sector (Chikuta, 2015).

Wildlife and nature preserves, therefore, constitute a type of natural heritage deeply ensconced in the tourism industry, especially in Africa. In Africa, heritage tourism is becoming increasingly popular (Hall & Seyfi, 2021; Novelli, Adu-Ampong, & Ribeiro, 2021; Timothy & Nyaupane, 2009). As such, heritage plays both a pull and a push role in travel decision-making (Lwoga & Adu-Ampong, 2021).

DOI: 10.4324/9781003153955-16

Heritage refers to our inheritances, what we have, and what we leave on to coming generations. It includes natural and cultural elements, living and architectural manifestations of culture, and physical and intangible characteristics that have cultural meaning or value (Timothy & Nyaupane, 2009). Around 40% of all foreign tourist travel activities are related to heritage tourism (UNWTO, 2018).

A wildlife preserve is a land- or water-based protected natural area dedicated to the conservation and scientific study of specific wild animals, flora, or geological features. Wildlife preserves may protect a wide range of species, including common, threatened, and endangered species, and can include a variety of habitats, such as forests, mountains, wetlands, and rivers. Natural heritage refers to natural sites or demarcated areas with natural elements such as physical and biological formations or groups of such formations that have exceptional universal significance from an aesthetic or scientific standpoint (Harrison & O'Donnell, 2010). Plants, animals, natural landscapes and landforms, oceans, and water bodies are all examples of natural heritage. In common with cultural heritage, natural heritage is often divided into tangible (e.g., plants, animals, and landforms) and intangible (e.g., aesthetic characteristics and contributions to biodiversity) components (Harrison, 2010).

The Role of Tourism in Developing Nature Preserves

In the early Renaissance, kings and other national rulers in Europe set aside protected territories as royal hunting reserves (Eagles, McCool, & Haynes, 2002). With the passage of time, these places became accessible to the general public, laying the groundwork for tourism. The first park was Yellowstone in the United States, which was established in 1872, and dedicated as a park to the benefit and amusement of the public (Eagles, McCool, & Haynes, 2002). In the nineteenth century, countries including the United States, New Zealand, Australia, and Canada began to create protected places, such as Queen Victoria Park at Niagara Falls, Canada, in 1885 and Tongariro National Park, New Zealand, in 1894 (Bishop, 2004). In the twentieth century, the idea of PAs flourished all over the world (Mutanga, 2017; Risso, 2017).

In Africa, the colonial era changed local communities' common access into a segregative policy known as 'fortress conservation', which gave rise to the eviction and dislodgment of native communities from their inherited land and prevented them from accessing the resources on which they relied (Büscher & Dietz, 2005; Siurua, 2006). During Africa's colonial era, environmental protection areas were established free of human pressure and habitation, with strictly prohibited access to resources. In Europe, explorers, adventurers, and professional hunter-traders armed with high-precision guns were permitted to trade in trophy animals (Jones, 2006; Murombedzi, 2003). However, European settlers and hunters in Africa engaged in the mass slaughtering of animals as a result of their predatory attitude toward African wildlife, endangering some species' survival. For example, in the 1890s, the emergence of colonial imperial authority corresponded with the concern to protect wildlife resources by the Kenyan government. Western tourists came to the newly founded British East Africa Protectorate to take part in enormous hunting safaris, resulting in severe wildlife slaughter (Honey, 1999). The British created legislation through the years to meet their recreational demands and to facilitate hunting safaris (Honey, 1999). The colonial government authorised the establishment of wildlife game reserves in 1896 and the National Parks Ordinance was passed in 1945, paving the way for founding more PAs, such as Nairobi Royal Park and Aberdare Royal Park, which were created not only to protect wildlife but also to provide exclusive recreation to European settlers (Chongwa, 2012). Kenya has enacted policies that have steadily increased its appeal as a tourist destination worldwide, starting from the initial times of wildlife hunting safaris. In Kenya, wildlife-based tourism has therefore become the country's main source of revenue (Chongwa, 2012).

In reaction to the mass killing of animals by European settlers and hunters, the Game Law was enacted in 1896 in Zimbabwe, with the goal of controlling consumption and to prevent the commercialisation of wildlife (Muposhi et al., 2016). More sanctuary areas and game reserves were established as a result of the necessity to conserve wildlife, culminating in the current Parks and Wildlife Estate (Muposhi et al., 2016), which reinforces the scientific, economic, aesthetic, and moral justification for wildlife conservation through sound management that endeavours to conserve the heritage of Zimbabwe for the good of current and future generations and stakeholders.

In Uganda, the British had interests in the Nile State with its immense resources; therefore, royal efforts to preserve nature were vital. As British interests in the region developed, the need to provide hunting safaris and leisure options to the region's growing settler population also grew (Honey, 1999). In South Africa, however, the formation of two game reserves, Sabie and Shingwedzi, which created Kruger National Park officially in 1926, was motivated by racial prejudice. The National Parks Act formally established land in South Africa for White men to utilise for hunting game and protecting other flora and fauna. Several forest reserves were established in many African countries from the late nineteenth century to the twentieth century as regions where not only the fauna would be conserved, but also tourism would be promoted swiftly (Chongwa, 2012; Mutanga, 2017).

Tourism and recreation in PAs are recognised by the International Union for Conservation of Nature (IUCN, 1994), among other activities (Tables 16.1 and 16.2). While PAs are managed for a number of reasons (Table 16.2), the importance of tourism and recreation in PAs is clear because they are allowed in all IUCN PA categories, except Ia: Strict Nature Reserve) (Table 16.2).

Africa's colonial conservation legacy was first a disjointed collection of wildlife reserves and parks created to meet the recreational and economic demands of expatriate Whites, settler populations, and foreign tourists (Chongwa, 2012). The requirements of local people and their environments were not considered when many of these PAs were established (Risso, 2017). The PA movement was linked to the forcible relocation of local inhabitants and denying them access to resources like meat, grazing lands, and firewood (Fischer, Muchapondwa, & Sterner, 2011). The evolution of pragmatic wildlife management methods into a refined conservation

TABLE 16.1 IUCN management categories of PAs

Category	*Description*
I	Strict Nature Reserve: PAs managed mainly for science or wilderness protection.
Ia	Strict Nature Reserve: PAs managed mainly for science.
Ib	Wilderness Area: PAs managed mainly for wilderness protection.
II	National Park: PAs managed mainly for ecosystem protection and recreation.
III	Natural Monument: PAs managed mainly for the conservation of specific natural features.
IV	Habitat/Species Management Area: PAs managed mainly for conservation through management intervention.
V	Protected Landscape/Seascape: PAs managed mainly for landscape/seascape conservation and recreation.
VI	Managed Resource Protected Area: PAs managed mainly for the sustainable use of natural ecosystems.

Source: IUCN (1994).

TABLE 16.2 The matrix of management objectives and IUCN PA management categories

Management objective	*Ia*	*Ib*	*II*	*III*	*IV*	*V*	*VI*
Scientific research	1	3	2	2	2	2	3
Wilderness protection	2	1	2	3	3	–	2
Preservation of species and genetic diversity (biodiversity)	1	2	1	1	1	2	1
Maintenance of environmental services	2	1	1	–	1	2	1
Protection of specific natural/cultural features	–	–	2	1	3	1	3
Tourism and recreation	–	2	1	1	3	1	3
Education	–	–	2	2	2	2	3
Sustainable use of resources from natural ecosystems	–	3	3	–	2	2	1
Maintenance of cultural/traditional attributes	–	–	–	–	–	1	2

Source: IUCN (1994).
Key: 1 = primary objective; 2 = secondary objective; 3 = potential applicable objective; – = not applicable.

science during the nineteenth century is represented in the histories of the PA systems of South Africa, Namibia, and Tanzania (Carruthers, 2017). With the arrival of ecologists as a primary research support for PAs in the 1950s, conservation prescriptions and management approaches in several African countries began to shift. In Kruger National Park, South Africa, this was seen as a principal project after the state's creation of PAs, while in Kenya, wildlife conservation was undertaken as a way of reclaiming the country's natural heritage after independence (Barratt Heitmann, 2018). Other countries, such as Angola, have invested very little in studying their preserves until the twenty-first century (Huntley et al., 2019).

To counter the negative effects of colonial practices on African communities, such as reducing social and economic costs of protectionist or fortress conservation approaches, and to ensure that conservation objectives were met, several integrated conservation and development projects (ICDPs) and community-based natural resource management (CBNRM) programmes were implemented in the region (Dressler et al., 2010). These include Living in a Finite Environment (LIFE) in Namibia, the Communal Areas Management Programme for Indigenous Resources (CAMPFIRE) in Zimbabwe, Tchuma Tchato in Mozambique, and the Administrative Management Design for Game Management Areas (ADMADE) in Zambia (Shereni & Saarinen, 2021). CBNRM has enhanced the foundations of conservation and its benefits to local communities by managing natural resources, such as land, forests, animals, and water, by collective local institutions (Roe, Nelson, & Sandbrook, 2009). Preserves are frequently protected against activities like land development, industrial usage, hunting, and motorised leisure to encourage wildlife conservation and to preserve Africa's natural heritage. The business sector and non-government organisations (NGOs) have been involved in natural resource conservation to supplement government efforts in formulating environmental policies, organising public support for environmental conservation, and safeguarding endangered forest and animal species (Pegas & Castley, 2014). The operations of NGOs and community-based organisations include promotion, orientation, and public education on a variety of conservation measures, such as the development and maintenance of wildlife and nature preserves, national parks, and other PAs.

The Role of Tourism in Conserving Wildlife and Nature Preserves

Tourism provides financial motivation for governments to undertake environmental legislation and conservation strategies. An example of such laws is the establishment of PAs such as national parks and nature reserves. This is because tourism in PAs (both consumptive and non-consumptive) has undoubtedly provided significant revenue for national governments (Stone & Nyaupane, 2016). Many African countries rely on natural landscape- and ecosystems-based tourism to generate revenue; hence, they recognise the necessity of safeguarding their coastlines, beaches, natural landscapes, and wildlife. For example, conservation initiatives emanating from tourism have benefited mountain gorillas in the Democratic Republic of the Congo, Uganda, and Rwanda. In Uganda, the economic worth of gorilla tourism is estimated to be around US$34.3 million, with gorilla trekking permits starting at US$600 (IGCP, 2022). This gave rise to policies and regulations that enhance conservation, including community conservation projects, veterinary programmes, transboundary collaboration, as well as stringent law enforcement measures. In fact, several governments have made critical efforts to realise nature-based tourism's revolutionary potential. Only 1,000 tourists visited Rwanda's national parks in 1999, for example. However, after the establishment of a tourism plan by the government, over 94,000 tourists visit every year. Moreover, about 410,000 people are employed in the tourism sector, while 15% of the income from tourism goes to community development around the national parks.

Tourism also contributes to the funding of conservation efforts in wildlife and nature reserves. Wildlife heritage is the only principal revenue driver for tourism in Africa. Tourism is an important source of income for land and wildlife conservation in Africa, with most preserves relying on revenue from tourism and other wildlife economies to operate. Revenue from Africa's 8,400 PAs amounts to about USD $48 billion. This indicates that financial opportunities can be available to African governments to conserve and develop their natural heritage for tourism. The World Tourism Organization (UNWTO) has estimated that wildlife watching constitutes about 80% of trips to Africa. In 2015, in Sub-Saharan Africa, it was estimated that about $142m per year was realised from entrance fees charged in different PAs in 14 countries. In a number of countries in Africa, each tourist on wildlife photographic tours spent about $433 every day. Tourism is vital to many PAs, which rely on visitors and tourism for financial support (Bushell & Bricker, 2017; Muhanna, 2006). As a result, biodiversity conservation and tourism are intertwined, with nature preserves providing a desirable tourism product, as well as a source of revenue for government agencies that can help protect biodiversity. Eagles, McCool, and Haynes (2002) attest that protected area visitation and tourism are the primary sources of funding for most PAs. Many protected places around the world are underfunded, with the global budget deficit for efficiently administering these sites estimated to be as high as US$440 billion per year (UNDP, 2019). By offering an extra source of money, tourism plays a critical role in filling this gap (Mutanga et al., 2021). Funds generated from tourism such as operating licenses, entrance fees, guiding services, and lodging can help pay wages for PA employees. Funds from tourism can also finance other essential PA activities like environmental education programs, ecosystem monitoring, invasive species eradication, and anti-poaching patrols (Higginbottom, Northrope, & Green, 2001).

Tourism also aids in ecosystem monitoring (Higginbottom, Northrope, & Green, 2001). While it is the duty of park rangers to protect wildlife and the natural environment, it is often difficult for them to do that alone. As such, tourists in natural regions can help to watch and safeguard animals. Game drive vehicles and passengers dissuade people from poaching in PAs, thereby assisting in the protection of endangered animals such as rhinos. Due to the COVID-19 pandemic, the importance of tourism in discouraging illegal hunters and providing sanctuaries for animals has become apparent. This is because, as tourism has dwindled, rhino poaching has apparently surged in tourist areas (Roth, 2020). Moreover, businesses in the tourism industry may be able to assist researchers in gathering environmental data.

Tourism helps increase community support for conservation (Mutanga et al., 2015). Nature-based tourism has grown in popularity over the previous decade. Wildlife tourism today employs approximately 22 million people worldwide and generates more than US$120 billion in revenue (WTTC, 2019). The increasing attention that wildlife tourism is receiving and its associated financial opportunities have the potential to shift community attitudes toward conservation.

Local populations may perceive wildlife as a threat if tourism is not available, and natural resources may only be valued for food. But when wildlife and other natural resources bring benefits to their community, residents may be more aware of the significance of preserving their natural heritage. Benefits from community-based tourism, for example, are pushing communities in Namibia to conserve wildlife through an expanding landscape of collective conservation (Silva & Mosimane, 2013). As a result, tourism can help minimise environmental degradation by offering alternative means of support for local communities, in addition to changing mindsets. Jobs such as housekeepers, cooks, and guides provide a substitute to ecologically damaging activities including slash-and-burn agriculture, illegal hunting, quarrying, and logging (Le Bel et al., 2011).

Finally, tourism can also raise environmental awareness among tourists. Tourists have numerous possibilities to learn about and experience the natural environment through tourism (Higginbottom, Northrope, & Green, 2001). As a result, visitors who connect with the environment during their travels may develop a greater appreciation for it and a desire to safeguard it. Wildlife interpretation methods such as guided tours, exhibitions, and brochures are used in PAs to enthuse conservation learning.

Nature and Wildlife as Heritage in the Tourism Industry

In Africa, there are around 8,400 PAs under state, private, and communal ownership, or a combination of these, covering up to 4.3 million km^2 and accounting for 13.8% of the land area on the continent (Vause, 2019). On the UNESCO World Heritage List, Africa now has 95 World Heritage Sites (WHSs), 38 of which are natural WHSs (Lwoga & Adu-Ampong, 2021). The global network of biosphere reserves, on the other hand, has 738 sites in 134 nations, including 90 sites in 33 African countries (UNESCO, 2022).

Heritage is a community asset that is regarded as a legacy passed down from ancestors (Harrison, 2010). As a result, the natural heritage legacy supplies society with resources that have been handed down from earlier generations, developed in the present, and bequeathed for the benefit of upcoming generations. Nature preserves are thus a natural heritage that preserves a community's biological and evolutionary origins for future generations. Just like cultural heritage, nature has been bestowed to society by its forebears, and currently, societies must take seriously their responsibility to protect it and pass it on to future generations in the best possible form.

Natural heritage has the potential to make natural diversity more accessible and enjoyable. As a result, natural heritage is basically an economic activity (Harrison, 2010). This is because the state and other organisations' involvement in natural heritage is mostly motivated by the economic potential of natural heritage and its links to tourism. As a result, tourism and natural heritage are inextricably linked on a fundamental economic level. At state, regional or local levels, tourism is required to pay for the advertising and conservation of natural heritage, while natural heritage brings in the tourism that consumes services and promotes the country's 'brand' (Harrison, 2010).

Natural heritage has become extremely valuable to the tourism industry in many nations, while also posing new conservation challenges. As a result, efficient natural heritage management necessitates striking the right balance between enjoying nature now and protecting its value for the future. The 1973 Washington Convention on International Trade in Endangered Species of Wild Fauna and Flora (CITES) declares that wild species are a part of natural heritage that cannot be replaced and that must be conserved for current and forthcoming generations, and that people and countries should be the best protectors of the wildlife. This principle is reiterated in the 1979 Bonn Convention on Migratory Species of Wild Animals, which states that each generation of people retains Earth's resources for future generations and has the responsibility to ensure that this heritage is preserved through judicious use and wise stewardship (Lyster, 1989). According to the Bern Convention for the Conservation of European Wildlife and Natural Habitats of 1979, wildlife is a natural heritage of aesthetic, scientific, cultural, recreational, economic, and intrinsic worth that must be maintained and passed on to future generations (Jen, 1999).

Biodiversity is encouraged through wildlife and environment preserves. The term 'biodiversity' refers to the wide range of species and living organisms on Earth, as well as their impact on the ecosystem. As a result, biodiversity is a form of intangible natural heritage made up of interactions between different species and the natural world. This is because it relates to living beings that are continually changing and interacting with one another, as well as their surroundings, as opposed to objects of heritage such as works of art or cultural artefacts. Because many animals and plants survive through symbiotic or predatory connections with other species, wildlife preserves encourage biodiversity by safeguarding natural areas and the species that inhabit them.

Local heritage resources are frequently used to attract tourists (Madden & Shipley, 2012; Timothy, 2021). Nature's everyday applications for leisure, exercise, and psychological regeneration go back well beyond written history, and have been a primary motivator for the construction of PAs (Stolton et al., 2015). Natural heritage-based tourist attractions are an intrinsic component of the tourism offered in many nations, making heritage a vital player in the tourism sector (Leask, Fyall, & Garrod, 2002). Africa's natural heritage and tourist attractions include its wildlife, mountains, waterfalls, rapids, beaches, and coral reefs. Prominent examples include Mt Kilimanjaro in Tanzania, Mt Kenya in Kenya, Victoria Falls on the border of Zambia and Zimbabwe, Zanzibar's beaches in Tanzania, and the wildebeest migration between the Maasai Mara in Kenya and the Serengeti in Tanzania (Lovei, 2017). The use of natural heritage as a tourism attraction is reflected in many forms of tourism, for example, nature-based tourism, ecotourism, wildlife tourism, and adventure tourism (Table 16.3).

TABLE 16.3 Forms of tourism and how they utilise natural heritage

Form of tourism	*Definition*	*How the form of tourism utilises natural heritage*	*Examples of activities involved*	*Sources*
Nature-based tourism	Travelling with the main aim of visiting a natural destination	Based on nature's attractions and on all the activities that are held in the natural surroundings: - Depends on the natural landscape or natural resources either as the setting for activities or where the land or resources are themselves the central component of the tourist activity - incorporates all forms of tourism (e.g., mass tourism, adventure tourism, low-impact tourism, and ecotourism) which use natural resources in an undeveloped form, including species, habitat, landscape, scenery, and salt and freshwater features	Wildlife viewing and Photography, fishing, downhill skiing, hunting, ecotourism, swimming, diving, hiking, campfire, picnic, learning, friends, or spending time with friends and/ or families	Kuenzi and McNeely (2008); Goodwin (1996); Suta, Hrnjic, and Banda (2017)

(*Continued*)

TABLE 16.3 (Continued)

Form of tourism	*Definition*	*How the form of tourism utilises natural heritage*	*Examples of activities involved*	*Sources*
Ecotourism	Travelling to relatively undisturbed natural areas with the particular aim of studying and appreciating the scenery and its flora and fauna, as well as any existing cultural manifestations (both past and present) found in these areas. It is responsible travel to natural areas that conserves the environment and sustains the livelihoods of local people	- The main tourist motivation is the observation and appreciation of nature, as well as the traditional cultures prevailing in natural areas - Ecotourism minimises negative impacts on the natural and socio-cultural environment and supports the conservation of natural areas by generating economic benefits for host communities, organisations, and authorities that are responsible for conserving the natural areas - It also increases awareness both among the local communities and tourists about the need to conserve natural and cultural assets	Nature tours, observation of flora and fauna, cultural and local heritage tourism, volcano tourism; fishing tourism, bird watching tourism, kayaking tourism, bushwalking, hiking tourism, cold water tourism, hunting; visiting parks, scenic driving, photography tourism, camping	Mihalic (2006); Goodwin (1996); UNWTO (2001)

Wildlife tourism	Based on encounters with wild animals in either the animals' natural environment or in captivity. It includes non-consumptive activities as well as those that involve killing or capturing animals	- Occurs in natural, aesthetically pleasing environments - Utilises nature in two ways: (i) Consumptive tourism - Includes harvesting of wildlife populations for recreation or commercial purposes - Mainly covers hunting tourism and fishing (ii) Non-consumptive tourism - Does not include killing or trapping of wildlife populations - The purpose of non-consumptive tourism rests on education, recreation, and research - Encompasses segments such as ecotourism, safari tours, and mountain tourism wildlife watching among others	Safaris, bird watching, whale watching, hunting, fishing, diving, playing with lions and tigers, elephant riding, gorilla trekking, monkey forests, ostrich riding	Curtin and Kragh (2014)
Adventure tourism	Nature tourism with a degree of risk-taking and physical endurance	- Explores new experiences often involving perceived risk or controlled danger associated with personal challenges in a natural environment or exotic settings - Adventure activities usually take place in enclosed areas within which flora and fauna are protected and preserved for sustainable tourism - Nature is an important ingredient in many adventure activities where contact with nature can bring the perceived risks and dangers, and also the excitement and transcendence of connecting with nature and wildlife	Hiking, walking, mountain climbing, white water rafting, or kayaking	Mihalic (2006); Beaver (2002); Swarbrooke et al. (2003)

Conclusion

This chapter elucidates the role of tourism in the development of wildlife and nature preserves in the African tourism landscapes. An analysis of the history of the establishment of nature reserves indicates that tourism has helped develop a region-wide system of nature preserves in two ways: the development of nature reserves for recreation purposes, and the provision of incentives for the conservation of wildlife and nature reserves. The chapter also illuminates the role of nature and wildlife as a form of heritage that is used by the tourism industry. Nature and wildlife are simultaneously intangible and tangible natural heritage composed of the interactions between various wildlife species and the natural world. The legacy of wildlife as a form of heritage is preserved through a system of PAs including natural WHSs and biosphere reserves. The use of natural heritage as a tourism asset is reflected in many forms of tourism that include nature-based tourism which is founded on nature's attractions; ecotourism which also entails observing and appreciating nature, as well as the traditional cultures in natural areas; wildlife tourism, which utilises nature through consumptive and non-consumptive activities; and adventure tourism, which involves perceived risk or controlled danger associated with personal challenges in a natural environment. Wildlife preserves, therefore, not only act as tourist attractions but have, through conservation, become a form of heritage that is used by the tourism industry. In many African countries, natural heritage is now a spearhead and inspiration for both conservation and tourism.

References

Barratt Heitmann, J.R. (2018). *National parks and development in Sub-Saharan Africa: An examination of social and economic Program effectiveness in the development paradigm*. Senior Project, Division of Social Studies, Bard College, New York.

Beaver, A. (2002). *A dictionary of travel and tourism terminology*. Wallingford: CABI.

Bishop, K. (2004). *Speaking a common language: The uses and performance of the IUCN system of management categories for protected areas*. Gland, Switzerland: IUCN.

Brito, B.R. (2020). Tourism and nature in São Tomé and Príncipe: Opportunities for the internationalisation of a small island state. In M. Novelli, E.A. Adu-Ampong, & M.A. Ribeiro (Eds.), *Routledge handbook of tourism in Africa*, pp. 342–353. London: Routledge.

Büscher, B., & Dietz, T. (2005). Conjunctions of governance: The state and the conservation-development nexus in Southern Africa. *Journal of Transdisciplinary Environmental Studies*, *4*(2), 1–15.

Bushell, R., & Bricker, K. (2017). Tourism in protected areas: Developing meaningful standards. *Tourism and Hospitality Research*, *17*(1), 106–120.

Carruthers, V. (2017). *The wildlife of southern Africa*. Cape Town: Penguin Random House.

CBD (2011). *Protected areas and the CBD: Convention on Biological Diversity (CBD)*. Montreal: Secretariat of the Convention on Biological Diversity.

Chikuta, O. (2015). *The development of a universal accessibility framework for national parks in South Africa and Zimbabwe*. Unpublished Doctoral dissertation, North-West University, South Africa.

Chongwa, M.B. (2012). The history and evolution of national parks in Kenya. *George Wright Forum*, *29*(1), 39–42.

Curtin, S., & Kragh, G. (2014). Wildlife tourism: Reconnecting people with nature. *Human Dimensions of Wildlife, 19*(6), 545–554.

Dressler, W., Büscher, B., Schoon, M., Brockington, D.A.N., Hayes, T., Kull, C.A., McCarthy, J., & Shrestha, K. (2010). From hope to crisis and back again? A critical history of the global CBNRM narrative. *Environmental conservation, 37*(1), 5–15.

Dudley, N., & Parish, J. (2006). *Closing the gap. Creating ecologically representative protected area systems: A guide to conducting the gap assessments of protected area systems for the Convention on Biological Diversity*. Montreal: Secretariat of the Convention on Biological Diversity.

Eagles, P.F.J., McCool, S.F., & Haynes, C.D.A. (2002). *Sustainable tourism in protected areas: Guidelines for planning and management*. Gland, Switzerland: IUCN.

Fischer, C., Muchapondwa, E., & Sterner, T. (2011). Bioeconomic model of community incentives for wildlife management before and after CAMPFIRE. *Environmental and Resource Economics, 48*, 303–319.

Forje, G.W., Tchamba, M.N., & Eno-Nku, M. (2021). Determinants of ecotourism development in and around protected areas: The case of Campo Ma'an National Park in Cameroon. *Scientific African, 11*, e00663.

Goodwin, H. (1996). In pursuit of ecotourism. *Biodiversity & Conservation, 5*(3), 277–291.

Harrison, R. (2010). What is heritage? In R. Harrison (Ed.), *Understanding the politics of heritage*, pp. 5–42. Manchester: Manchester University Press.

Harrison, R., & O'Donnell, D. (2010). Natural heritage. In S. West (Ed.), *Understanding heritage in practice*, pp. 88–126. Manchester: Manchester University Press.

Hall, C.M., & Seyfi, S. (Eds.) (2021). *Cultural and heritage tourism in the Middle East and North Africa: Complexities, management and practices*. London: Routledge.

Higginbottom, K., Northrope, C., & Green, R. (2001). *Positive effects of wildlife tourism on wildlife*. Gold Coast, QLD: CRC for Sustainable Tourism.

Honey, M.S. (1999). Treading lightly? Ecotourism's impact on the environment. *Environment: Science and Policy for Sustainable Development, 41*(5), 4–9.

Huntley, B.J., Beja, P., Pinto, P.V., Russo, V., Veríssimo, L., & Morais, M. (2019). Biodiversity conservation: History, protected areas and hotspots. In B.J. Huntley, V. Russo, F. Lages, & N. Ferrand (Eds.), *Biodiversity of Angola. Science and conservation: A modern synthesis*, pp. 495–512. Cham, Switzerland: Springer.

IGCP (2022). International Gorilla Conservation Programme. Online: https://igcp.org/ Accessed 12 January 2022.

IUCN (1994). *Guidelines for protected area management categories*. Gland, Switzerland: IUCN.

Jen, S. (1999). *The Convention on the Conservation of European Wildlife and Natural Habitats: Procedures of application in practice*. Bern: Council of Europe.

Job, H., Becken, S., & Lane, B. (2017). Protected areas in a neoliberal world and the role of tourism in supporting conservation and sustainable development: An assessment of strategic planning, zoning, impact monitoring, and tourism management at natural World Heritage Sites. *Journal of Sustainable Tourism, 25*(12), 1697–1718.

Jones, S. (2006). A political ecology of wildlife conservation in Africa. *Review of African Political Economy, 33*(109), 483–495.

Kuenzi, C., & McNeely, J. (2008). Nature-based tourism. In O. Renn & K. Walker (Eds.), *Global risk governance: Concept and practice using the IRGC framework*, pp. 155–178. Dordrecht: Springer.

Leask, A., Fyall, A., & Garrod, B. (2002). Heritage visitor attractions: Managing revenue in the new millennium. *International Journal of Heritage Studies, 8*(3), 247–265.

Le Bel, S.B., Murwira, A., Mukamuri, B., Czudek, R., Taylor, R., & La Grange, M. (2011). Human wildlife conflicts in southern Africa: Riding the whirlwind in Mozambique and in Zimbabwe. In J. López-Pujol (Ed.), *The importance of biological interactions in the study of biodiversity*, pp. 283–322. Rijeka, Croatia: InTech.

Lovei, M. (2017). Africa can benefit from nature-based tourism in a sustainable manner. World Bank, 27 September 2017. Online https://blogs.worldbank.org/nasikiliza/africa-can-benefit-from-nature-based-tourism-in-a-sustainable-manner#:~:text=Nature%2Dbased%20tourism%20can%20offer,on%2C%20September%2028%2C%202017. Accessed 18 February 2021.

Lwoga, N.B., & Adu-Ampong, E.A. (2021). Heritage and heritage tourism in Africa: Historical overview and contemporary issues. In M. Novelli, E.A. Adu-Ampong, & M.A. Ribeiro (Eds.), *Routledge handbook of tourism in Africa*, pp. 69–84. London: Routledge.

Lyster, S. (1989). The Convention on the Conservation of Migratory Species of Wild Animals (The Bonn Convention). *Natural Resources Journal, 29*, 979.

Madden, M., & Shipley, R. (2012). An analysis of the literature at the nexus of heritage, tourism, and local economic development. *Journal of Heritage Tourism*, 7(2), 103–112.

Mihalic, T. (2006). Nature-based products, ecotourism and adventure tourism. In D. Buhalis & C. Costa (Eds.), *Tourism business frontiers: Consumers, products and industry*, pp. 111–117. Oxford: Elsevier.

Muhanna, E. (2006). Sustainable tourism development and environmental management for developing countries. *Problems and Perspectives in Management, 4*(2), 14–30.

Muposhi, V.K., Gandiwa, E., Bartels, P., & Makuza, S.M. (2016). Trophy hunting, conservation, and rural development in Zimbabwe: Issues, options, and implications. *International Journal of Biodiversity, 2016*, 8763980.

Murombedzi, J. (2003). *Pre-colonial and colonial conservation practices in southern Africa and their legacy today*. Gland, Switzerland: IUCN.

Mutanga, C.N. (2017). *Tourism and wildlife conservation: Protected area-community relationships and nature tourism dynamics in Zimbabwe*. Unpublished doctoral thesis, Chinhoyi University of Technology, Zimbabwe.

Mutanga, C.N., Chikuta, O., Muboko, N., Gandiwa, E., Kabote, F., & Kaswaurere, T.W. (2021). Wildlife tourism, conservation and community benefits in Zimbabwe. In M. Novelli, E.A. Adu-Ampong, & M.A. Ribeiro (Eds.), *Routledge handbook of tourism in Africa*, pp. 459–473. London: Routledge.

Mutanga, C.N., Vengesayi, S., Gandiwa, E., & Muboko, N. (2015). Community perceptions of wildlife conservation and tourism: A case study of communities adjacent to four protected areas in Zimbabwe. *Tropical Conservation Science, 8*(2), 564–582.

Naidoo, R., Weaver, L.C., Diggle, R.W., Matongo, G., Stuart-Hill, G., & Thouless, C. (2016). Complementary benefits of tourism and hunting to communal conservancies in Namibia. *Conservation Biology, 30*(3), 628–638.

Novelli, M., Adu-Ampong, E.A., & Ribeiro, M.A. (Eds.) (2021). *Routledge handbook of tourism in Africa*. London: Routledge.

Pegas, F.D.V., & Castley, J.G. (2014). Ecotourism as a conservation tool and its adoption by private protected areas in Brazil. *Journal of Sustainable Tourism, 22*(4), 604–625.

Risso, L.C. (2017). Discourses of the environmental conservation in the protected areas and the insertion of the indigenous and traditional communities in this debate. *Brazilian Geographical Journal: Geosciences and Humanities Research Medium*, 7(1), 7–18.

Roe, D., Nelson, F., & Sandbrook, C. (Eds.) (2009). *Community management of natural resources in Africa: Impacts, experiences and future directions*. London: International Institute for Environment and Development.

Roth, R. (2020). Poachers kill more rhinos as coronavirus halts tourism to Africa. *The New York Times*, 6 July 2020. Online: https://www.nytimes.com/2020/04/08/science/coronavirus-poaching-rhinos.html Accessed 10 July 2021.

Shereni, N.C., & Saarinen, J. (2021). Community perceptions on the benefits and challenges of community-based natural resources management in Zimbabwe. *Development Southern Africa, 38*(6), 879–895.

Silva, J. A., & Mosimane, A.W. (2013). Conservation-based rural development in Namibia: A mixed-methods assessment of economic benefits. *The Journal of Environment & Development, 22*(1), 25–50.

Siurua, H. (2006). Nature above people: Rolston and "fortress" conservation in the South. *Ethics and the Environment, 11*(1), 71–96.

Stolton, S., Dudley, N., Avcıoğlu Çokçalışkan, B., Hunter, D., Ivanić, K.-Z., Kanga, E., Kettunen, M., Kumagai, Y., Maxted, N., Senior, J., Wong, M., Keenleyside, K., Mulrooney, D., & Waithaka, J. (2015). Values and benefits of protected areas. In G.L. Worboys, M. Lockwood, A. Kothari, S. Feary, & I. Pulsford (Eds.), *Protected area governance and management*, pp. 145–168. Canberra: Australian National University Press.

Stone, M.T., & Nyaupane, G.P. (2016). Protected areas, tourism and community livelihoods linkages: A comprehensive analysis approach. *Journal of Sustainable Tourism*, 24(5), 673–693.

Suta, N., Hrnjic, A., & Banda, A. (2017). Natural resources management in tourism: Dimensions and impact of tourist offer in the southeastern Europe national parks. In M.H. Bilgin, H. Danis, E. Demir, & U. Can (Eds.), *Financial environment and business development*, pp. 537–562. Gland, Switzerland: Springer.

Swarbrooke, J., Beard, C., Leckie, S., & Pomfret, G. (2003). *Adventure tourism: The new frontier*. Oxford: Butterworth Heinemann.

Timothy, D.J. (2021). *Cultural heritage and tourism: An introduction*, 2nd Edn. Bristol: Channel View Publications.

Timothy, D.J., & Nyaupane, G.P. (2009). Introduction: Heritage tourism and the less-developed world. In D.J. Timothy & G.P. Nyaupane (Eds.), *Cultural heritage and tourism in the developing world: A regional perspective*, pp. 3–19. London: Routledge.

Twining-Ward, L., Li, W., Bhammar, H., & Wright, E. (2018). *Supporting sustainable livelihoods through wildlife tourism*. Washington, DC: World Bank Group.

UNESCO (2022). World Network of Biosphere Reserves. Online: https://en.unesco.org/biosphere/wnbr.

UNDP (2019). *The Biodiversity Finance Initiative (BIOFIN), Finance for nature, a commitment of the whole society to the sustainability of the planet – July 2019*. New York: UNDP.

UNWTO (2001). *The British ecotourism market*. Madrid: World Tourism Organization.

UNWTO (2018). *Tourism and culture synergies*. Madrid: World Tourism Organization.

Vause, J. (2019). *Building a wildlife economy: Developing nature-based tourism in Africa's state protected areas*. Nairobi: United Nations Environment Programme (UNEP).

WTTC (2019). *The economic impact of global wildlife tourism: Travel & tourism as an economic tool for the protection of wildlife*. London: World Travel and Tourism Council.

17

CLIMATE CHANGE, CULTURAL HERITAGE, AND TOURISM IN AFRICA

Critical Issues and Changes

Jarkko Saarinen and Naomi Moswete

Introduction

Change characterizes tourism and its relationships with social and physical environments. Indeed, the tourism phenomenon, its various patterns, segments, and products, are in a constant state of change and the industry's operations have diverse impacts on destinations and wider sociocultural, economic, and ecological systems (Hall & Lew, 2009). This applies also to cultural heritage tourism, which is often considered static and stable rather than dynamic, transformative, and impactful (Richards, 2021; Smith, 2016). Like many other forms of modern tourism, however, cultural heritage tourism has been increasingly used for local and regional development, community empowerment, restoration of degraded sites, and economic diversification (Butler, 2017; Keitumetse, 2016; Moswete & Lacey, 2015; Saarinen, Moswete, & Monare, 2014; Sharpley, 2002; Zeppel & Hall, 1992). Consequently, cultural heritage-based tourism has brought about changes to local communities and their environments (Keitumetse & Nthoi, 2009; Manwa, Moswete, & Saarinen, 2016; Mbaiwa & Sakuze, 2009). Furthermore, well-managed tourism is seen as a strong potential instrument for the sustainable use of local natural and cultural heritage resources (Lowenthal, 2006; Rogerson, 2012; Stone, Lenao, & Moswete, 2020; Timothy & Nyaupane, 2009), which also creates potential changes for localities and their livelihoods structures, strategies, and environments.

Tourism as an agent of change is visible in many parts of Africa (Ahebwa, Aporu, & Nyakaana, 2016; Keitumetse & Nthoi, 2009; Lenao & Saarinen, 2015; Nhamo, 2018; Patel, 2000; Rogerson & Visser, 2020), which serves as an overall geographical context for this chapter. In addition to the impacts of tourism development on surrounding natural environments, cultural heritage tourism and its operations and attractions are transformed by various internal and external drivers (Bryson, 2018; Burnes, 2004). In general, internal drivers refer to elements within a regional and

DOI: 10.4324/9781003153955-17

local environment and tourism sector, many of which are controlled or guided by the industry and its main stakeholders. External drivers relate to conditions that originate outside the local tourism industry and local destination. Thus, external drivers are usually beyond the control of the local destination. In this respect, tourism is often seen as a highly vulnerable sector because it is dependent on, and impacted by, multiple external forces, such as geopolitics, the global economy, financial and political crises, pandemics, and changes in social and ecological systems and environments.

This chapter focuses on the role of one key external driver of change to contemporary and future heritage tourism: global climate change (see Gössling & Hall, 2006; Scott & Becken, 2010). As noted by Hall (2016), there is very little specific literature on the relationship between climate change and cultural heritage-based tourism. Indeed, the recent focus on climate change and heritage relations has emphasized much more natural heritage than cultural heritage (Brooks et al., 2020; Hall & Ram, 2016). Here, the aim is not to analyse the nature and detailed processes of climate change and its impacts on cultural heritage tourism in Africa. Obviously, the continent and its cultural assets and attractions are much too large, diverse, and vibrant for that sort of analysis. Therefore, the chapter focuses on heritage tourism on a general level, specifically on the potential impacts of climate change on cultural heritage attractions and sites. The chapter first discusses the idea of cultural heritage-based tourism and its changing nature on a conceptual level, followed by a general and critical view of the cultural tourism resources and attractions in the African context, highlighting World Heritage Sites (WHSs) as a demonstrative example. This is followed by a discussion on climate change and its potential impacts on cultural heritage tourism in the African context.

Cultural Heritage Tourism

The literature is full of categories and classifications of heritage tourism (see Timothy & Boyd, 2003). There are also many related niches, such as cultural tourism, ethnic tourism, and indigenous tourism, all of which overlap with, or are part of, heritage tourism (Manwa, Moswete, & Saarinen, 2016; Richards, 2021; Timothy, 2021). In general, heritage can be divided into natural and cultural resources. Although a strict distinction between nature and culture may be questionable in local heritage contexts and attractions because the two perspectives are essentially inseparable in heritage spaces (Brooks et al., 2020; Hall et al., 2016), we focus specifically on cultural heritage in this chapter. Regardless of the nature–culture divide, the core idea of heritage refers to the use of the past for present-day purposes, one of which is tourism. For example, the National Trust for Historic Preservation (2015) in the United States describes heritage tourism as travelling to experience the places, artefacts, and activities that authentically represent the stories and people of the past.

However, heritage is not based solely on history and is not always rooted in factual authenticity. Instead, heritage is selective, based on cultural features and traditions from the past that we value in the present and desire to pass on to future generations by maintaining and managing them (Hewison, 1987; Timothy & Boyd, 2006). Thus, what we see and value as heritage is socially determined in the present; heritage is "the contemporary uses of the past" (Ashworth & Tunbridge, 1999, p. 105). Therefore, as human values, ideologies, and frameworks are constantly, albeit slowly, transforming, our constructions and representations of heritage also change through time. Richards (2021, pp. 2–10), for example, interprets the evolution and changing characteristics of cultural tourism based on different "turns" in academic research, including the cultural, mobilities, performative, creative, and curatorial turns.

A key element that defines what is valued as heritage, or not, is power relations (Desmond, 1999; Hall, 1992, 1997; Picard, 1996). As Cheong and Miller (2000, p. 372) have indicated, power is "everywhere in tourism" (see also Squire, 1994) and its social and cultural exchanges, practices, and host–guest relations (van Beek & Schmidt, 2012). This connection between cultural heritage-based tourism and power issues is a highly valid point when evaluating heritage resources, their use, and management in environments with a colonial past, for example (Nash, 1989; van der Merwe & Patel, 2005). In this respect, Chirikure, Ndoro, and Deacon (2018) note that the management of heritage in Africa is largely based on Western ideas usually without considering local contexts, circumstances, and knowledge systems. For Chirikure, Ndoro, and Deacon (2018), the root cause of the marginalization of local views and voices is the legacy of colonialism and the inherited administrative structures from the colonial systems of African states after they achieved independence.

Besides nature and culture, heritage tourism can also be classified based on the materiality or immateriality of its products and assets. Tangible heritage can be either immovable or movable (Timothy & Boyd, 2003). The former refers to buildings and other physical structures, such as the correctional facilities on Robben Island (Cape Town, South Africa) and Constitution Hill (Johannesburg, South Africa) (Strange & Kempa, 2003; van der Merwe, 2013) (Figure 17.1). Movable relics include objects and artefacts in museums and exhibitions. Immaterial or intangible heritage is usually manifested in social traditions, values, beliefs, norms, and ways of living, including festivals and events. Like the nature–culture division of heritage resources, the tangible-intangible distinction is somewhat artificial and blurred in heritage tourism contexts. Musical performances, for example, entail a combination of instruments and oral traditions. The inseparability of material–immaterial and nature–culture heritage forms is very typical in Africa's cultural tourism (Brooks et al., 2020; Chirikure, Ndoro, & Deacon, 2018; Manwa, Moswete, & Saarinen, 2016; Ndoro, 2015; Sarmento, 2010).

Different classifications and sub-categories of heritage tourism make the phenomenon complex in theory and practice for sustainable management. Moreover, the reach and influence of heritage may include personal, local, national, and global scales (Timothy & Boyd, 2003), which has an impact on how certain resources can attract visitors and different market segments. As a result, the definitions and

FIGURE 17.1 Constitution Hill in Johannesburg, South Africa. The place is a former prison and military fort that imprisoned Mahatma Gandhi and Nelson Mandela. As such, the site represents dark heritage in tourism.

Photo: J. Saarinen.

understandings of heritage and heritage tourism are widely debated in the tourism research literature (see Ashworth & Tunbridge, 1999; Garrod & Fyall, 2001; Poria, Butler, & Airey, 2003; Timothy, 2021). In this chapter, we understand cultural heritage-based tourism broadly as a form of tourism that engages cultural heritage resources that stakeholders wish to protect, manage and use in the present and the future. Although this definition begs further analysis, such as the meaning of "stakeholder", it is also open to a realistic understanding that the valuation of cultural heritage can be conflicting or ambivalent in tourism planning and development contexts.

Cultural Heritage Tourism Resources and Attractions: An African Perspective

World Heritage Sites: Global Cultural Heritage Conservation in Local Contexts

The range of African cultural heritage is rich and varied. In general, heritage in Africa refers to a "range of a society's inherited traditions, monuments, objects, places, and culture, as well as the contemporary activities, knowledge, meanings, and behaviors that are drawn from them" (Brooks et al., 2020, p, 299). As indicated earlier, it is impossible to fully capture the continent's cultural heritage resources for tourism in a single book chapter. Instead, we try to provide a locally relevant view of cultural heritage tourism resources, their use, and development. This relates to the critical questions in the local–non-local nexus regarding who decides what is valuable heritage and what are acceptable ways of using and managing those resources for tourism and local development.

UNESCO's WHSs provide a demonstrative example of the complexities of local heritage resources and their utilization in a global tourism and development context. UNESCO has dedicated itself to identifying and protecting manifestations of cultural and natural heritage around the world that is considered to be of outstanding value to humanity. The world's heritage is increasingly threatened by the destruction caused by socioeconomic growth, overuse, and environmental degradation, which led to the initiation of the World Heritage List in 1972. At the time of writing, the list includes 1,154 individual sites based on their cultural and/or natural heritage values; there are 897 cultural, 218 natural, and 39 combined natural and cultural heritage sites listed and regarded as universally valuable places (UNESCO, 2022). UNESCO also maintains the List of World Heritage in Danger, which presently includes some 52 WHSs. The List of World Heritage in Danger aims to highlight local conditions that threaten the sites and emphasizes their management solutions and more sustainable uses (UNESCO, 2022).

Most of the currently listed WHSs in Africa are concentrated in Algeria, Egypt, Ethiopia, Morocco, Senegal, South Africa, and Tunisia (Nhamo, 2018). Notably, UNESCO separates North Africa from the rest of Africa and includes that region with the Middle East (Southwest Asia). Of the total number of WHSs in Africa, there were 89 cultural WHSs and 6 mixed sites in 35 different African countries in 2021 (Table 17.1). Of Africa's cultural heritage sites, ten are included on the List of World Heritage in Danger. Apart from the Middle East, including countries such as Iraq and Syria, Africa has significantly more WHSs in danger than any other region. For example, of the more than 464 cultural and mixed WHSs in Europe and North America, only three of them are on the danger list (UNESCO, 2022).

Does this mean that Africa's heritage sites are endemically under acute threat? Not necessarily, but according to Brooks et al. (2020, p. 298; see also Anag et al., 2002; Brooks, 2005; Chirikure, Ndoro & Deacon, 2018).

> heritage in Africa is facing multiple threats, including economic development, agricultural expansion, urbanization, the direct and indirect effects of conflict, hydro-meteorological hazards, mineral, and fossil fuel exploration, looting and illicit trade in cultural materials, the impacts of tourism, vandalism, low prioritization and neglect.

TABLE 17.1 UNESCO Cultural and Mixed World Heritage Sites in Africa by country in 2021

Algeria	Al Qal'a of Beni Hammad
	Djémila
	M'Zab Valley
	Tassili n'Ajjer (mixed)
	Timgad
	Tipasa
	Kasbah of Algiers
Angola	Mbanza Kongo, Vestiges of the Capital of the former Kingdom of Kongo
Benin	Royal Palaces of Abomey
Botswana	Tsodilo
Burkina Faso	Ruins of Loropéni
	Ancient Ferrous Metallurgy Sites of Burkina Faso
Cabo Verde	Cidade Velha, Historic Centre of Ribeira Grande
Chad	Ennedi Massif: Natural and Cultural Landscape (mixed)
Côte d'Ivoire	Historic Town of Grand-Bassam
Egypt	Abu Mena
	Ancient Thebes with its Necropolis
	Historic Cairo
	Memphis and its Necropolis – the Pyramid Fields from Giza to Dahshur
	Nubian Monuments from Abu Simbel to Philae
	Saint Catherine Area
Eritrea	Asmara: A Modernist African City
Ethiopia	Rock-Hewn Churches, Lalibela
	Fasil Ghebbi, Gondar Region
	Aksum
	Lower Valley of the Awash
	Lower Valley of the Omo
	Tiya
	Harar Jugol, the Fortified Historic Town
	Konso Cultural Landscape
Gabon	Ecosystem and Relict Cultural Landscape of Lopé-Okanda (mixed)
Gambia (the)	Kunta Kinteh Island and Related Sites
	Stone Circles of Senegambia*
Ghana	Forts and Castles, Volta, Greater Accra, Central and Western Regions
	Asante Traditional Buildings
Kenya	Lamu Old Town
	Sacred Mijikenda Kaya Forests
	Fort Jesus, Mombasa
	Thimlich Ohinga Archaeological Site
Lesotho	Maloti-Drakensberg Park*
Libya	Archaeological Site of Cyrene
	Archaeological Site of Leptis Magna
	Archaeological Site of Sabratha
	Rock-Art Sites of Tadrart Acacus
	Old Town of Ghadamès
Madagascar	Royal Hill of Ambohimanga

(*Continued*)

TABLE 17.1 (Continued)

Malawi	Chongoni Rock-Art Area
Mali	Old Towns of Djenné
	Timbuktu
	Cliff of Bandiagara (Land of the Dogons) (mixed)
	Tomb of Askia
Mauritania	Ancient Ksour of Ouadane, Chinguetti, Tichitt and Oualata
Mauritius	Aapravasi Ghat
	Le Morne Cultural Landscape
Morocco	Medina of Fez
	Medina of Marrakesh
	Ksar of Ait-Ben-Haddou
	Historic City of Meknes
	Archaeological Site of Volubilis
	Medina of Tétouan (formerly known as Titawin)
	Medina of Essaouira (formerly Mogador)
	Portuguese City of Mazagan (El Jadida)
	Rabat, Modern Capital and Historic City: a Shared Heritage
Mozambique	Island of Mozambique
Namibia	Twyfelfontein or/Ui-//aes
Niger	Historic Centre of Agadez
Nigeria	Sukur Cultural Landscape
	Osun-Osogbo Sacred Grove
Senegal	Island of Gorée
	Island of Saint-Louis
	Stone Circles of Senegambia*
	Saloum Delta
	Bassari Country: Bassari, Fula and Bedik Cultural Landscapes
South Africa	Fossil Hominid Sites of South Africa
	Robben Island
	Maloti-Drakensberg Park (mixed)*
	Mapungubwe Cultural Landscape
	Richtersveld Cultural and Botanical Landscape
	ǂKhomani Cultural Landscape
Sudan	Gebel Barkal and the Sites of the Napatan Region
	Archaeological Sites of the Island of Meroe
Togo	Koutammakou, the Land of the Batammariba
Tunisia	Amphitheatre of El Jem
	Archaeological Site of Carthage
	Medina of Tunis
	Punic Town of Kerkuane and its Necropolis
	Kairouan
	Medina of Sousse
	Dougga/Thugga
Uganda	Tombs of Buganda Kings at Kasubi
United Republic of Tanzania	Ngorongoro Conservation Area 1 (mixed)
	Ruins of Kilwa Kisiwani and Ruins of Songo Mnara

TABLE 17.1 (Continued)

	Stone Town of Zanzibar
	Kondoa Rock-Art Sites
Zimbabwe	Great Zimbabwe National Monument
	Khami Ruins National Monument
	Matobo Hills

(*Source*: UNESCO, 2022)
* indicates a binational WHS

This is a long list of threats, and recently armed conflicts have intensified challenges in heritage conservation, management, and human wellbeing, in general, in the Sahara–Sahel region (Nhamo, 2018; Santarém, Saarinen, & Brito, 2021). From a critical perspective, however, the extant conflicts do not perhaps fully explain the overall long-term pattern of the List of World Heritage in Danger with an emphasis on African heritage.

According to Chirikure, Ndoro, and Deacon (2018), the underlying reason for Africa's "in danger" evaluations is the region's legacy of colonialism and Westernized heritage management models. Because of this, Western ideas have been directly transferred to Africa "without considering the local situations", which "has suffocated local knowledge systems" that have historically provided contextual and outcome-based solutions to local heritage conservation (Chirikure, Ndoro, & Deacon, 2018, p. 1; see also Robbins, 2012). Perhaps more than many other regions of the world, Africa's heritage is predominantly an active, living resource (Chirikure, Ndoro, & Deacon, 2018, pp. 4–5) that "is continuously 'being made'" (Brooks et al., 2020, p. 298). If the region's living heritage, which is constantly being made and remade, is not fully understood, the result may be externally driven conservation and management that lacks "opportunit[ies] for local people to become meaningfully involved in decision-making about heritage places" (Chirikure, Ndoro, & Deacon, 2018, p. 1). In this respect, the management principles, practices, and evaluation protocols of WHSs can be critically regarded as a continuation of the legacy of colonialism and may be incongruous with local needs, social mores, and values.

A different kind of criticism towards WHS management has emerged based on the current neoliberal conservation paradigm that has also been introduced to African cultural heritage settings. Subsequently, the tourism industry has become an increasingly typical development tool used in the neoliberal governance of cultural heritage conservation in Africa (Mowforth & Munt, 1998). Related to this, Büscher (2013) depicts tourism as a magical "Holy Grail" that provides all the goods needed for destination residents by integrating different goals of contemporary heritage management. Based on the expected growth of cultural heritage tourism at a given site, various management needs and local impacts are easier to justify for different stakeholders such as community members and entrepreneurs (Duffy, 2002; Mowforth & Munt, 1998). However, previous research indicates that there are many potential challenges to equating tourism to community development and wellbeing (Lapeyre, 2010; Mbaiwa, 2005; Ramutsindela, 2007; Rogerson & Rogerson, 2011).

In this respect, questions about how community benefits are created and how they can be optimized in heritage tourism planning and development are important to consider (Harilal, Tichaawa, & Saarinen, 2021; Hunt et al., 2015; Saarinen, 2019a). Indeed, the industry has the capacity to create benefits for communities if it is planned and developed in a sustainable way that is guided by sufficient grassroots participation and empowerment (Scheyvens, 2002, 2011). This perspective highlights the positive impacts of tourism with an emphasis on tourism as a tool for positive change. Recently, this development focus has been linked to the United Nations Sustainable Development Goals (SDGs) (Hughes & Scheyvens, 2016; Rasoolimanesh et al., in press; Saarinen, 2020). While good governance and sustainable tourism management are extremely important internal drivers, there are also major external drivers that will challenge and potentially have serious impacts on cultural heritage management and tourism, and eventually, for tourism-dependent communities.

Cultural Heritage-Based Tourism in Times of Change in Africa

"This Changes Everything": Global Climate Change

Naomi Klein's (2015) book title, *This Changes Everything: Capitalism vs. the Climate*, says it all. Whether the current hegemonic political economy agrees or not, global climate change will create major impacts for all aspects of life, environments, and economies. Indeed, as the Intergovernmental Panel on Climate Change (IPCC) (2013, p, 13) states, "Human influence on the climate system is clear. This is evident from the increasing greenhouse gas concentrations in the atmosphere, positive radiative forcing, observed warming, and understanding of the climate system". Thus, as Hall et al. (2015) have concluded, the reality of climate change is not open to any major disagreement in academia.

According to the recent World Meteorological Organization (2020) report, increasing temperatures and rising sea levels, changing precipitation patterns, and extreme weather conditions, such as heavy rainfall, storms, floods, droughts, and heat waves, are threatening socioeconomic development and human wellbeing in Africa. Similarly, the IPCC's report notes that warming will have devastating effects on livelihoods, specifically in agriculture, which represents the core livelihood across the continent (Intergovernmental Panel on Climate Change, 2014). Furthermore, temperatures in Africa are projected to rise faster than the global average increase during the 21st century (James & Washington, 2013; Joshi et al., 2011; Sanderson, Hemming, & Betts, 2011). "Africa is therefore an exposure and vulnerability 'hot spot' for climate variability and change impacts" (World Meteorological Organization, 2020, p. 3).

African cultural heritage resources, attractions, and heritage tourism will also be impacted by climate change. There is a growing body of scholarship on the effects of climate change on tourism in Africa, specifically the southern African region (Saarinen, Fitchett, & Hoogendoorn, 2022), and countries such as Botswana, South

Africa, and Zambia have been increasingly studied in tourism-climate change research (Dube, 2003; Dube & Nhamo, 2018a, 2018b; Dube & Nhamo, 2018b; Fitchett, Grant, & Hoogendoorn, 2016; Hambira et al., 2021; Hambira, Saarinen, & Moses, 2020; Mmopelwa & Blignaut, 2006; Moswete & Dube, 2013; Saarinen et al., 2020). The rest of the continent, however, has received far less research attention (Hambira & Mbaiwa, 2020; Hoogendoorn et al., 2021), and the estimated impacts of climate change on heritage and heritage tourism are understudied. Fitchett and Roshan (2021) recently noted that much of the world's cultural heritage is protected indoors, in climate-controlled environments. Thus, the weather and climate-related elements are not always considered crucial issues for the built environment and tangible heritage. However, these conditions do not usually apply to much of the Global South, where cultural heritage attractions are more often outdoors, with the outdoor environments forming a greater part of the cultural tourism experience. Therefore, it is easy to agree with Fitchett and Roshan (2021) that there is an urgent need to evaluate the threats of climate change to cultural heritage in the Global South.

The Impacts of Climate Change on Africa's Heritage and Heritage Tourism

Hall et al. (2016) have reviewed some of the main issues concerning climate change and the protection and management of cultural heritage in the Anthropocene. The idea of the Anthropocene refers to increasing anthropogenic influences and presence in (former) natural systems that have progressed to a state in which humans are now one of the "great forces of/in nature" (Moore, 2015; Morton, 2012; Robbins & Moore, 2013). As an idea, it firmly connects the global to the local (Saarinen, 2019b). Due to globalization processes linked with the Anthropocene, distant cultural and natural heritage environments are seen to be increasingly impacted by changes caused somewhere else (Mbaiwa et al., 2019). In this context, Hall et al. (2016) discuss the climate change-related evidence of threats and observed changes in cultural landscapes, the built environment, buried archaeology, and parks and gardens. They conclude that climate change poses a major threat to these kinds of cultural heritage sites and related tourism, and that failure to respond to local climate change impacts means "that more and more irreplaceable cultural heritage will be lost" (Hall et al., 2016, p. 19).

Brooks et al. (2020) describe the potential consequences of a changing climate for different forms of heritage resources in Africa. They categorized the changes into two types: the effects of sea-level rise and coastal erosion, and climate variability (i.e., variations in mean climatic conditions and temperatures). With regard to the first type of change, they examine the effects of sea level rise and erosion, as well as specific heritage forms in relation to climate variability throughout the African continent. Brooks and his colleagues note that traditional coastal economies are a crucial component of heritage in Africa and are now largely at the mercy of rising seas. Likewise, much of the continent's cultural heritage assets are located in coastal areas, because this is where the majority of colonial, seafaring, transportation, and general urban heritage developed in ancient days and in more modern times. Many of Africa's coastal areas are exceedingly vulnerable to natural processes, including shoreline retreat and flooding. Besides the dangers of the floodwaters themselves, flooding also deposits silt and other debris in sensitive cultural zones. In some coastal areas, flood-caused sedimentation is known to have buried many archaeological remains and other historic relics (Hall et al., 2016). Thus, coastal heritage is increasingly threatened by coastal erosion and flooding, and by mid-century many coastal areas are expected to face significant threats due to climate change-induced sea level rises (Brits, 2022; Kamal et al., 2021; Vousdoukas et al., 2020, 2022).

North Africa has the most vulnerable sites, including the North Sinai Archaeological Site Zone, Egypt; the ancient Punic and Roman ruins in Tipasa, Algeria; Carthage, Tunisia; and Qaitbay Citadel, Egypt (Kamal et al., 2021; Vousdoukas et al., 2022). Other particularly vulnerable cultural sites include Kuntah Kinteh Island, The Gambia; the Aného–Glidji conglomeration, Togo; Stone Town in Zanzibar City, Tanzania; and numerous other low-lying heritage sites in Cameroon, Cape Verde, Mauritania, Mozambique, Namibia, Republic of the Congo, Côte d'Ivoire, Djibouti, Senegal, Sudan, and Tanzania (Brits, 2022; Vousdoukas et al., 2022). A recent study by Vousdoukas and his colleagues (2022, p. 256) finds that approximately 20% of all coastal heritage sites in Africa are at risk from a 1-in-100-year extreme sea level event, and that number is expected to nearly quadruple by the end of the current century, depending upon CO_2 and emissions levels. The same study found that nearly one-third (30%) of Africa's cultural heritage sites are exposed to a 100-year coastal extreme event, equalling some 419 km^2 of exposed area of cultural heritage significance.

Cultural heritage is more susceptible to flooding than to erosion, whereas natural heritage is more prone to erosion, and natural areas are more able to recover from, or adapt to, episodic flooding (Vousdoukas et al., 2022). In contrast, "archaeological remains are a non-renewable resource. Once they are gone, they are gone forever" (Timothy & Tahan, 2020, p. 213). Vousdoukas et al. (2022, p. 260) raise the question about how much area and artefact a cultural heritage locality can lose to flooding and erosion before it loses its economic, cultural, and ecological value and suggest that studies to determine this must be done on a site-by-site basis. Hard protection measures, such as seawalls and breakwaters may be part of the short-term solution. Many destinations have started to implement these barriers to erosion and flooding, but these require a great deal of maintenance and may not be a good long-term solution.

According to the second perspective, climate variability can affect both tangible and intangible cultural heritage. According to Brooks et al. (2020, p. 303), "changes in seasonal conditions such as the timing of rainfall can affect livelihoods, while extremes in weather and climate can affect physical aspects of heritage and communities and populations who act as repositories of knowledge, skills and practice". Brooks and his co-authors examine the effects of climate change variability on tangible heritage in the form of rock art (e.g., Tsodilo Hills, Botswana), and intangible heritage in the form of pastoralism. Rock art sites, for example, are subject to many weathering mechanisms, while because of increased droughts, pastoralism today is having a hard time surviving without being combined with sedentary farming. As noted earlier, agriculture is under a major climate change threat in Africa (Intergovernmental Panel on Climate Change, 2014), especially in the Sahelian region, which severely endangers the future of pastoralism as a way of life (Kirkbride, 2008). Other implications of climate change and climate variability affect mixed tangible and intangible heritage, such as the management of sacred groves and traditional vernacular architecture and building materials (e.g., adobe structures). Sandstone structures (North Africa) and mud-brick architecture (Sub-Saharan Africa) are among the most iconic forms of African heritage and are under severe threat of extreme weather events brought on by climate change (Brooks et al., 2020; Obafemi & Kurt, 2016). Likewise, environmental pollution and excess precipitation in some areas of Africa have resulted in the decay of historic built environments (Friesem et al., 2011; Rovella et al., 2021), and the desertification process in the Sahelian region, hastened by global warming, continues to afflict the prevailing pastoralist cultures and distress the built environment of the Sahelian areas of Mauritania, Mali, Niger, Chad, Senegal, Burkina Faso, Nigeria, Sudan, and Eritrea (Brimblecombe et al., 2011; Siriman & Wang, 2021; World Bank, 2020).

While the relationships between tourism and climate have been increasingly studied in Africa (Saarinen, Fitchett, & Hoogendoorn, 2022), there is a major dearth of concrete research-based cases on the impacts of climate change on cultural heritage and the tourism it attracts. One of the few recent examples is living museums in Namibia (Green & Saarinen, 2022). Living museums are exhibitionary museological sites that rely on people's ordinary lives and everyday cultural practices, such as weaving, hunting, animal husbandry, gardening, food preparation, music, dance, and storytelling. These intangible elements of culture, together with representative built structures (e.g., homes, religious shrines, and communal buildings) are explained and interpreted through guided tours and live presentations (Living Culture Foundation, 2021; Naumova, 2015). Green and Saarinen (2022) and Saarinen, Fitchett, & Hoogendoorn, 2022 studied the Ovahimba Living Museum in northern Namibia, which is one of the driest regions in southern Africa. The area has faced a consistent and recurring drought since 2013, and 2019 was reported to be the driest year in over 90 years of recorded history in Namibia (Keja-Kaereho & Tjizu, 2019). Many African ethnic groups have traditionally adapted to changing environmental conditions. For example, the Ovahimba have relied on their livestock growing and maize and vegetable gardening. Owing to increased drought conditions, however, the group faces considerable challenges in practising their traditional livelihoods (Keja-Kaereho & Tjizu, 2019); the growing tourism industry has offered an additional income source and livelihood option for Ovahimba communities (Saarinen, 2012), but it cannot replace their traditional knowledge and customary practices of hunting and horticulture.

The increasingly persistent droughts and resulting environmental variabilities have directly affected both the Ovahimbas' traditional livelihoods, as well as their ability to demonstrate their living culture to tourists. This has led to a certain degree of cultural adaptation for tourism purposes. For example, local community members no longer (or much less often) demonstrate activities that are directly related to the natural environment, such as their hunting practices and their agricultural successes. Instead, they demonstrate more innocuous activities, such as craft-making and explaining their rituals and traditions (Green & Saarinen, 2022). Thus, due to environmental change, the Ovahimba have adapted and altered their everyday behaviour in their own lives, as well as in the lives they choose to depict for tourists. This has transformed the living museum ideal of displaying "normal" living and ordinary routines (van Veuren, 2001) into staged performances that take the place of everyday life and are tourism centric by nature (Saarinen, 2007). Green and Saarinen (2022) and Saarinen, Fitchett, and Hoogendoorn (2022) found that climate change has contributed to local communities becoming more tourism-dependent as they can no longer live their traditional way of life because of increasingly limited natural resources.

The ways in which the supplementary tourism-focused activities will change the Ovahimbas' identity and cultural heritage are as yet unknown, but as noted by Robbins (2012, p. 23), new environmental conditions "lead to new kinds of people" with alternative opportunities, constraints, and imperatives. Indeed, as many authors have indicated, cultural heritage is neither fixed nor unchanging, but rather socially produced, conditioned, and constantly negotiated and renegotiated (Park, 2013). Tourism has been one of the most influential forces for utilizing, commercializing, and transforming cultural heritage resources in Africa (Akama & Sterry, 2002; Brooks, 2005; Keitumetse, 2016; Manwa, Moswete, & Saarinen, 2016; Mbaiwa, 2005; Mbaiwa & Sakuze, 2009), and throughout the world (Richards, 2021; Smith, 2016; Timothy, 2014, 2021).

Conclusion

Global climate change is increasingly recognized as a major threat to cultural heritage in Africa, which is extremely problematic for tourism because the industry's sustained success is based on those very heritage resources. It is possible that large portions of the continent's most important cultural heritage assets will disappear or greatly diminish as a result of the expected impacts of climate change by the end of the century, and the impacts will already be discernible in the next few decades. Although there are nascent efforts to understand the impacts of climate change on cultural heritage in other parts of the world, despite acknowledged concerns and well-understood threats, there has been limited research interest in climate change threats to tangible and intangible cultural heritage and heritage tourism in Africa. For Brooks et al. (2020, p. 315), these threats and projected impacts on cultural heritage and the local communities that depend on those resources raise critical questions about climate justice, which highlights the position that "those least responsible for anthropogenic climate change are often those most affected by its impacts".

Indeed, the impacts of global climate change on Africa's cultural heritage milieu are deeply connected to the need for sustainable development and social wellbeing. For Hall et al. (2016), this emphasizes that cultural heritage and heritage tourism issues cannot be, and should not be, viewed differently from broader societal and environmental concerns. There is a real need to consider and understand how a constantly evolving tourism industry operates in cultural heritage settings and destination communities and how it might be steered onto a path of sustainable development. Without this sort of sound sustainable management approach, tourism continues to represent simply an additional threat to the protection of cultural heritage sites. Climate change threats and the growing need to manage sustainable cultural heritage-based tourism also highlight the urgency of better cultural resource governance that would strengthen the adaptative capacity of heritage conservation and management, as well as the resilience of local communities whose heritage is being shared, towards the direct and indirect impacts of external and internal drivers of change.

References

Ahebwa, W., Aporu, J.P., & Nyakaana, J.B. (2016). Bridging community livelihoods and cultural conservation through tourism. *Tourism and Hospitality Research, 16*(2), 103–115.

Akama, J., & Sterry, P. (Eds.) (2002). *Cultural tourism in Africa: Strategies for the new millennium*. Arnhem, Netherlands: Association for Tourism and Leisure Education.

Anag, G., Cremaschi, M., de Lernia, S., & Liverani, M. (2002). Environment, archaeology, and oil: The Messak Settafet Rescue Operation (Libyan Sahara). *African Archaeological Review, 19*, 67–73.

Ashworth, G., & Tunbridge, J. (1999). Old cities, new pasts: Heritage planning in selected cities of Central Europe. *GeoJournal, 49*, 105–116.

Brimblecombe, P., Bonazza, A., Brooks, N., Sampedro, G., Harris, I., & Sabbioni, C. (2011). Impact of climate change on earthen buildings. In L. Rainer, A. Bass Rivera, & D. Gandreau (Eds.), *Proceedings, Terra 2008: The 10th International Conference on the Study and Conservation of Earthen Architectural Heritage*, pp. 278–282. Los Angeles: Getty.

Brits, E. (2022). Africa coastal heritage under climate threat: Urgent measures required for protection. *Nature Africa*, 7 June, 2022. Online: https://www.nature.com/articles/d44148-022-00078-9#:~:text=Some%20of%20the%20most%20notable,to%20Africa's%20slave%20trade%20history

Brooks, N. (2005). Cultural heritage and conflict: The threatened archaeology of Western Sahara. *Journal of North African Studies, 10*, 413–439.

Brooks, N., Clarke, J., Ngaruiya, W.G., & Wangui, E.E. (2020). African heritage in a changing climate. *Azania: Archaeological Research in Africa, 55*(3), 297–328.

Bryson, J. M. (2018). *Strategic planning for public and non-profit organizations: A guide to strengthening and sustaining organizational achievement*, 4th Edn. New York: Wiley.

Burnes, B. (2004). *Managing change: A strategic approach to organisational dynamics*. London: Prentice Hall.

Büscher, B. (2013). *Transforming the frontier: Peace parks and the politics of neoliberal conservation is in southern Africa*. Durham: Duke University Press.

Butler, G. (2017). Fostering community empowerment and capacity building through tourism: Perspectives from Dullstroom, South Africa. *Journal of Tourism and Cultural Change, 15*(3), 199–212.

Cheong, S.M., & Miller, M.L. (2000). Power and tourism: A Foucauldian observation. *Annals of Tourism Research, 27*(2), 371–390.
Chirikure, S., Ndoro, W., & Deacon, J. (2018). Approaches and trends in African heritage management and conservation. In W. Ndoro, S. Chirikure, & J. Deacon (Eds.), *Managing heritage in Africa: Who cares?*, pp.1–21. London: Routledge.
Desmond, J.C. (1999). *Staging tourism: Bodies on display from Waikiki to Sea World*. Chicago: University of Chicago Press.
Dube, K., & Nhamo, G. (2018a). Climate variability, change and potential impacts on tourism: Evidence from the Zambian side of the Victoria Falls. *Environmental Science and Policy, 84*, 113–123.
Dube, K. & Nhamo, G. (2018b). Climate variability, change and potential impacts on tourism: Evidence from the Zimbabwean side of the Victoria Falls. *Environment Development Sustainability, 21*(4), 2025–2041.
Dube, O.P. (2003). Impact of climate change, vulnerability and adaptation options: Exploring the case of Botswana through Southern Africa: A review. *Botswana Notes and Records, 35*, 147–168.
Duffy, R. (2002). *A trip too far: Ecotourism, politics and exploitation*. London: Earthscan.
Fitchett, J.M., Grant, B., & Hoogendoorn, G. (2016). Climate change threats to two low-lying South African coastal towns: Risks versus perceptions. *South African Journal of Science, 112*(5/6), 86–94.
Fitchett, J.M., & Roshan, G. (2021). Climate change threats to cultural and heritage tourism in Iran. In S. Seyfi & C.M. Hall (Eds.), *Cultural and heritage tourism in the Middle East and North Africa: Complexities, management and practices*, pp. 218–238. London: Routledge.
Friesem, D., Boaretto, E., Eliyahu-Behar, A., & Shahack-Gross, R. (2011). Degradation of mud brick houses in an arid environment: A geoarchaeological model. *Journal of Archaeological Science, 38*(5), 1135–1147.
Garrod, B., & Fyall, A. (2001). Heritage tourism: A question of definition. *Annals of Tourism Research, 28*(4), 1049–1052.
Gössling, S., & Hall, C. M. (2006). An introduction to tourism and global environmental change. In S. Gössling & C.M. Hall (Eds.), *Tourism and global environmental change: Ecological, economic, social and political interrelationships*, pp. 1–33. London: Routledge.
Green, I., & Saarinen, J. (2022). Changing environment and the political ecology of authenticity in heritage tourism: A case of the Ovahimba and the Ju/'Hoansi-San Living Museums in Namibia. In J. Saarinen, N. Moswete, & B. Lubbe (Eds.), *Southern African perspectives on sustainable tourism development: Tourism and changing Localities*, pp. 139–152. Cham, Switzerland: Springer.
Hall, C.M. (2016). Heritage, heritage tourism and climate change. *Journal of Heritage Tourism, 11*(1), 1–9.
Hall, C.M., Amelung, B., Cohen, S., Eijgelaar, E., Gössling, S., Higham, J., Leemans, R., Peeters, P., Ram, Y., & Scott, D. (2015). On climate change scepticism and denial in tourism. *Journal of Sustainable Tourism, 23*(1), 4–25.
Hall, C.M., Baird, T., James, M., & Ram, Y. (2016). Climate change and cultural heritage: Conservation and heritage tourism in the Anthropocene. *Journal of Heritage Tourism, 11*(1), 10–24.
Hall, C.M., & Lew, A. (2009). *Understanding and managing tourism impacts: An integrated approach*. London: Routledge.
Hall, C.M., & Ram, Y. (2016). Heritage in the Intergovernmental Panel on Climate Change (IPCC)assessment reports: A lexical assessment. *Journal of Heritage Tourism, 11*(1), 96–104.
Hall, S. (1992). The west and the rest: Discourse and power. In S. Hall & B. Gieber (Eds.), *Formations of modernity*, pp. 85–95. Cambridge: Polity Press & The Open University.

Hall, S. (1997). The work of representation. In S. Hall (Ed.), *Representation: Cultural representations and signifying practices*, pp. 13–64. Cambridge: The Open University & Sage.

Hambira, W., & Mbaiwa, J. (2020). Tourism and climate change in Africa. In M. Novelli, E. Akwasi Adu-Ampong, & M.A. Ribeiro (Eds.), *Routledge handbook of tourism in Africa*, pp. 98–116. London: Routledge.

Hambira, W., Saarinen, J., Atlhopheng, J., & Manwa, H. (2021). Climate change, tourism and community development: Perceptions of Maun residents, Botswana. *Tourism Review International, 25*(2–3), 105–117.

Hambira, W., Saarinen, J. & Moses, O. (2020). Climate change policy in a world of uncertainty: Changing environment, knowledge and tourism in Botswana. *African Geographical Review, 39*(3), 252–266.

Harilal, V., Tichaawa, T., & Saarinen, J. (2021). The impacts of ecotourism and conservation measures in protected areas on local communities in Cameroon. *Tourism Review International, 25* (2–3), 89–103.

Hewison, R. (1987). *The heritage industry: Britain in a climate of decline*. London: Methuen.

Hoogendoorn, G., Stockigt, L., Saarinen, J., & Fitchett, J.M. (2021). Adapting to climate change: The case of snow-based tourism in Afriski, Lesotho. *African Geographical Review, 40*(1), 92–104.

Hughes, E., & Scheyvens, R. (2016). Corporate social responsibility in tourism post-2015: A Development First approach. *Tourism Geographies, 18*(5), 469–482.

Hunt, C., Durham, W., Driscoll, L., & Honey, M. (2015). Can ecotourism deliver real economic, social, and environmental benefits? A study of the Osa Peninsula, Costa Rica. *Journal of Sustainable Tourism, 23*(3), 339–357.

Intergovernmental Panel on Climate Change (2013). Summary for policymakers. In *Climate change 2013: The physical science basis, contribution of working group I to the fifth assessment report of the Intergovernmental Panel on Climate Change*. Cambridge: Cambridge University Press.

Intergovernmental Panel on Climate Change (2014). *Climate change: Impacts, adaptation, and vulnerability*. Cambridge: Cambridge University Press.

James, R., & Washington, R. (2013). Changes in African temperature and precipitation associated with degrees of global warming. *Climatic Change, 117*(4), 859–872.

Joshi, M., Hawkins, E., Sutton, R., Lowe, J., & Frame, D. (2011). Projections of when temperature change will exceed 2°C above pre-industrial levels. *Nature Climate Change, 1*(8), 407–412.

Kamal, I., Fekri, M., Abou El-Magd, I., & Soliman, N. (2021). Mapping the impacts of projected sea-level rise on cultural heritage sites in Egypt: Case study (Alexandria). *Journal of the Faculty of Tourism and Hotels—University of Sadat City, 5*(1/2), 1–20.

Keitumetse, S., & Nthoi, O. (2009). Investigating the impact of World Heritage Site tourism on the intangible heritage of a community: Tsodilo Hills World Heritage Site, Botswana. *International Journal of Intangible Heritage, 4*, 144–151.

Keitumetse, S.O. (2016). *African cultural heritage conservation and management: Theory and practice from southern Africa*. Cham, Switzerland: Springer.

Keja-Kaereho, C., & Tjizu, B. (2019). Climate change and global warming in Namibia: Environmental disasters vs human life and the economy. *Management and Economics Research Journal, 5*(1), 1–11.

Kirkbride, M. (2008). *Survival of the fittest: Pastoralism and climate change in East Africa*. Oxford: Oxfam International.

Klein, N. (2015). *This changes everything: Capitalism vs. the climate*. New York: Simon & Schuster.

Lapeyre, R. (2010). Community-based tourism as a sustainable solution to maximise impacts locally? The Tsiseb Conservancy case, Namibia. *Development Southern Africa, 27*, 757–772.

Lenao, M., & Saarinen, J. (2015). Integrated rural tourism as a tool for community tourism development: Exploring culture and heritage projects in the North-East District of Botswana. *South African Geographical Journal, 97*(2), 203–216.

Living Culture Foundation (2021, May 1). Traditional cultures in Namibia. Online: https://www.lcfn.info/

Lowenthal, G. (2006). Natural and cultural heritage. *International Journal of Heritage Studies, 11*(1), 81–92.

Manwa, H., Moswete, N., & Saarinen, J. (Eds.) (2016). *Cultural tourism in southern Africa.* Bristol: Channel View Publications.

Mbaiwa, J.E. (2005). Enclave tourism and its socio-economic impacts in the Okavango Delta, Botswana. *Tourism Management, 26*(2), 157–172.

Mbaiwa, J.E., Mogomotsi, P.K., Mbaiwa, T., & Siphambe, G.B. (2019). Globalisation, tourism, and ecosystems management. In D.J. Timothy (Ed.), *Handbook of globalization and tourism*, pp. 198–212. Cheltenham: Edward Elgar.

Mbaiwa, J.E., & Sakuze, L. (2009). Cultural tourism and livelihood diversification: The case of Gcwihaba Caves and XaiXai village in the Okavango Delta, Botswana. *Journal of Tourism & Cultural Change*, 7(1), 61–75.

Mmopelwa, G., & Blignaut, J.N. (2006). The Okavango Delta: The value of tourism. *Southern African Journal of Economics & Management Sciences, 9*(1), 113–127.

Moore, A. (2015). Tourism in the Anthropocene Park? New analytic possibilities. *International Journal of Tourism Anthropology, 4*(2), 186–200.

Morton, T. (2012). From modernity to the Anthropocene: Ecology and art in the age of asymmetry. *International Journal of Social Science, 63*(207–208), 39–51.

Moswete, N., & Dube, O.P. (2013). Wildlife-based tourism and climate: Potential opportunities and challenges for Botswana. In L. D'Amore & P. Kalifungwa (Eds.), *Meeting the challenges of climate change to tourism: Case studies of best practice*, pp. 395–416. Cambridge: Cambridge Scholars.

Moswete, N., & Lacey, G. (2015). "Women cannot lead": Empowering women through cultural tourism in Botswana. *Journal of Sustainable Tourism, 23*(4), 600–617.

Mowforth, M., & Munt, I. (1998). *Tourism and sustainability: A new tourism in the Third World.* London: Routledge.

Nash, D. (1989). Tourism as a form of imperialism. In V.L. Smith (Ed.), *Host and guests: The anthropology of tourism*, pp. 37–47. Philadelphia: University of Pennsylvania Press.

National Trust for Historic Preservation (2015, June 17). Today's word: Heritage tourism. Online: https://savingplaces.org/stories/preservation-glossary-todays-word-heritage-tourism#.YNCF4udRU2w

Naumova, A. (2015). Touching the past: Investigating lived experiences of heritage in living history museums. *The International Journal of the Inclusive Museum*, 7(3–4), 1–9.

Ndoro, W. (2015). World Heritage Sites in Africa: What are the benefits of nomination and inscription? In W. Logan, M. Nic Craith, & U. Kockel (Eds.), *A companion to heritage studies*, pp. 392–409. New York: Wiley.

Nhamo, A. (2018). Burning images: A critical review of rock art conservation in Zimbabwe. *Conservation and Management of Archaeological Sites, 20*, 58–75.

Obafemi, A.P.O., & Kurt, S. (2016). Environmental impacts of adobe as a building material: The north Cyprus traditional building case. *Case Studies in Construction Materials, 4*, 32–41.

Park, H. (2013). *Heritage tourism*. London: Routledge.

Patel, Z., (2000). Rethinking sustainable development in the post-apartheid reconstruction of South African cities. *Local Environment, 5*(4), 383–399.

Picard, M. (1996). *Bali: Cultural tourism and touristic culture.* New York: Archipelago Press.
Poria, Y., Butler, R., & Airey, D. (2003). The core of heritage tourism. *Annals of Tourism Research, 30*(1), 238–254.
Ramutsindela, M. (2007). *Transfrontier conservation in Africa: At the confluence of capital, politics and nature.* Wallingford: CABI.
Rasoolimanesh, S.M., Ramakrishna, S., Hall, C.M., Esfandiar, K., & Seyfi, S. (in press). A systematic scoping review of sustainable tourism indicators in relation to the sustainable development goals. *Journal of Sustainable Tourism.* doi:10.1080/09669582.2020.1775621.
Richards, G. (2021). *Rethinking cultural tourism.* Cheltenham: Edward Elgar.
Robbins, P. (2012). *Political ecology: A critical introduction.* Oxford: Wiley-Blackwell.
Robbins, P., & Moore, S.A. (2013). Ecological anxiety disorder: Diagnosing the politics of the Anthropocene. *Cultural Geographies, 20*(1), 3–19.
Rogerson, C.M. (2012). Tourism – agriculture linkages in rural South Africa: Evidence from the accommodation sector. *Journal of Sustainable Tourism, 20,* 477–495.
Rogerson, C.M., & Rogerson. J. (2011). Tourism research within the Southern African Development Community. *Tourism Review International 15,* 213–224.
Rogerson, J.M., & Visser, G. (Eds.), (2020). *New directions in South African tourism geographies.* Cham, Switzerland: Springer.
Rovella, N., Aly, N., Comite, V., Randazzo, L., Fermo, P., Barca, D., Alvarez de Buergo, M., & La Russa, M.F. (2021). The environmental impact of air pollution on the built heritage of historic Cairo (Egypt). *Science of the Total Environment, 764,* 142905.
Saarinen, J. (2007). Cultural tourism, local communities and representations of authenticity: The case of Lesedi and Swazi cultural villages in Southern Africa. In B. Wishitemi, A. Spenceley, & H. Wels (Eds.), *Culture and community: Tourism studies in eastern and southern Africa,* pp. 140–154. Amsterdam: Rozenberg.
Saarinen, J. (2012). Tourism development and local communities: The direct benefits of tourism to OvaHimba communities in the Kaokoland, North-West Namibia. *Tourism Review International, 15,* 149–157.
Saarinen, J. (2019a). Communities and sustainable tourism development: Community impacts and local benefit creation tourism. In S.F. McCool & K. Bosak (Eds.), *A research agenda for sustainable tourism,* pp. 206–222. Cheltenham: Edward Elgar.
Saarinen, J. (2019b). What are wilderness areas for? Tourism and political ecologies of wilderness uses and management in the Anthropocene. *Journal of Sustainable Tourism 27*(4), 472–487.
Saarinen, J. (2020). Tourism and sustainable development goals: Research on sustainable tourism geographies. In J. Saarinen (Ed.), *Tourism and sustainable development goals: Research on sustainable tourism geographies,* pp. 1–10. London: Routledge.
Saarinen, J., Fitchett, J., & Hoogendoorn, G. (2022). *Climate change and tourism in southern Africa.* Abington: Routledge.
Saarinen, J., Moswete, N., Atlhopheng, J., & Hambira, W. (2020). Changing socio-ecologies of Kalahari: Local perceptions towards environmental change and tourism in Kgalagadi, Botswana. *Development Southern Africa, 37*(5), 855–870.
Saarinen, J., Moswete, N., & Monare, M.J. (2014). Cultural tourism: New opportunities for diversifying the tourism industry in Botswana. *Bulletin of Geography: Socio-economic Series, 26,* 7–18.
Sanderson, M.G., Hemming, D.L., & Betts, R.A. (2011). Regional temperature and precipitation changes under warming. *Philosophical Transactions of the Royal Society A, 369*(1934), 85–98.
Santarém, F., Saarinen, J., & Brito, J. C. (2021). Assessment and prioritization of cultural ecosystem services in the Sahara-Sahelian region. *Science of the Total Environment, 777,* 146053.

Sarmento, J. (2010). Fort Jesus: Guiding the past and contesting the present in Kenya. *Tourism Geographies, 12*(2), 246–263.
Scheyvens, R. (2002). *Tourism for development: Empowering communities*. Harlow: Prentice Hall.
Scheyvens, R. (2011). *Tourism and poverty*. London: Routledge.
Scott, D., & Becken, S. (2010). Adapting to climate change and climate policy: Progress, problems and potentials. *Journal of Sustainable Tourism, 18*(3), 283–295.
Sharpley, R. (2002). Rural tourism and the challenge of tourism diversification: The case of Cyprus. *Tourism Management, 23*(3), 233–244.
Siriman, D., & Wang, Y. (2021). The survival of earthen architecture in Malian Sahel, case study: The Historic City of Djenné. *Current Urban Studies, 9*(1), 83–106.
Smith, M. (2016). *Issues in cultural tourism studies*. London: Routledge.
Squire, S.J. (1994). Accounting for cultural meanings: The interface between geography and tourism studies re-examined. *Progress in Human Geography, 18*(1), 1–16.
Stone, M., Lenao, M., & Moswete, N. (Eds.) (2020). *Natural resources, tourism and community livelihoods in southern Africa*. London: Routledge.
Strange, C., & Kempa, M. (2003). Shades of dark tourism: Alcatraz and Robben Island. *Annals of Tourism Research, 30*(2), 386–405.
Timothy, D.J. (2014). Contemporary cultural heritage and tourism: Development issues and emerging trends. *Journal of Public Archaeology, 13*(1–3), 30–47.
Timothy, D.J. (2021). *Cultural heritage and tourism: An introduction*, 2nd Edn. Bristol: Channel View Publications.
Timothy, D.J., & Boyd, S.W. (2003). *Heritage tourism*. Harlow: Prentice Hall.
Timothy, D.J., & Boyd, S.W. (2006). Heritage tourism in the 21st century: Valued traditions and new perspectives. *Journal of Heritage Tourism, 1*(1), 1–17.
Timothy, D.J., & Nyaupane, G.P. (Eds.) (2009). *Cultural heritage and tourism in the developing world: A regional perspective*. London: Routledge.
Timothy, D.J., & Tahan, L.G. (2020). Understanding perspectives on archaeology and tourism. In D.J. Timothy & L.G. Tahan (Eds.), *Archaeology and tourism: Touring the past*, pp. 205–223. Bristol: Channel View Publications.
UNESCO (2022). World Heritage List. Online: https://whc.unesco.org/en/list/
van Beek, W., & Schmidt, A. (Eds.) (2012). *African hosts & their guests: Cultural dynamics of tourism*. Rochester, NY: James Currey.
van der Merwe, C.D. (2013). The limits of urban heritage tourism in South Africa: The case of Constitution Hill, Johannesburg. *Urban Forum, 24*, 573–588.
van der Merwe, C.D., & Patel, Z. (2005). Understandings of urban regeneration, heritage and environmental justice at Constitution Hill, Johannesburg. *Urban Forum, 16*, 244–258.
van Veuren, E.J. (2001). Transforming cultural villages in the spatial development initiatives of South Africa. *South African Geographical Journal, 83*(2), 137–148.
Vousdoukas, M.I., Clarke, J., Ranasinghe, R., Reimann, L., Khalaf, N., Minh Duong, T., Ouweneel, B., Sabour, S., Iles, C.E., Trisos, C.H., Feyen, L., Mentaschi, L., & Simpson, N.P. (2022). African heritage sites threatened as sea-level rise accelerates. *Nature Climate Change, 12*, 256–265.
Vousdoukas, M.I., Ranasinghe, R., Mentaschi, L., Plomaritis, T.A., Athanasiou, P., Luijendijk, A., & Feyen, L. (2020). Sandy coastlines under threat of erosion. *Nature Climate Change, 10*, 260–263.
World Bank (2020). Where climate change is reality: Supporting Africa's Sahel pastoralists to secure a resilient future. Online: https://www.worldbank.org/en/news/immersive-story/2020/09/21/where-climate-change-is-reality-supporting-africas-sahel-pastoralists-secure-a-resilient-future

World Meteorological Organization. (2020). *State of the climate in Africa 2019.* International conference, Geneva, Switzerland. Online: https://public.wmo.int/en/events/meetings/state-of-climate-africa-2019

Zeppel, H., & Hall, C.M. (1992). Arts and heritage tourism. In B. Weiler & C.M. Hall (Eds.), *Special interest tourism*, pp. 47–68. London: Belhaven Press.

18

HERITAGE TOURISM IN AFRICA

Where Do We Go from Here?

Dallen J. Timothy

The contributions in this book together constitute a wide-reaching treatise on the growing importance of heritage tourism in Africa—a region whose heritage tourism sector has not received academic attention commensurate with its importance, despite its richness, variety, and diversity (Butler & Ivanovic, 2016; Lwoga & Adu-Ampong, 2020; Ndoro et al., 2018; Rogerson & Rogerson, 2011; Rogerson & Visser 2011, 2014). The lack of scholarly attention is particularly acute in the least developed countries of the realm. A great deal has been written about cultural tourism in the most visited countries, but there is much more to learn about the unique characteristics and management challenges and opportunities in countries beyond the normative and pervasive studies undertaken in Egypt, Morocco, Ghana, Kenya, Tanzania, Tunisia, South Africa, and Zimbabwe.

The aim of this book is to shed light on as many manifestations of cultural heritage in Africa as possible and to provide an understanding of the opportunities and challenges that face the cultural heritage sector there. Yet, obviously within the limitations of a single book, not every element of heritage tourism or unique cultural experience can possibly be included, and not every country is examined in detail. Nevertheless, the effort has succeeded in elucidating current patterns and trends, opportunities and challenges, and the socioeconomic and political vagaries of Africa's heritagescapes and the tourisms they induce.

Although much has been covered in the pages of this volume, there remains a dearth of knowledge about the dynamics of heritage tourism in this part of the world, which has so much potential for sustainable tourism growth into the future to help alleviate poverty and protect the tangible and intangible patrimony of the entire continent. This concluding chapter highlights several future directions that might fruitfully be brought to bear on additional research that would be useful in helping scholars and industry stakeholders better understand the value of cultural heritage and its tourism manifestations in Africa.

DOI: 10.4324/9781003153955-18

Issues to Consider

Most work on heritage tourism in Africa is overly descriptive, which means that more theoretical and analytical work is needed to understand the unique dynamics and performances of sustainable heritage tourism development. Balancing heritage use and protection is a critical part of sustainable tourism everywhere, but nowhere is this truer than in Africa (Baillie & Sørensen, 2021; Breen, 2007; Ndoro et al., 2018).

Sustainable Heritage Tourism

Sustainability is the primary tourism paradigm of the 21st century and is defined, operationalized, and studied from a wide range of perspectives, most notably with regard to the natural environment, ecosystem management, and climate change. In the Global South, these are critical issues, but just as important are sustainability principles that are geared toward social, cultural, and economic environments (Timothy & Nyaupane, 2009).

Local control of tourism and its resources is a key principle of sustainable tourism development and may take different forms. This includes grassroots decision-making and the empowerment of native peoples to decide which elements of their cultural heritage should be shared with tourists and which should remain private and protected for their sacrosanct and sociocultural values (Buzinde, 2019; Monaheng, 2016). When communities have the authority to initiate, support, or reject tourism development, they become more politically empowered, and their acceptance of tourism can result in greater economic growth through employment, entrepreneurial growth, and monetary equality between men and women and among different ethnic groups (Ramakau-Ntene, 2006). Another perspective on empowerment is the increasingly critical concept of indigenous rights, including land rights, resource rights, and intellectual or cultural property rights. These are all extremely important in the context of heritage tourism.

Unfortunately, heritage protection efforts (natural and cultural) in many parts of Africa have been made without considering participatory processes. This has denied people access to their heritage and disempowered them through actions that infringe on their cultural and place-based rights. Traditional practices and ways of managing local environments have been sidelined in favor of Western practices directed by non-local decision-makers (Ndoro, 2004). The May–June 2022 forced removal of thousands of Maasai people from their ancestral lands in Tanzania, ostensibly to help protect the area from overuse, reflects a severe lack of indigenous rights. In these efforts, the government has denied access and dismissed this group's ownership of its own heritage in order to 'protect' the area's heritage and to facilitate trophy-hunting and safari tourism. Despite unilateral government moves such as this, with more appropriate participatory planning, heritage tourism has the potential to empower communities, particularly when indigenous knowledge, indigenous rights, and grassroots leadership underpin tourism development efforts (Chauma & Ngwira, 2022; Shabalala & Ezeuduji, 2022; Timothy, 2015).

In this context, the role of host communities and descendent communities as stewards of cultural resources becomes increasingly important. Destination residents are important stakeholders in simultaneously protecting cultural heritage and promoting heritage tourism. Grassroots participation in protecting the past and selling it for its tourism-induced benefits is a delicate process. By understanding their role as good stewards of their inheritance from previous generations, residents in cultural destinations can become more effective caretakers. This requires them to know about the workings of tourism and how they can best be responsible stakeholders in its development (Ashley et al., 2015; Lwoga, 2017). Residents are the best interpreters of their own heritage, decision-makers, and disseminators of indigenous knowledge. As they act as protectors, curators, and custodians of their cultural heritage, they also become part of the attractiveness of a well-cared-for heritage tourism destination.

The importance of indigenous cultures and intangible heritage cannot be overstated. Given the lack of tangible remains of the past in so many parts of Africa, intangible, living heritage takes on an increasingly important role in tourism and in the protection of the region's exceptional cultural patrimony (Ololo & Dieke, 2022; Togola & Al Karjousli, 2014). Local knowledge and traditional ways of valuing cultural inheritances are key in protecting the past and in decision-making for the future. Cultural assets must be evaluated more holistically to understand their connections with local social mores and tourists' interests, and careful planning is necessary to protect and utilize heritage simultaneously while achieving economic development goals, especially poverty alleviation. Future research must examine the triple bottom line in heritage tourism development, particularly with regard to indigenous knowledge and indigenous rights, local leadership, stakeholder empowerment, and what these mean for the people most affected by tourism. Future research needs to address these problems and challenges. We need to assess heritage in all its forms and what it means to different communities and how they might best be good stewards, stakeholders, and decision-makers. Sustainable heritage tourism has the potential to help African states achieve the UN Sustainable Development Goals within their borders, particularly those related to poverty alleviation, wellbeing, equality, and effective partnerships.

Heritage, Nation-Building, and Identity

Also part of empowered communities is a common heritage identity, shared by members of a community, region, or country. Heritage, whether tangible or intangible, is extremely political. It is frequently manipulated, interpreted, and commodified as a propaganda tool to uphold official state narratives and to support the positions of parties in power. Heritage tourism is known to be a tool for enhancing social solidarity, patriotism, and identity as a community, and national pride grows when people have opportunities to share what is important to them with outsiders (Richards, 2007; Timothy, 2021a; Timothy & Boyd, 2003). This is especially important in settler societies, as well as multinational countries that have a strong colonial history. In such countries, ethnicity, race, and cultural background often divide people, whereas heritage tourism often functions as an adhesive for building a sense of nationhood and a common historical memory, including their shared struggles against colonial control. Many of the wars of independence in Africa have functioned as cohesive tools and builders of new nationhood following independence in the latter part of the 20th century, leading many states to commemorate the post-colonial heritage over that of the colonial period (Birmingham, 1995; Chirikure et al., 2021; Herbst, 1990; Marschall, 2007; Teye, 2009).

The socio-political role of heritage and heritage tourism in nation-building, re-writing history, empowering communities through social solidarity, and unfortunately even disempowering certain minority groups, is a topic worthy of additional research in Africa. This work would have salient implications for social sustainability, heritage interpretation and management, tourism decision-making, and the differential power dynamics at play at the state and community levels.

Extraordinary Heritage versus Ordinary Heritage

Although the tendency of the global tourism system is to recognize, commemorate, and market the grandest and most opulent heritage (e.g., castles, fortresses, cathedrals, and temples), including those of colonial elites (Timothy, 2014a, 2014b), Africa is home to unlimited manifestations of ordinary heritage that may be closer to the hearts of the people than faraway colonial buildings in a capital city that is far detached from the ordinary lives of ordinary people and which may represent eras of colonial repression. UNESCO has been criticized in the past for its Western focus on an extraordinary heritage of 'universal value', usually ignoring the ordinary vernacular heritage of the majority of the world's inhabitants (Sullivan, 2004; Tucker & Carnegie, 2014). Fortunately, however, changes in the agency during the past three decades have acknowledged that even commonplace cultural heritage is important to protect and learn about, and it too has universal value. In 1992, UNESCO officially recognized the importance of cultural landscapes and modified its legal instruments to codify this into the World Heritage List. Further in this direction, UNESCO approved the Convention for the Safeguarding of the Intangible Cultural Heritage in 2003, as a natural progression toward appreciating, protecting, and advocating for the world's immaterial and living cultural heritage. At the time of writing, 121 cultural landscapes and 630 elements of intangible heritage had been listed, including a growing number in Africa. These include the Congolese rumba (dance in DR Congo and Congo), Malagasy kabary (poetic oration in Madagascar), Aixan/Gana/Ob#ANS TSI//Khasigu (ancestral musical sound knowledge and skills in Namibia), Enkipaata, Eunoto, and Olng'esherr male rites of passage among the Maasai of Kenya, and traditional hand puppetry in Egypt. Likewise, the Ennedi Massif and Natural and Cultural Landscape (Chad), Konso Cultural Landscape (Ethiopia), Le Morne Cultural Landscape (Mauritius), and the Matobo Hills (Zimbabwe) have all been inscribed recently as critical cultural landscapes that exhibit the priceless ordinary heritage of ordinary people. Despite this growing recognition of the importance of ordinary heritage, tourism overall continues to focus on the largest-scale, most ancient, most tangible, most colorful, and grandest manifestations of human heritage, but this is changing.

Thousands of elements of ordinary heritage, built and living, tangible and intangible, are found throughout Africa, and in an effort to de-colonize the region's heritage, a greater emphasis has been placed on local heritage and a growing pride in what existed before (and after) colonialism (Pwiti & Ndoro, 1999). While it may be unreasonable to suggest that all ordinary heritage can be preserved for its heritage value or for tourism, representative examples certainly should be something that heritage stakeholders consider before it is too late, as so much heritage is being lost to intentional destruction (e.g., looting), climate change, and overdevelopment (Timothy, 2021a; Wahome, 2013). Such efforts should take center stage among empowered communities who choose to protect their ordinary heritage. However, the main challenge to this remains the same: many cultures do not see their mundane lives and living spaces as heritage to be valued and valorized. Thus, coalescing Western views of heritage, which includes the traditional focus on extraordinary heritage as well as the newer focus on ordinary heritage, with traditional views seems incompatible, but it does not have to be (Baillie & Sørensen, 2021; Timothy, 2014b, 2018, 2021b, 2021c; Togola & Al Karjousli, 2014).

Part of that incompatibility is the Western emphasis on excavating a site, restoring it, and museumifying it in order to learn and teach about past societies, whereas in many traditional societies, people learn about the past and their forebears through oral traditions, legends, and performances. Among some traditional societies, remnants of ancient sites are considered gifts passed down from ancestors that can be used for sustenance, such as selling artifacts to collectors or utilizing archaeological pieces for new construction projects. Other traditional societies see ancient artifacts differently—as something to be left undisturbed, unexcavated, and certainly not museumified out of respect for the ancients and because the items do not belong to modern-day descendants. Both of these perspectives are, of course, incompatible with Western paradigms of heritage protection and use, which has previously stoked conflict in heritage management in Africa (Timothy, 2021b).

Africa as a whole may have less globally recognized grandiose heritage when compared to other regions, but it certainly does not lack a remarkable patrimony (Timothy, 2021a). A greater recognition of what the continent offers, especially in Sub-Saharan Africa, is warranted in scholarly research and in tourism practice. This is particularly true in the case of native cultures and knowledge systems, such as traditional medicine, beliefs, agricultural practices, food systems, and handicrafts, many of which are on the verge of disappearing (Akendengue, 1992; Alabi, 1994). More work is needed to encourage the preservation of indigenous knowledge and ordinary heritage, most of which is yet to be discovered and explored the way it has been in other parts of the world (Carabelli, 2016; Mawere, 2014; Wahome & Gathungu, 2021).

Such an effort also raises important issues related to authenticity—authentic places, people, material culture, living culture, and tourist experiences (Boonzaaier & Wels, 2018; Green & Saarinen, 2022; Manwa, 2014; Phelan, 2015; Saarinen, 2007;

Sonwabile Henama et al., 2016). Authenticity in the realm of African tourism is lacking in research attention, despite its importance in heritage tourism discourse more generally (Chhabra, 2021). Commemorating ordinary heritage can help authenticate the African experience for tourists and residents as residents seek to share what is important to them and what perspectives of daily life they wish to impart. On the contrary, it can also lead to certain inauthenticities as traditions, rituals, and places are altered to meet the needs and satisfy the tastes of tourists (Hannam & Offeh, 2012; Reid, 2002).

Perspectives on Diasporas

Perhaps the most researched theme in African heritage tourism is the legacy of slavery. Much has been written about the transatlantic slave trade and its tourism manifestations in West Africa and other parts of the world (e.g., the Caribbean, South America, and North America) (Engmann, 2021; Hannam & Offeh, 2012; Magnani, 2014; Richards, 2005; Schramm, 2007; Teye & Timothy, 2004; Yankholmes et al., 2009; Yankholmes & McKercher, 2015; Yankholmes & Timothy, 2017), but little, if any, of the East Africa slave trade from the Swahili Coast to South Asia and the Middle East has been examined from a heritage perspective.

Much has been written about the experiences of the West African slave diaspora on return journeys to the motherland, their treatment by host communities, and the adapted narratives they often consume while visiting slave sites in Ghana (Araujo, 2010; Gijanto, 2011; Mensah, 2015; Parry, 2018; Teye et al., 2011; Timothy & Teye, 2004). Likewise, there is a growing body of literature on return travel by various African diasporas of more recent vintage (Arnone, 2011; Graf, 2016; Marques da Silva & Pinheiro, 2021; Marschall, 2015, 2017; Otoo et al., 2021), reflecting new modes of involvement beyond the well-established experiences of the descendants of slaves.

Despite the growing body of work on other African diasporas and migration patterns, there is a need to understand diasporas and large-scale migrations beyond the forced migration of slaves more generally and how these other migration patterns manifest in tourism. Tourism manifestations include the return travel of first and second-generation emigrants; ancestral, roots, and genealogy tourism; clan reunions; and cultural tours that familiarize diasporic populations and emigres with the cultures and peoples of their ancestral homelands, often in an effort to build global solidarity with various causes in the homeland. Several African countries have found success in catering to their diasporic markets abroad, and tourism is often seen as a way that emigres can support and invest in the development of their motherlands beyond only sending remittances (Arnone, 2011), and creates a sense of belonging among immigrants and their children who might not feel rooted in either their adopted land or in their original homeland (Graf, 2016).

For the industry, more work is needed to understand the nuances of successful or challenging diaspora marketing. From a theoretical perspective, it is incumbent upon scholars to shed light on solidarity-building and identity affirmation through return travel, as well as the role of colonial history on different diasporic identities. For example, even different ethnic groups from the same country, such as the Ovimbundu, Ambundu, and Bokongo people from Angola living in Portugal may exhibit different cultural or national identities abroad in some instances or exhibit state identities (Angolan) in other situations. They might have different cultural practices or shared cultural experiences under the broader umbrella of Angolan immigrants, while maintaining their own tribal or ethnic identities separately from those of the modern state from which they came. The intertwined concepts of migration and diaspora are critical in understanding people's connections to their ancestral homelands from a tourism perspective. There is much to be learned about the cultural and national experiences of different diasporic groups and the roles that colonialism and ethnic identity play in identity formation and connections to place.

Niches for Future Focus: Food Tourism, Agritourism, and Religious Tourism

One of the biggest untapped cultural tourism assets in Africa is its culinary heritage and agricultural landscapes. Traditional foods in Africa have been heavily influenced by physical geography, tribal practices, and contact with the outside world, or lack thereof in some cases. The diversity of gustatory traditions, recipes, flavors, eating practices, and farming and hunting methods together manifest a valuable yet underutilized potential resource. Regional foods that depict the characteristics of place, culture, and tradition are in high demand among some tourist cohorts who desire to have more authentic experiences with places that are off the normative mass tourism grid. Likewise, food can help form a bond between diasporic peoples and their original homeland, which also manifests in return travel. Foodie tourism (food tourism, culinary tourism, wine tourism, and others) is growing significantly around the world as a form of special interest tourism, and some niche markets are willing to pay a lot of money to have authentic food experiences.

Part of this is the connection to agritourism, which is a form of heritage tourism that focuses on the agricultural traditions, patterns, and products of destinations. Africa is home to many specialty products, such as coffee, tea, cocoa, and groundnuts, which could provide a foundation for rural tourism development throughout the continent as it has in many other destinations (Jolliffe, 2007, 2010). Most of Africa's tourism takes place in urban areas. The growth of culinary tourism and agritourism has strong potential to help diversify the region's heritage tourism product and bring jobs and community development opportunities into rural areas that have few other economic options.

Much of Africa's tangible and intangible culture is founded upon spiritual and religious traditions, both traditional religion and elements of faith brought in by Arab and European colonizers. Religious tourism, including pilgrimage, while not as well recognized in Africa as it is in places such as Europe, Asia, and the Middle East, is an increasingly salient manifestation of cultural heritage and has the potential to provide development opportunities in underprivileged regions (Ali & Dajen, 2013). Much more work is needed to realize the full potential of religious tourism in all its forms on the continent—in Islam, Christianity, and traditional African religions, as long as it respects the values and sacrosanct spaces of worshippers and pilgrims and enables residents to benefit socially and economically.

Food tourism, agritourism, and religious tourism are only three examples of often overlooked types of heritage tourism that have a great deal of potential for growth in Africa. Sacred places of local renown have begun to emerge as global spaces of faith and religion, especially through social media. Likewise, globalization processes and various media influences have started to spread the word about the health and cultural benefits of African culinary traditions, with Ethiopian, Ghanaian, and Kenyan restaurants cropping up in Europe and North America, with the assistance of their diasporic groups abroad. More research needs to address the 'other', lesser-known types of tourism in order to highlight what is unique about Africa and to stress the expanding opportunities beyond the normative heritage tourism sites that continue to draw the largest crowds of cultural tourists to the continent. This can help spread the positive impacts of tourism to a larger number of communities and regions and expose examples of ordinary heritage to the outside world.

A Heritage of Poverty

Western travelers are increasingly attracted to the voyeuristic act of touring poverty (Chhabra & Chowdhury, 2012; Dovey & King, 2012; Freire-Medeiros, 2013; Giddy & Hoogendoorn, 2018; Selinger, 2009; Timothy, 2022; Tzanelli, 2018), because of the contrast and sense of otherness poverty provides from their own lives of relative affluence and comfort, and because it is an ostensible representation of 'authentic life' in traditionally poor countries (Giddy & Hoogendoorn, 2018). Although Africa customarily has been viewed as a 'continent of wars, famines and entrenched poverty…there is now a focus on "Africa rising"' (Beegle et al., 2016, p. 3). Nevertheless, the prevailing perception of Africa in most Western societies is one of corruption, conflict, disease, and poverty—defining characteristics that are perpetuated by the media and, according to Poncian (2015), how Africans often view and portray themselves to the world.

Touring poor communities, squatter settlements, and downtrodden urban neighborhoods has become a popular undertaking in certain tourist destinations since the 1990s and in fact has become standard fare in urban tourism in the Global South (Dürr, 2012; Freire-Medeiros, 2013; Frenzel, 2016; Frenzel et al., 2015; Frisch, 2012; Giddy & Hoogendoorn, 2018; Rogerson & Visser, 2014). Although many poor communities and squatter settlement dwellers might not consider their habitats to be worthy of preservation, commemoration, or commoditization for the tourist gaze, these areas in fact have become part of the identity and image of many tourist destinations, and the very act of touring has heritagized poverty, shantytowns, and squatter settlements, with residents, guides, and even destination management organizations being key actors in this heritagization process (Rogerson &

Visser, 2014; Yagi & Frenzel, 2022). With regard to the heritagization of slums, Yagi and Frenzel (2022, p. 1) assert that 'tourism is not merely a practice in which heritage is consumed. Tourism also produces heritage'. Through tourism's heritagization effect and based upon the long-time view of Africa as a home to poverty and urban decay and the media narratives that feed this worldview, poverty and slums have become a part of the heritage of African urban life (Rogerson & Visser, 2014). Although this niche sector generally arrived later than it did in South America and Asia, slum tourism, or tourists 'peeking on poverty', is now a thriving activity in Cape Town and Johannesburg, South Africa (Booyens & Rogerson, 2019b; Burgold & Rolfes, 2013; Frenzel, 2014); Soweto Township, South Africa (Booyens & Rogerson, 2019a; Briedenhann & Ramchander, 2006; Hoogendoorn et al., 2020; Rogerson, 2014; Rogerson & Mthombeni, 2015); Nairobi, Kenya (Chege & Mwisukha, 2013; Chege & Waweru, 2014; Frenzel & Blakeman, 2015; Kieti & Magio, 2013; Yagi & Frenzel, 2022); Windhoek, Namibia (Steinbrink et al., 2016); and Cairo, Egypt (Hendawy & Madi, 2016; Mekawy, 2012).

In addition to gazing on poverty fleetingly, tourists at times have been able to fully immerse themselves by paying to pretend to be poor in the relative safety of a fabricated shanty town. A 2013 media headline announced that 'Well-off tourists can now spend a night in a fake South African slum' (Macha, 2013, n.p.), referring to the opening of a shanty town at the Emoya Luxury Hotel and Spa in Bloemfontein. There, tourists were able to spend a night in a made-to-order tin shack (complete with water, electricity, and internet). This special lodging arrangement, which meets the definition of Roy's (2004) 'aestheticization of poverty', received widespread condemnation from local residents and national and foreign commentators who argued that the operation made a mockery of people who live in poverty by no choice of their own, and it was frequently referred to as 'poverty porn'. This experiment in the commercial arrogating of poverty— 'paying to be poor'—was short-lived and is no longer operational (Mashishi, 2021).

The touristification and heritagization of slums in Africa raise many questions about place attachment, formal versus informal action and living spaces, sustainable livelihoods, authentic versus inauthentic tourist experiences, profiteering from people's misfortune, and other perspectives on the commodification of poverty for tourist consumption. More academic attention is needed to understand tourism niches that are unique to Africa and how they might either suppress the poor or enhance their economic and social empowerment through entrepreneurial opportunities, employment, and improved social services.

Image Problems

As previously noted, Africa suffers from negative images of conflict, war, terrorism, poverty, disease, corruption, inaccess, crime, and low levels of economic development (Beegle et al., 2016; Poncian, 2015). However, perhaps an even greater problem for the region in tourism marketing terms is its lack of image altogether. Certain countries, such as Egypt, Tunisia, Morocco, South Africa, Ghana, Kenya,

Botswana, and Tanzania, have a generally positive tourism image abroad, yet the majority of countries in the region are relatively unknown to the outside world. This is largely a result of lack of media coverage (or inaccurate media information), the low visibility of trade and diplomatic relations between African states and Western countries, Africa's position in the heart of the Global South, and general geographical ignorance outside of Africa.

Africa is a vast region with many individual countries and cultures, but it is often uncritically lumped together in the Western psyche and in the tourism marketplace without any sort of unpacking of regional or national differences. People from the Global North often do not even know about the individual countries of Africa and that each one has its own history, heritage, and identity. To illustrate this anecdotally, several years ago, one of the author's acquaintances participated in a cruise along the east coast of Africa. When asked which countries she visited, she simply replied, 'I don't know. It was just Africa'. The author asked if her cruise stopped in Kenya and Tanzania, to which she replied 'I think so. That sounds familiar, but to me it was just Africa'. This illustrates not only geographic ignorance (de Blij, 2007; Stock, 2013) but even the broader tendency for Westerners to treat Africa as a single place, undifferentiated in its regions, states, or cultures (Jaksch & Nieves, 2014). Many Americans, for example, believe that Africa is a country rather than a continent with more than 50 countries (Randolph & DeMulder, 2008), and many Westerners would likely not be aware of regional differences in culture and national heritage.

Routes and Trails

Cultural routes and trails have become increasingly important in many Western societies as venues for heritage tourism. There are far fewer heritage trails in the Global South, but they are an emerging product in many developing countries. Trails or routes are linear corridors that can be traversed by foot, bicycle, horseback, or car. They may be organic or purposive—organic trails forming through time in a natural way as walking paths, hunting trails, or transportation corridors develop into tourism trails. Purposive trails are those designed intentionally to become tourist routes and usually are planned to draw attention to certain themes that link individual nodes (attractions) together into a linear tourist resource, which is more valuable in its totality than the individual nodes on their own (Timothy & Boyd, 2015). Routes of both origins can play valuable roles in protecting cultural heritage as they disperse crowds and provide natural corridors for funneling tourists in and out of an area. They are also key in developing tourism as they increase collaboration between sites and destinations, as well as connect nodal points for a greater economic impact that can be shared within a region.

Heritage trails of many scales (urban neighborhood; destination area; regional, national, and cross-border) have become increasingly popular in Africa, as some countries have begun to realize their ability to advance heritage tourism (O'Brien & Curtis, 1979; Proos & Hattingh, 2020), promote conservation, and stimulate economic development, as well as their ability to engage communities in efforts toward greater tourism participation and public activism (Ahebwa et al., 2016; Glen & Roets, 2015; Snowball & Courtney, 2010). The Resistance and Liberation Heritage Route in South Africa, for example, is regarded as a tool for protecting the legacy of South Africa's struggle for freedom and a provider of jobs and tourism revenue. The trail, which connects more than a dozen sites throughout South Africa, focuses on commemorating and celebrating the country's independence and freedom from apartheid (Bialostocka, 2014). The Liberation Heritage Route was listed on UNESCO's Tentative List of World Heritage in 2009, indicating the universal value of its theme and the sites it connects to the concepts of freedom, liberty, democracy, and human rights (Weiss, 2014). The trail aims to raise awareness of the country's history and contribute to the local economy in places where each site is located (Bialostocka, 2014), and to develop social cohesion, national solidarity, and an ethos of human rights (Kotze & Zuma, 2017; Weiss, 2014). Several multinational trails and routes have been in the works since the early 2000s, focusing on the cultural and natural heritage of South Africa, Lesotho, Eswatini, Zimbabwe, Malawi, Tanzania, Botswana, Mozambique, and Namibia (Briedenhann & Wickens, 2004), although their fruition has been hampered by border restrictions and other geopolitical forces, as well as differing levels of development and enthusiasm on opposite sides of state boundaries.

Some of the main challenges to developing short- or long-distance cultural trails in many parts of Africa are problems of accessibility and ineffective public transportation, security concerns, lack of awareness among community members, lack of funding, and a lack of will to collaborate (Glen & Roets, 2015). This is particularly the case in developing countries, where heritage sites are spread apart in rural regions and may not include the most remarkable monuments and historic sites. However, the principle of heritage routes is to connect 'small sites of mainly local significance into a cultural heritage route and market them as a package while also improving the management and conservation of heritage assets' (Snowball & Courtney, 2010, p. 563). Such a justification often overcomes a lack of will and encourages heritage community buy-in (Timothy & Boyd, 2015). Additional research is sorely needed to gauge the potential of other heritage trails in Africa and to emphasize the role of trails in connecting individual places of ordinary heritage value that come become constituent parts of extraordinary heritage through collaboration and careful planning.

Concluding Comments

Africa is one of the least visited regions of the world, yet it is one of the most culturally rich and diverse areas on the planet with much 'undiscovered' tourism potential (Drummond et al., 2021; Wahome & Gathungu, 2021). Likewise, tourism in most of Africa is under-researched and possesses significant potential for academic assessments. Most of Africa represents an underutilized laboratory for understanding unique relationships between cultural heritage and tourism. Most of the research heretofore published on heritage tourism in Africa is concentrated in less than a quarter of the countries on the continent; thus, there is much geographic scope for additional work that lays bare the intricacies, mysteries, and manifestations of heritage tourism in Africa. Living indigenous cultures, ancient civilizations and the remains of our oldest forebears, urban built environments, rural landscapes and livelihoods, colonial and industrial complexes, religious practices and events, diasporas and migrations, food trends, nature preserves, and places of dark history exemplify the range of heritage tourism opportunities that exist in Africa, but there is so much more yet to explore. The discussions contained herein only begin to scratch the surface of what Africa has to offer global tourism. The continent is a clean slate for collecting data and testing paradigms, theories, and concepts so that we can better understand what this so-called 'marginal' region of the world will yet offer the heritage tourism experience. Misperceptions and stereotypes of Africa repel many tourists, yet simultaneously and oxymoronically, Africa's mystique attracts many others.

The diverging histories of colonial overlords and local people are frequently a cause of contention in Africa. The historic environment there is a mix of indigenous and colonial cultures, and both perspectives must work in harmony to create a holistic and balanced approach to the past. While colonial heritage is often celebrated in Africa over that of the native people, as modernization and globalization continue to unfold it will become increasingly important for the cultural heritage of Africans to be elevated so that it has a far less chance of disappearing in the midst of modernization and the frequent over-emphasis on European pasts.

There are thousands of directions heritage tourism could go in Africa. This chapter has highlighted only a few to start the discussion moving forward. There are many unique dynamics, histories, and human relations that make Africa what it is today, worthy of meaningful research attention. It is my sincere hope that this book is a step in the right direction and has raised more questions than it has answered.

References

Ahebwa, W.M., Aporu, J.P., & Nyakaana, J.B. (2016). Bridging community livelihoods and cultural conservation through tourism: Case study of Kabaka Heritage Trail in Uganda. *Tourism and Hospitality Research*, *16*(2), 103–115.

Akendengue, B. (1992). Medicinal plants used by the Fang traditional healers in Equatorial Guinea. *Journal of Ethnopharmacology*, *37*(2), 165–173.

Alabi, S. (1994). Nigerian tradition, culture and environment: Opportunities for heritage tourism development. *Tourism Recreation Research*, *19*(2), 69–77.

Ali, M., & Dajen, A. (2013). A geographical analysis of cultural tourism in North Ethiopia: A case study. *Radix International Journal of Research in Social Science*, *2*(1), 1–17.

Araujo, A. (2010). Welcome the diaspora: Slave trade heritage tourism and the public memory of slavery. *Ethnologies*, *32*(2), 145–178.

Arnone, A. (2011). Tourism and the Eritrean diaspora. *Journal of Contemporary African Studies*, *29*(4), 441–454.

Ashley, K.S., Osmani, M., Emmitt, S., Mallinson, M., & Mallinson, H. (2015). Assessing stakeholders' perspectives towards the conservation of the built heritage of Suakin, Sudan. *International Journal of Heritage Studies*, *21*(7), 674–697.

Baillie, B., & Sørensen, M.L.S. (Eds.) (2021). *African heritage challenges: Communities and sustainable development*. Singapore: Springer.
Beegle, K., Christiaensen, L., Dabalen, A., & Gaddis, I. (2016). *Poverty in a rising Africa*. Washington, DC: World Bank Publications.
Bialostocka, O. (2014). Using the past to build the future: A critical review of the Liberation Heritage Route (LHR) Project of South Africa. *Africa Insight, 44*(2), 94–107.
Birmingham, D. (1995). *The decolonization of Africa*. London: University College London.
Boonzaaier, C., & Wels, H. (2018). Authenticity lost? The significance of cultural villages in the conservation of heritage in South Africa. *Journal of Heritage Tourism, 13*(2), 181–193.
Booyens, I., & Rogerson, C.M. (2019a). Creative tourism: South African township explorations. *Tourism Review, 74*(2), 256–267.
Booyens, I., & Rogerson, C.M. (2019b). Re-creating slum tourism: Perspectives from South Africa. *Urbani Izziv, 30*, 52–63.
Breen, C. (2007). Advocacy, international development and World Heritage Sites in Sub-Saharan Africa. *World Archaeology, 39*(3), 355–370.
Briedenhann, J., & Ramchander, P. (2006). Township tourism: Blessing or blight? The case of Soweto in South Africa. In M.K. Smith & M. Robinson (Eds.), *Cultural tourism in a changing world: Politics, participation and (re) presentation*, pp. 104–123. Bristol: Channel View Publications.
Briedenhann, J., & Wickens, E. (2004). Tourism routes as a tool for the economic development of rural areas—vibrant hope or impossible dream? *Tourism Management, 25*(1), 71–79.
Burgold, J., & Rolfes, M. (2013). Of voyeuristic safari tours and responsible tourism with educational value: Observing moral communication in slum and township tourism in Cape Town and Mumbai. *Die Erde–Journal of the Geographical Society of Berlin, 144*(2), 161–174.
Butler, G., & Ivanovic, M. (2016). Cultural heritage tourism development in post-apartheid South Africa: Critical issues and challenges. In H. Manwa, N. Moswete, & J. Saarinen (Eds.), *Cultural tourism in Southern Africa*, pp. 58–75. Bristol: Channel View Publications.
Buzinde, C.N. (2019). Tourism and indigenous communities: Linking reterritorialisation and decolonization in North Africa. In D.J. Timothy (Ed.), *Routledge handbook on tourism in the Middle East and North Africa*, pp. 95–105. London: Routledge.
Carabelli, R. (2016). Heritage preservation and contemporary production in northern Africa towns. In C.N. Silva (Ed.), *Urban planning in North Africa*, pp. 147–158. London: Routledge.
Chauma, E.C., & Ngwira, C. (2022). Managing a World Heritage Site in Malawi: Do residents' sentiments matter? *Journal of Heritage Tourism, 17*(2), 142–157.
Chege, P.W., & Mwisukha, A. (2013). Benefits of slum tourism in Kibera slum in Nairobi, Kenya. *International Journal of Arts and Commerce, 2*(4), 94–102.
Chege, P.W., & Waweru, F.K. (2014). Assessment of Status, Challenges and Viability of Slum Tourism: Case Study of Kibera Slum in Nairobi, Kenya. *Assessment, 4*(6), 38–48.
Chhabra, D. (2021). *Resilience, authenticity, and digital heritage tourism*. London: Routledge.
Chhabra, D., & Chowdhury, A. (2012). Slum tourism: Ethical or voyeuristic. *Tourism Review International, 16*(1), 69–73.
Chirikure, S., Ndoro, W., Bugarin, F.T., di Lernia, S., Ichumbaki, E.B., & Lwoga, N.B. (2021). Usable pasts forum: UNESCO and heritage tourism in Africa. *African Archaeological Review, 38*(3), 513–533.
de Blij, H. (2007). *Why geography matters. Three challenges facing America: Climate change, the rise of China and global terrorism*. Oxford: Oxford University Press.
Dovey, K., & King, R. (2012). Informal urbanism and the taste for slums. *Tourism Geographies, 14*(2), 275–293.

Drummond, J., Drummond, F., & Rogerson, C.M. (2021). Latent opportunities for heritage tourism in South Africa: Evidence from Mahikeng and surrounds. *African Journal of Hospitality, Tourism and Leisure, 10*(5), 1591–1609.

Dürr, E. (2012). Urban poverty, spatial representation and mobility: Touring a slum in Mexico. *International Journal of Urban and Regional Research, 36*(4), 706–724.

Engmann, R.A.A. (2021). Coups, castles, and cultural heritage: Conversations with Flight Lieutenant Jerry John Rawlings, former President of Ghana. *Journal of Heritage Tourism, 16*(6), 722–737.

Freire-Medeiros, B. (2013). *Touring poverty*. London: Routledge.

Frenzel, F. (2014). Slum tourism and urban regeneration: Touring inner Johannesburg. *Urban Forum, 25*(4), 431–447.

Frenzel, F. (2016). *Slumming it: The tourist valorization of urban poverty*. London: Zed Books.

Frenzel, F., & Blakeman, S. (2015). Making slums into attractions: The role of tour guiding in the slum tourism development in Kibera and Dharavi. *Tourism Review International, 19*(1–2), 87–100.

Frenzel, F., Koens, K., Steinbrink, M., & Rogerson, C.M. (2015). Slum tourism: State of the art. *Tourism Review International, 18*(4), 237–252.

Frisch, T. (2012). Glimpses of another world: The favela as a tourist attraction. *Tourism Geographies, 14*(2), 320–338.

Giddy, J.K., & Hoogendoorn, G. (2018). Ethical concerns around inner city walking tours. *Urban Geography, 39*(9), 1293–1299.

Gijanto, L.A. (2011). Competing narratives: Tensions between diaspora tourism and the Atlantic past in The Gambia. *Journal of Heritage Tourism, 6*(3), 227–243.

Glen, N., & Roets, M. (2015). Creating a cultural-environmental themed route in the Alfred Nzo District of the Eastern Cape in South Africa. In World Tourism Organization (Ed.), *Global Report on Cultural Routes and Itineraries*, pp. 85–87. Madrid: UNWTO.

Graf, S. (2016). Diaspora tourism and the negotiation of belonging: Journeys of young second-generation Eritreans to Eritrea. *Ethnic and Racial Studies, 40*(15), 2710–2727.

Green, I., & Saarinen, J. (2022). Changing environment and the political ecology of authenticity in heritage tourism: A case of the Ovahimba and the Ju/'Hoansi-San living museums in Namibia. In J. Saarinen, B. Lubbe, & N.N. Moswete (Eds.), *Southern African perspectives on sustainable tourism Management: Tourism and changing localities*, pp. 139–152. Cham, Switzerland: Springer.

Hannam, K., & Offeh, F. (2012). The institutional commodification of heritage tourism in Ghana. *Africa Insight, 42*(2), 18–27.

Hendawy, M., & Madi, B. (2016). Slum tourism: A catalyst for urban development? Reflections from Cairo's Ashwa'iyat (informal areas). In S. Attia, S. Shabka, Z. Shafik, & A. Ibrahim (Eds.), *Dynamics and resilience of informal areas: International perspectives*, pp. 57–72. Cham, Switzerland: Springer.

Herbst, J. (1990). War and the state in Africa. *International Security, 14*(4), 117–139.

Hoogendoorn, G., Letsatsi, N., Malleka, T., & Booyens, I. (2020). Tourist and resident perspectives on 'slum tourism': The case of the Vilakazi precinct, Soweto. *GeoJournal, 85*. 1133–1149.

Jaksch, M.L., & Nieves, A.D. (2014). Africa is a country? Digital diasporas, ICTs, and heritage development strategies for social justice. *The Journal of Interactive Technology and Pedagogy, 6*, 1–13.

Jolliffe, L. (ed.) (2007). *Tea and tourism: Tourists, traditions and transformations*. Bristol: Channel View Publications.

Jolliffe, L. (ed.) (2010). *Coffee culture, destinations and tourism*. Bristol: Channel View Publications.

Kieti, D.M., & Magio, K.O. (2013). The ethical and local resident perspectives of slum tourism in Kenya. *Advances in Hospitality and Tourism Research, 1*(1), 37–57.

Kotze, S., & Zuma, B. (2017). Navigating freedom: Social cohesion policies and the democratic cultural landscape of Durban's Liberation Heritage Route. *Museum International, 69*(1–2), 98–113.

Lwoga, N.B. (2017). Factors influencing local residents' intentions to conserve the built heritage in Tanzania. *Journal of Heritage Tourism, 12*(4), 394–409.

Lwoga, N.B., & Adu-Ampong, E.A. (2020). Heritage and heritage tourism in Africa: Historical overview and contemporary issues. In M. Novelli, E.A. Adu-Ampong, & M.A. Ribeiro (Eds.), *Routledge handbook of tourism in Africa*, pp. 69–84. London: Routledge.

Macha, N. (2013). Well-off tourists can now spend a night in a fake South African slum. *Global Voices Online*, 1 December 2013. Online: https://theworld.org/stories/2013-12-01/well-tourists-can-now-spend-night-fake-south-african-slum

Magnani, E. (2014). Culture and tourism? Limits and potential of sustainable tourist development in Gorée, Senegal and Ilha de Moçambique, Mozambique. *Via Tourism Review, 4/5*, n.p. (Online journal).

Manwa, H. (2014). Authenticity of the Basotho Hat: Perceptions of weavers and vendors. *Mediterranean Journal of Social Sciences, 5*(20), 981–989.

Marques da Silva, N., & Pinheiro, T.R. (2021). Back to the roots: Quilombola travel in Brazil and West Africa. *World Leisure Journal, 63*(1), 23–39.

Marschall, S. (2007). Constructing identity through cultural and heritage tourism in South Africa. In. B. Wishitemi, A. Spenceley, & H. Wels (Eds.), *Culture and community: Tourism studies in Eastern and Southern Africa*, pp. 155–172. Amsterdam: Rozenberg.

Marschall, S. (2015). 'Homesick tourism': Memory, identity and (be)longing. *Current Issues in Tourism, 18*(9), 876–892.

Marschall, S. (2017). Transnational migrant home visits as identity practice: The case of African migrants in South Africa. *Annals of Tourism Research, 63*, 140–150.

Mashishi, N. (2021). South African hotel did create fake 'shantytown' but deluxe 'poverty experience' no longer on offer. *Africa Check*, 13 April 2021. Online: https://africacheck.org/fact-checks/meta-programme-fact-checks/south-african-hotel-did-create-fake-shantytown-delux-poverty

Mawere, M. (2014). *Culture, indigenous knowledge and development in Africa: Reviving interconnections for sustainable development*. Bamenda, Cameroon: Langaa Research and Publishing.

Mekawy, M.A. (2012). Responsible slum tourism: Egyptian experience. *Annals of Tourism Research, 39*(4), 2092–2113.

Mensah, I. (2015). The roots tourism experience of diaspora Africans: A focus on the Cape Coast and Elmina castles. *Journal of Heritage Tourism, 10*(3), 213–232.

Monaheng, T. (2016). Integrating Indigenous knowledge in the development of cultural tourism in Lesotho. In H. Manwa, N. Moswete, & J. Saarinen (Eds.), *Cultural tourism in Southern Africa*, pp. 31–46. Bristol: Channel View Publications.

Ndoro, W. (2004). Traditional and customary heritage systems: Nostalgia or reality? The implications of managing heritage sites in Africa. In de Merode, E., Smeets, R., & Westrik, C. (Eds.), *Linking universal and local values: Managing a sustainable future for World Heritage*, pp. 81–84. Paris: UNESCO.

Ndoro, W., Chirikure, S., & Deacon, J. (Eds.) (2018). *Managing heritage in Africa: Who cares?* London: Routledge.

O'Brien, R., & Curtis, C. (1979). The Donkin Heritage Trail, Port Elizabeth. *Planning and Building Developments, 40*, 13–26.

Ololo, N.G., & Dieke, P.U.C. (2022). Change and continuity in event tourism development in Nigeria: Examples from Ekpe Cultural Festival in Abia State. *Tourism Review International, 26*(1), 57–71.

Otoo, F.E., Kim, S.S., & King, B. (2021). African diaspora tourism-How motivations shape experiences. *Journal of Destination Marketing & Management, 20*, 100565.

Parry, T.D. (2018). 'What is Africa to me' now? African-American heritage tourism in Senegambia. *Journal of Contemporary African Studies, 36*(2), 245–263.

Phelan, K.V. (2015). Staged authenticity and identity conflicts: Cultural tourism in Africa. In G.R. Ricci (Ed.), *Travel, tourism and identity: Culture & civilization*, pp. 163–174. London: Routledge.

Poncian, J. (2015). The persistence of Western negative perceptions about Africa: Factoring in the role of Africans. *Journal of African Studies and Development*, 7(3), 72–80.

Proos, E., & Hattingh, J. (2020). Advancing heritage tourism in the central Karoo: The South African War Battlefields Route. *Development Southern Africa, 37*(4), 678–693.

Pwiti, G., & Ndoro, W. (1999). The legacy of colonialism: Perceptions of the cultural heritage in Southern Africa, with special reference to Zimbabwe. *The African Archaeological Review, 16*(3), 143–153.

Ramakau-Ntene, M. (2006). *Exploring the tourism potential of heritage sites: The case of the Botha Bothe Plateau in Lesotho*. Unpublished doctoral dissertation, University of Cape Town.

Randolph, B., & DeMulder, E. (2008). I didn't know there were cities in Africa! Challenging children's–and adults'—misperceptions about the African continent. *Teaching Tolerance, 34*, 1–5.

Reid, D. (2002). Development of cultural tourism in Africa: A community-based approach. In J. Akama & P. Sterry (Eds.), *Cultural tourism in Africa: Strategies for the new millennium*, pp. 25–34. Arnhem, Netherlands: Association for Tourism and Leisure Education.

Richards, G. (2007). Introduction: Global trends in cultural tourism. In G. Richards (Ed.), *Cultural tourism: Global and local perspectives*, pp. 1–24. London: Psychology Press.

Richards, S.L. (2005). What is to be remembered? Tourism to Ghana's slave castle-dungeons. *Theatre Journal, 57*(4), 617–637.

Rogerson, C.M. (2014). Rethinking slum tourism: Tourism in South Africa's rural slumlands. *Bulletin of Geography, Socio-economic Series, 26*(26), 19–34.

Rogerson, C.M., & Mthombeni, T. (2015). From slum tourism to slum tourists: Township resident mobilities in South Africa. *Nordic Journal of African Studies, 24*(3–4), 319–338.

Rogerson, C.M., & Rogerson, J.M. (2011). Tourism research within the Southern African Development Community: Production and consumption in academic journals, 2000–2010. *Tourism Review International, 15*(1–2), 213–224.

Rogerson, C.M., & Visser, G. (2011). African tourism geographies: Existing paths and new directions. *Tijdschrift voor Economische en Sociale Geografie, 102*(3), 251–259.

Rogerson, C.M., & Visser, G. (2014). A decade of progress in African urban tourism scholarship. *Urban Forum, 25*(4), 407–417.

Roy, A. (2004). Transnational trespassings. In A. Roy & N. Alsayyad (Eds.), *Urban informality: Transnational perspectives from the Middle East, Latin America, and South Asia*, pp. 289–318. Lanham, MD: Lexington Books.

Saarinen, J. (2007). Cultural tourism, local communities and representations of authenticity: The case of Lesedi and Swazi cultural villages in Southern Africa. In B. Wishitemi, A. Spenceley, & H. Wels (Eds.), *Culture and community: Tourism studies in Eastern and Southern Africa*, pp. 140–154. Amsterdam: Rozenberg.

Schramm, K. (2007). Slave route projects: Tracing the heritage of slavery in Ghana. In F. de Jong & M. Rowlands (Eds.), *Reclaiming heritage: Alternative imaginaries of memory in West Africa*, pp. 71–98. New York: Routledge.

Selinger, E. (2009). Ethics and poverty tours. *Philosophy and Public Policy Quarterly, 29*(1/2), 2–7.

Shabalala, L.P., & Ezeuduji, I.O. (2022). Community involvement and participation in sustainable development of cultural heritage tourism: A case from South Africa. In E. Christou & A. Fotiadis (Eds.), *Reviving tourism, in the post-pandemic era*, pp. 422–435. Sindos, Greece: School of Economics and Business, International Hellenic University.

Snowball, J.D., & Courtney, S. (2010). Cultural heritage routes in South Africa: Effective tools for heritage conservation and local economic development? *Development Southern Africa, 27*(4), 563–576.

Sonwabile Henama, U., van Rensburg, M.J., & Nicolaides, A. (2016). Nkandla: The unexplored frontier for heritage tourism to Zululand, South Africa. *African Journal of Hospitality, Tourism and Leisure, 5*(2), 1–16.

Steinbrink, M., Buning, M., Legant, M., Schauwinhold, B., & Süßenguth, T. (2016). *Touring Katutura!: Poverty, Tourism, and Poverty Tourism in Windhoek, Namibia.* Potsdam: Universitätsverlag Potsdam.

Stock, R. (2013). *Africa south of the Sahara: A geographical interpretation.* New York: Guilford.

Sullivan, S. (2004). Local involvement and traditional practices in the World Heritage system. In E. de Merode, R. Smeets, & C. Westrik (Eds.), *Linking universal and local values: Managing a sustainable future for World Heritage*, pp. 49–55. Paris: UNESCO.

Teye, V.B. (2009). Tourism and Africa's tripartite cultural past. In D.J. Timothy & G.P. Nyaupane (Eds.), *Cultural heritage and tourism in the developing world: A regional perspective*, pp. 165–185. London: Routledge.

Teye, V.B., & Timothy, D.J. (2004). The varied colors of slave heritage in West Africa: White American stakeholders. *Space and Culture*, 7(2), 145–155.

Teye, V.B., Turk, E., & Sönmez, S. (2011). Heritage tourism in Africa: Residents' perceptions of African-American and White tourists. *Tourism Analysis, 16*(2), 169–185.

Timothy, D.J. (2014a). Contemporary cultural heritage and tourism: Development issues and emerging trends. *Public Archaeology*, 13(3), 30–47.

Timothy, D.J. (2014b). Views of the vernacular: Tourism and heritage of the ordinary. In J. Kaminski, A. Benson, & D. Arnold (Eds.), *Contemporary issues in cultural heritage tourism*, pp. 32–44. London: Routledge.

Timothy, D.J. (2015). Cultural heritage, tourism and socio-economic development. In R. Sharpley & D.J. Telfer (Eds.), *Tourism and development: Concepts and issues*, 2nd Edn, pp. 237–249. Bristol: Channel View Publications.

Timothy, D.J. (2018) Making sense of heritage tourism: Research trends in a maturing field of study. *Tourism Management Perspectives, 25*, 177–180.

Timothy, D.J. (2021a). *Cultural heritage and tourism: An introduction*, 2nd Edn. Bristol: Channel View Publications.

Timothy, D.J. (2021b). Democratising the cultural past: Western values, the Global South and cross-cultural perspectives in heritage tourism. In J. Saarinen & J.M. Rogerson (Eds.), *Tourism, change and the Global South*, pp. 33–47. London: Routledge.

Timothy, D.J. (2021c). Heritage and tourism: Alternative perspectives from South Asia. *South Asian Journal of Tourism and Hospitality*, 1(1), 35–57.

Timothy, D.J. (2022). Volunteer tourism, poverty, and people on the margins: Altruism, ego-enhancement, and voyeurism in the Himalaya. In G.P. Nyaupane, & D.J. Timothy (eds) *Tourism and development in the Himalaya: Social, environmental, and economic forces*, pp. 170–183. London: Routledge.

Timothy, D.J., & Boyd, S.W. (2003). *Heritage tourism.* Harlow: Prentice Hall.

Timothy, D.J., & Boyd, S.W. (2015). *Tourism and trails: Cultural, ecological and management issues*. Bristol: Channel View Publications.

Timothy, D.J., & Nyaupane, G.P. (Eds.) (2009). *Cultural heritage and tourism in the developing world: A regional perspective*. London: Routledge.

Timothy, D.J. & Teye, V.B. (2004). American children of the African diaspora: Journeys to the motherland. In T. Coles & D.J. Timothy (Eds.), *Tourism, diasporas and space*, pp. 111–123. London: Routledge.

Togola, D.C., & Al Karjousli, S. (2014). Capture the wealth of heritage in a holy city of Islam: The problem of developing tourism in Djenné in Mali. *Via Tourism Review, 4/5*, n.p. (Online journal).

Tucker, H., & Carnegie, E. (2014). World Heritage and the contradictions of 'universal value'. *Annals of Tourism Research, 47*, 63–76.

Tzanelli, R. (2018). Slum tourism: A review of state-of-the-art scholarship. *Tourism Culture & Communication, 18*(2), 149–155.

Wahome, E.W. (2013). An assessment of the challenges associated with the loss of tangible heritage in Kenya. *International Journal of Arts and Commerce, 2*(9), 77–92.

Wahome, E.W., & Gathungu, J.J.W. (2021). Redefining sustainability in the conservation and promotion of the cultural heritage tourism product in Kenya. In V.G.B. Gowreesunkar, S.W. Maingi, H. Roy, & R. Micera (Eds.), *Tourism destination management in a post-pandemic context*, pp. 257–271. Bingley: Emerald.

Weiss, L.M. (2014). The historical narration of rights in South Africa: Past, present, and future tense. *Heritage & Society*, 7(2), 121–138.

Yagi, T., & Frenzel, F. (2022). Tourism and urban heritage in Kibera. *Annals of Tourism Research, 92*, 103325.

Yankholmes, A., Akyeampong, O.A., & Dei, L.A. (2009). Residents' perceptions of Transatlantic slave trade attractions for heritage tourism in Danish-Osu, Ghana. *Journal of Heritage Tourism, 4*(4), 315–329.

Yankholmes, A., & McKercher, B. (2015). Rethinking slavery heritage tourism. *Journal of Heritage Tourism, 10*(3), 233–247.

Yankholmes, A., & Timothy, D.J. (2017). Social distance between local residents and African-American expatriates in the context of Ghana's slavery-based heritage tourism. *International Journal of Tourism Research, 19*(5), 486–495.

INDEX

For Product Safety Concerns and Information please contact our EU representative GPSR@taylorandfrancis.com
Taylor & Francis Verlag GmbH, Kaufingerstraße 24, 80331 München, Germany

www.ingramcontent.com/pod-product-compliance
Ingram Content Group UK Ltd.
Pitfield, Milton Keynes, MK11 3LW, UK
UKHW020145200926
475702UK00005B/929

* 9 7 8 0 3 6 7 7 2 2 2 3 4 *